CYBER SECURITY SOLUTIONS FOR PROTECTING AND BUILDING THE FUTURE SMART GRID

CYBER SECURITY SOLUTIONS FOR PROTECTING AND BUILDING THE FUTURE SMART GRID

Edited by

DIVYA ASIJA

Department of Electrical and Electronics Engineering,
Amity School of Engineering & Technology,
Amity University Uttar Pradesh, Noida, India

R.K. VIRAL

Department of Electrical and Electronics Engineering,
Amity School of Engineering & Technology,
Amity University Uttar Pradesh, Noida, India

RESUL DAŞ

Software Engineering, Firat University, Elazığ, Türkiye

GURKAN TUNA

Computer Technologies, Trakya University, Edirne, Türkiye

ELSEVIER

Elsevier
Radarweg 29, PO Box 211, 1000 AE Amsterdam, Netherlands
125 London Wall, London EC2Y 5AS, United Kingdom
50 Hampshire Street, 5th Floor, Cambridge, MA 02139, United States

ISBN: 978-0-443-14066-2

For information on all Elsevier publications visit our website at https://www.elsevier.com/books-and-journals

Publisher: Megan Ball
Acquisitions Editor: Rachel Pomery
Editorial Project Manager: Mason Malloy
Production Project Manager: Prem Kumar Kaliamoorthi
Cover Designer: Miles Hitchen

Typeset by TNQ Technologies

Contents

Contributors

Ipek Abasıkeleş-Turgut
Computer Engineering, Faculty of Engineering and Life Sciences, Iskenderun Technical University, Hatay

Shyam V. Aradhye
Department of Computer Science and Engineering, School of Computational Sciences, COEP Technological University, Pune, Maharashtra, India

Divya Asija
Department of Electrical and Electronics Engineering, Amity School of Engineering and Technology, Amity University Uttar Pradesh, Noida, India

Sinan Atıcı
Department of Software, SECHARD Information Technologies Company, Istanbul, Türkiye

Mohit Bansal
Department of Electrical and Electronics Engineering, GL Bajaj Institute of Technology and Management, Noida, Uttar Pradesh, India

Jignesh G. Bhatt
Department of Instrumentation and Control Engineering, Faculty of Technology, Dharmsinh Desai University, Nadiad, Gujarat, India

Ved S. Bilaskar
Department of Computer Science and Engineering, School of Computational Sciences, COEP Technological University, Pune, Maharashtra, India

Resul Daş
Software Engineering, Technology Faculty, Firat University, Elazığ, Türkiye

Doygun Demirol
Department of Computer Science and Technology, Bingol University, Bingöl, Türkiye

Muhammed Zekeriya Gündüz
Department of Computer Science and Technology, Bingol University, Bingöl, Türkiye

Ruchika Gupta
Mangalmay Institute of Management & Technology, Greater Noida, Uttar Pradesh, India

Sweta Gupta
Department of Electrical Engineering, Netaji Subhas University of Technology (NSUT), Dwarka, Delhi, India

Kazım Hanbay
Software Engineering, Inonu University, Malatya, Türkiye

V.S.K.V. Harish
Department of Electrical Engineering, Netaji Subhas University of Technology (NSUT), Dwarka, Delhi, India

Deepak D. Kshirsagar
Department of Computer Science and Engineering, School of Computational Sciences, COEP Technological University, Pune, Maharashtra, India

Manoj Kumar
Department of Electrical and Electronics Engineering, Amity School of Engineering and Technology, Amity University Uttar Pradesh, Noida, India

Aman Kumar
Department of Electrical and Electronics Engineering, Amity School of Engineering and Technology, Amity University Uttar Pradesh, Noida, India

Tapsi Nagpal
Lingaya's Vidyapeeth, Faridabad, Haryana, India

Pushparaj R. Nimbalkar
Department of Computer Science and Engineering, School of Computational Sciences, COEP Technological University, Pune, Maharashtra, India

Prateema Sharma
Lingaya's Vidyapeeth, Faridabad, Haryana, India

Snehal S. Shinde
Department of Computer Science and Engineering, School of Computational Sciences, COEP Technological University, Pune, Maharashtra, India

Priyank Srivastava
Galgotias University, Greater Noida, Uttar Pradesh, India

Tushar Tomar
Department of Electrical and Electronics Engineering, Amity School of Engineering and Technology, Amity University Uttar Pradesh, Noida, India

Gurkan Tuna
Computer Technologies, Trakya University, Edirne, Türkiye

R.K. Viral
Department of Electrical and Electronics Engineering, Amity School of Engineering and Technology, Amity University Uttar Pradesh, Noida, India

PART I

Renewable Energy system security: Vulnerabilities, challenges and impacts

CHAPTER 1

Smart grids and cyber security

Resul Daş[1] and Muhammed Zekeriya Gündüz[2]
[1]Software Engineering, Technology Faculty, Firat University, Elazığ, Türkiye; [2]Department of Computer Science and Technology, Bingol University, Bingöl, Türkiye

1.1 Introduction

Technological developments aiming to serve humanity are increasing day by day. The acceleration of infrastructure and systems development has led to more effective use of the Internet. The use of the Internet for different objects and systems in daily life has led to the emergence of the concept of the Internet of Things (IoT) [1]. IoT means connecting everyday physical objects and devices to the Internet, allowing them to send and receive data. IoT has numerous applications across various industries, such as financial services, manufacturing, healthcare, transportation, industrial automation, agriculture, smart city, retail, and energy [2]. These are just a few examples, and as the technology continues to evolve, new applications will be emerge. Some of these applications are included in critical infrastructures. Fig. 1.1 shows major critical infrastructure sectors.

The smart grid is at the forefront of IoT-based energy applications. So, it is used as IoT-based smart grid in literature, too. IoT applications can be used to monitor and control energy usage, improve energy efficiency, enhance grid reliability, enable demand response, support renewable energy integration, and reduce costs for the smart grid [3]. IoT-based smart grid is a modern power grid that uses digital technology to monitor and manage the generation, transmission, distribution, and consumption of electricity. They have complex and bidirectional communication systems to optimize energy generation and consumption, reduce waste, and improve overall efficiency.

The smart grid includes advanced technologies to improve the electricity supply's efficiency, reliability, and sustainability. A smart grid system enables bidirectional communication and control between the grid and consumers, allowing for more efficient management of energy usage and integration of renewable energy sources [4]. The smart grid use sensors

Cyber Security Solutions for Protecting and Building the Future Smart Grid
ISBN 978-0-443-14066-2
https://doi.org/10.1016/B978-0-443-14066-2.00003-7

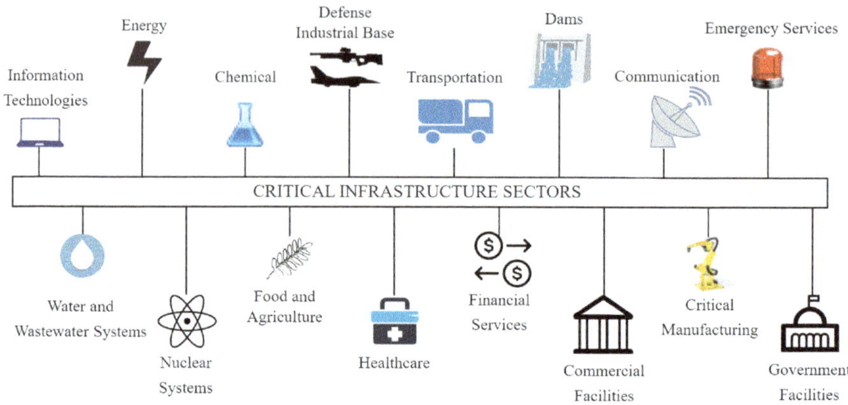

Figure 1.1 Major critical infrastructure sectors.

and smart meters to monitor energy usage in real time and to provide detailed information. This information allows for better management of energy demand and supply, which can reduce costs, increase efficiency, and improve grid stability [5]. In addition to improving the performance of the grid, the smart grid can also provide benefits to consumers. For example, it can enable consumers to monitor their energy usage in real time, allowing them to adjust their consumption patterns to save money and reduce their environmental footprint. The smart grid enables the balancing of electricity supply and demand, thereby improving grid reliability and reducing the risk of blackouts [6]. It also promotes the integration of renewable energy sources and mitigates the need for fossil fuel power plants, resulting in reduced greenhouse gas emissions [7]. Consequently, the smart grid plays a crucial role in the transition toward a more sustainable and efficient energy system. Fig. 1.2 shows a holistic view of the smart grid.

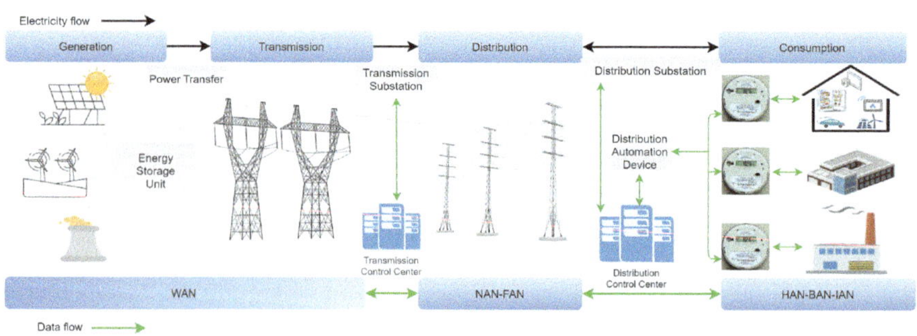

Figure 1.2 A holistic view of the smart grid [5].

Cyber security encompasses the protection of digital systems, networks, and information from unauthorized access, theft, or damage. It involves various techniques, procedures, and tools that aim to protect electronic devices, computer systems, networks, and data from potential cyber threats. The primary objective of cyber security is to provide the confidentiality, integrity, and availability of systems and data. Confidentiality, integrity, and availability are the main objectives of cyber security and called as CIA triad [8]. Integrating the essentials of cyber security into a system architecture from beginning to end is imperative, as cyber security is a process.

Cyber security is a critical concern for the smart grid because a cyberattack on the grid could have severe consequences for public safety, national security, and the economy. A cyberattack on a power system could disrupt the power supply, cause blackouts, equipment damage, or even physical harm to individuals. To mitigate the risks, the smart grid operators must implement robust cyber security measures to protect their systems from cyberattacks [9].

Cyber security in the smart grid applications is a collaborative responsibility shared by the power industry, government agencies, and consumers [8]. The power industry must undertake robust cyber security measures to secure its systems, while government agencies must provide regulatory frameworks and resources to encourage and enforce cyber security measures. Moreover, consumers must be aware of potential threats and take appropriate measures to strengthen their devices and networks against cyber threats.

Increasing confidence in the smart grids necessitates the urgent need to prioritize cyber security measures. As these systems are becoming increasingly integrated into our daily lives, it is crucial to adopt and implement robust cyber security protocols to counteract potential cyber threats. Therefore, ensuring the cyber security of the smart grid remains a top priority, and proactive measures must be taken to mitigate potential risks and ensure an uninterrupted and secure energy supply.

1.2 Cyber security: Why a growing threat?

The smart grid includes digital communication technologies to enhance efficiency, reliability, and security. However, the integration of advanced technologies in the grid also creates new cyber security challenges. The technologies that make the smart grid more efficient and effective also make them vulnerable to cyberattacks. Cyber security is a growing threat for the smart grid because of the potential consequences of a successful cyberattack on a smart grid system. It has become a growing threat for several reasons such as increased reliance on technology, expanding attack surface, sophisticated attacks, lack of cyber security skills, and human error [10]. Also, it is a critical concern for the smart grid because an attack on the grid could have severe consequences for public safety, national security, and the economy.

A cyberattack could disrupt the power supply, cause blackouts, equipment damage, or even physical harm to individuals. It could also compromise sensitive data related to energy generation, transmission, distribution, and consumption. Cyberattacks on the smart grid pose a significant threat because they can result in cascading failures that impact multiple systems and cause widespread disruption [11]. Increased complexity, high impact, lack of security standards, legacy systems, and interconnectedness are some prominent reasons why cyberattacks pose a significant risk for smart grids.

Moreover, the smart grid systems are often connected to other critical infrastructure systems, such as water and gas networks, making them even more attractive targets for cyberattacks. In addition, as more devices become connected to the smart grid through the IoT, there are more potential entry points for cyber criminals to exploit. Furthermore, using artificial intelligence (AI), machine learning (ML), and other cutting-edge technologies by cyber criminals leads to an increase in sophisticated cyberattacks [12].

Increasing cyber security concerns prevent the rapid development of the systems. Because cyber security threats are becoming much more complex. Eliminating these complex threats has become serious concern. This situation makes cyber security the biggest problem that is worrying.

The smart grid requires robust security measures to ensure the CIA triad for data and operations. To mitigate the risks, robust cyber security measures must be implemented against potential cyberattacks on the smart grid. The responsibility of ensuring cyber security in the smart grid is a shared responsibility between the power industry, government agencies, and consumers. The power industry must invest in and implement strong cyber security measures to protect its systems, while government agencies must provide regulatory frameworks and support to encourage and enforce cyber security measures [13]. Consumers must also be aware of the risks and take steps to secure their own devices and networks to prevent cyberattacks.

It is crucial that the smart grid operators and other stakeholders take proactive steps to prevent, detect, and respond to these threats to ensure the security and reliability of the grid.

1.3 Cyber security objectives and requirement

Developing of information and communication technologies (ICT) has played a crucial role in the evolution of the conventional electricity grid into the modern smart grid. Despite the numerous advantages offered by the smart grid, its immunity is not sufficient to cyber security issues. The concerns have the potential to obstruct the progress of smart grid applications. Nevertheless, it is anticipated that consistent advancements will enhance the smart grid experience in the forthcoming years.

The CIA triad is essential for secure communication infrastructures and the operation, and management of energy. The cyber security challenges of the smart grid revolve around the imperative task of ensuring the CIA triad of the control systems and ICT. The CIA triad is called cyber security objectives. Cyber security objectives aim to maintain digital assets and infrastructure against cyber threats. The protection of smart grid systems against cyberattacks is a crucial issue and requires the implementation of key security principles based on the CIA triad [8].

Confidentiality aims to prevent unauthorized access or disclosure of data. It means that access to information should only be granted to authorized individuals. In a smart grid network, home devices are connected to the power grid via bidirectional and real-time communication, thus making it vulnerable to cyberattacks. If attackers gain access to consumer information, they can misuse it to track their lifestyle, monitor their appliance usage, and determine their presence at home [14]. Confidentiality is also closely linked to privacy, which is a major concern for smart grid users.

Integrity is defined as the prevention of data from unauthorized alteration or destruction. Maintaining the truth of data is an integral part of integrity, and it helps provide a secure real-time monitoring system for the smart grid. It is essential to ensure that data are correct and nontampered and that it is not altered in an unauthorized or undetected manner. It is important to ensure that the consumption data recorded in the smart meter is correct and not tampered or changed without authorization or without being noticed [15].

Availability aims to protect the information system from breakdown and timely access to the information system. Availability attacks can take the form of corrupting, blocking, or delaying information. Therefore, it is essential to ensure that information is available to authorized parties in the smart grid without compromising its security. In a smart grid, availability is critical as the power system is expected to be available at all times. Preventing DDoS attacks that lead to blackouts is crucial to ensuring data availability. Availability is the most important security objective for a smart grid application [16].

Typical cyberattacks in the smart grid applications target at least one of the CIA triad principles. Malicious users or attackers can exploit information to gain an advantage or harm others. Confidentiality attacks aim to enable unauthorized access to information, while integrity attacks aim to manipulate or falsify data. Availability attacks aim to interrupt power delivery, delay, or break communication. Therefore, the implementation of the CIA triad principles is crucial for protecting smart grid systems against cyberattacks. The smart grid applications necessitate supplementary security requirements, apart from the CIA triad, to ensure a comprehensive cyber security approach [8].

Authentication is a critical component of the smart grid cyber security as it ensures that only authorized entities can access sensitive information and control systems. The authentication process involves verifying the identity of the user or device attempting to access the system by validating their credentials, such as usernames and passwords, digital certificates, or biometric data. By implementing strong authentication mechanisms, the smart grid operators can prevent unauthorized access to critical infrastructure and reduce the risk of cyberattacks [12].

Authenticity is often associated with the verification of user identity or the validity of data, messages, or transactions. Authenticity is necessary to verify that the parties involved are who they claim to be. This is an important aspect of the smart grid cyber security as it ensures that only authorized users can access and modify the grid's systems and data. Authenticity can be ensured through the use of encryption, digital signatures, and certificates to verify the identity and integrity of data and commands transmitted across the grid [5].

Authorization refers to the process of determining whether a user, application, or system has the necessary permissions and rights to access a particular resource or perform a specific action within a system or network. In other words, it involves defining and enforcing access control policies to ensure that only authorized entities can access or modify specific resources, such as files, databases, or applications. Authorization is an essential aspect of cyber security, as it helps prevent unauthorized access, modification, or theft of sensitive data and resources [9].

Accountability is an important aspect of cyber security for the smart grid. It refers to the ability to trace actions and activities in the smart grid system to their responsible parties. If a security problem emerges, accountability operations can determine who is responsible for it. The assets receiving the data cannot later deny it and this is called nonrepudiation. In the event of a cyberattack or other security incident, accountability allows for the identification and attribution of the responsible parties, which is essential for responding to the incident and taking appropriate action. Accountability can be achieved through the use of audit trails, logs, and other monitoring and reporting mechanisms within the smart grid system. Accountability is needed to secure the smart grid applications in terms of confidentiality, integrity, and privacy [5].

Auditing is a process of examining and analyzing the smart grid security and performance. It involves reviewing the system's log files and other data to detect potential security breaches or suspicious activities. Auditing helps to ensure that the system is operating as intended, and any deviations from normal operations can be detected and addressed promptly. It also assists to identify areas where security measures may need to be strengthened. Regular auditing of the smart grid is critical to maintaining the CIA triad of the data and operations. Auditability without authentication cannot ensure accountability. Auditability and authentication should be provided together to ensure exact accountability [8].

Privacy in the context of smart grids refers to the protection of personal and sensitive information of individuals that may be collected, stored, processed, and transmitted by the smart grid system. Privacy includes electricity usage patterns, device location, user behavior, etc. Privacy is a critical aspect of the smart grid security, as the unauthorized use or disclosure of this information could result in various negative consequences, including identity theft, fraud, and reputational damage. Therefore, it is essential to implement privacy measures, such as data encryption, access controls, and user consent, to protect sensitive information and ensure that the privacy rights of individuals are respected [17].

Dependability for smart grid refers to the ability of the grid to consistently deliver reliable and secure services, even in the face of failures, errors, or malicious attacks. It encompasses a range of attributes such as availability, reliability, safety, and resilience. To ensure dependability, smart grid systems must be designed with redundancy, fault tolerance, error detection, and self-healing capabilities, as well as robust cyber security measures to prevent and respond to cyberattacks [5].

Survivability for the smart grid refers to the ability of the system to continue operating and providing critical services in the face of cyberattacks, natural disasters, and other disruptive events. This includes the ability to detect and respond to cyber threats in real time, as well as the ability to recover and restore services in the event of a disruption. To ensure the survivability of the smart grid, robust cyber security measures must be in place such as disaster recovery plans, backup systems, and response procedures to mitigate the impact of an attack [5].

Safety criticality refers to the extent to which a system, process, or component is crucial to safeguarding people, the environment, or other valuable assets. In particular, safety-critical systems are those whose malfunction could lead to significant harm or damage, as seen in the instance of nuclear power plants or aviation systems. Safety criticality refers to systems that can potentially lead to severe outcomes due to the existence of some unexpected situations such as earthquakes, floods, and tsunamis may result in substantial physical damage, human injuries, or even deaths. It requires the implementation of appropriate cyber security measures and disaster recovery plans to prevent or mitigate the impact of potential failures or cyberattacks [9].

In addition to the CIA triad, the aforementioned cyber security requirements must be met strictly to ensure security against cyberattacks in the smart grid applications. Fig. 1.3 shows the smart grid security objectives and requirements.

1.4 Reported cyberattacks globally in power systems

Cyberattackers have improved their proficiency, while much of the world's vital infrastructure still relies on outdated technologies, making them vulnerable to simple cyberattacks. The use of IoT technologies in critical infrastructures and the increasing of organized cyber attacks present a worrying scenario. Recent cyberattacks have targeted SCADA control systems in various critical infrastructures. Stuxnet, Havex, BlackEnergy3, and Industroyer are the most notable examples. Malware targeting critical infrastructure, including water plants, gas plants, power plants, and transportation systems, is sophisticated and tailored to specific purposes. The integration of many IoT-based devices into critical infrastructure for efficient communication increases their exposure to cyberattacks.

IoT-based solutions are a leading technology for enhancing critical infrastructures. Any device with an IP address can connect to the Internet.

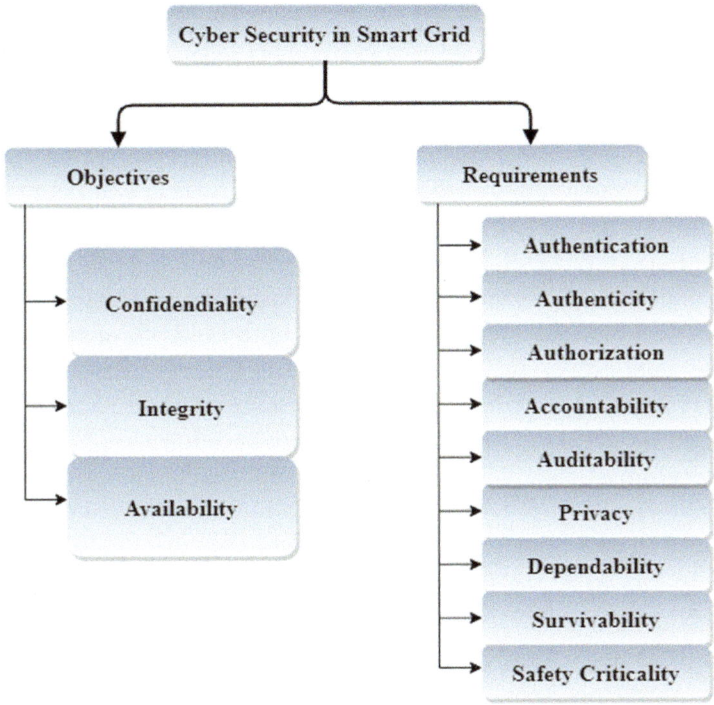

Figure 1.3 Smart grid security objectives and requirements.

So, IoT devices are vulnerable to almost all IP-based cyberattacks. The security vulnerabilities of the Internet make this technology prone to cyber threats. This section examines the most significant examples of IoT-based cyberattacks worldwide and their potential dangers to power systems. The attacks target the critical infrastructure of power systems and can have severe consequences, including power outages, service disruptions, and even physical damage to equipment.

The types of cyberattacks reported in power systems include malware attacks, phishing attacks, DDoS, ransomware attacks, advanced persistent threats (APTs), and insider attacks. Malware attacks involve the use of malicious software to gain unauthorized access to systems and disrupt their operations. Phishing attacks aim to obtain sensitive information such as passwords and login credentials from unsuspecting users. DDoS attacks flood systems with traffic, making them unavailable to legitimate users. Ransomware attacks involve the use of malware to encrypt systems and demand payment with bitcoin for decryption. Attackers use sophisticated techniques to gain long-term access to a system and steal sensitive data in an APT attack.

Insider attacks are carried out by individuals with authorized access to systems, who use their access to cause harm. There are just a few examples of power company hacking incidents that have occurred in recent years [2].

Stuxnet (2010): Stuxnet, identified as one of the most significant cyber weapons ever developed, attacked SCADA systems by targeting PLCs that enable the automation of electromechanical processes. Stuxnet is a sophisticated worm that targeted Iran's nuclear program and caused significant damage to its centrifuges. Stuxnet is thought to be a joint development of the US and Israel. The worm is highly sophisticated and utilizes a range of advanced techniques to infect and propagate through computer networks. Stuxnet is a milestone in cyber warfare, revealing the vulnerability of critical infrastructure to cyberattacks. Stuxnet used advanced techniques such as 0-day exploits, rootkit capabilities, and advanced encryption methods, demonstrating a high level of sophistication typically associated with nation-state actors [18].

The Havex cyberattack (2014): Also known as Energetic Bear or Dragonfly attack, was a state-sponsored malware campaign that targeted industrial control and SCADA systems in the energy sector primarily in Europe. It aimed to steal sensitive information and disrupt energy production. Attackers used spear-phishing emails and malicious software downloads to distribute the Havex malware, enabling them to gain control of critical infrastructure systems [19].

Turkish Power Company Attack (2015): The Turkish power grid was targeted by a cyberattack that caused a widespread power outage across several regions. The attack, which was believed to be coordinated and sophisticated, affected distribution management systems and caused disruption to the electricity supply for several hours. The attack was attributed to a group with ties to the Russian government, and it involved using malware to gain access to the power grid's control systems. The incident raised concerns about the vulnerability of critical infrastructure to cyberattacks and the need for more robust cyber security measures in the power industry [20].

Ukraine Power Grid Attack (2015 and 2016): Hackers targeted Ukraine's power grid twice, resulting in power outages that affected hundreds of thousands of people. The attacks were believed to have been carried out by state-sponsored Russian hackers. The malware used in the attack, known as BlackEnergy3, was a variant of the BlackEnergy malware family and was specifically designed to target industrial control systems (ICS) and SCADA systems in the energy sector [19].

Industroyer was another malware campaign that targeted critical infrastructure systems in Ukraine in 2016. This attack specifically targeted the power grid and caused significant power outages in the country. The malware was designed to target ICS and SCADA systems and was distributed through spear-phishing emails and malicious software downloads [21].

NotPetya Cyber Attack (2017): NotPetya was a destructive cyberattack is considered one of the most devastating cyberattacks in history. It started in Ukraine and quickly spread to other countries, affecting companies and organizations worldwide. NotPetya was a ransomware attack that used a variant of the Petya malware but with more advanced capabilities [22].

United States Power Grid Hack (2018): The US Department of Homeland Security reported a warning that Russian hackers had been targeting US power companies for at least a year. The attackers gained access to the power companies' networks and could potentially disrupt or disable the power grid, posing a significant threat to national security. The incident underscored the ongoing and increasing risk of cyberattacks targeting critical infrastructure systems in the United States and around the world.

South Korea Power Company Hack (2019): South Korea's nuclear power operator, Korea Hydro and Nuclear Power, suffered a data breach that exposed confidential information, including blueprints and manuals for two nuclear reactors. The hackers were reportedly from North Korea.

Australian Energy Company Hack (2019): It was reported that a Chinese hacking group had successfully infiltrated the computer systems of an Australian energy company. The group was able to steal confidential information, including technical details about the company's power generation and distribution systems.

Israel Power Grid Hack (2020): Israel's electricity authority was targeted by a cyberattack that disrupted its operations for several hours. The authorities said that the attack was carried out by a nation-state actor but did not disclose the identity of the perpetrator.

Indian Power Grid Attack (2020): The power distribution company of the Indian state of Maharashtra experienced a ransomware attack that affected the company's internal IT systems, communication networks, and email servers, resulting in significant operational disruption. The attackers demanded a ransom in exchange for restoring access to the company's systems and decrypting its data.

The incidents that highlight the increasing threat of cyberattacks targeting critical infrastructure systems underline the need for robust cyber security measures to protect against such attacks and increased investment in cyber security measures and greater awareness of the need to protect critical infrastructure from cyber threats.

1.5 Need of sustainable/renewable-integrated energy grids for cyber security

The integration of sustainable and renewable energy sources, such as wind and solar power, into energy grids is a growing trend with important implications for cyber security. As digital technologies, including IoT devices and cloud-based platforms, become more prevalent in the energy sector, the risk of cyberattacks also increases. The integration renewable energy sources into power grids can provide a number of benefits for cyber security [23].

The integration of sustainable and renewable energy sources into power grids can help to improve cyber security by reducing the risk of large-scale disruption caused by a single point of failure, increasing resilience and redundancy, promoting advanced monitoring and automation technologies, and improving cyber security standards. The prominent advantages renewable-integrated energy grids in cyber security are as follows [2–5], [8,11,17,24,25]:

Decentralization of power generation: By reducing reliance on centralized fossil-fuel-based power plants, which are more vulnerable to cyberattacks, the integration of renewable energy sources can help to improve cyber security in energy grids. Decentralized, renewable energy systems are often more resilient to cyberattacks because they are distributed and can continue to operate even if certain parts of the system are compromised. Renewable energy systems are often distributed and decentralized, with power generated from multiple sources located in different geographic locations. This reduces the risk of large-scale disruption caused by a single point of failure, such as a cyberattack on a centralized power plant.

Resilience and redundancy: Renewable energy systems can often be designed with built-in redundancy, which helps to ensure that power continues to be delivered even in the event of an attack or other disruption.

Increased monitoring and automation: Renewable energy systems often rely on advanced monitoring and automation technologies, which can help to detect and respond to cyber threats more quickly and effectively.

Improved cyber security standards: As renewable energy systems become more integrated into power grids, cyber security standards are likely to improve, as governments and companies recognize the need to protect these critical infrastructure assets.

Reduced dependence on fossil fuels: By promoting the use of renewable energy sources, sustainable energy grids can help to reduce dependence on fossil fuels, which are often located in politically unstable regions and are vulnerable to supply disruptions caused by cyberattacks or other factors.

To achieve the complete realization of the advantages of sustainable and renewable-integrated energy grids for cybersecurity, a collaborative effort among governments, energy companies, and cyber security experts is essential to invest in and develop the necessary technologies and infrastructure. Achieving this objective will necessitate interdisciplinary collaboration among various sectors, such as energy, cybersecurity, and information technology, and a dedicated effort to emphasize cyber security in the planning and execution of energy systems.

In order to tackle the challenges, it is crucial to enforce cyber security measures throughout the energy grid, encompassing both the renewable energy systems and the central control systems responsible for managing energy distribution. This can include measures such as encryption, intrusion detection systems with machine learning technics, and regular security audits and updates. It is also important to educate stakeholders about the risks of cyberattacks and the measures that can be taken to mitigate them, including training employees and working closely with government agencies to develop coordinated response plans in the event of an attack. Additionally, sustainable and renewable-integrated energy grids can also benefit from enhanced security measures, such as the use of blockchain technology to create secure, decentralized databases of energy transactions, and the implementation of secure communication protocols to prevent unauthorized access to critical infrastructure.

1.6 Risks across legacy power systems grid: Generation, transmission, and distribution

Legacy power systems, which are typically characterized by their centralized, top-down design and heavy reliance on fossil fuels, can carry risks across all stages of electricity generation, transmission, and distribution. Legacy power systems are susceptible to various risks such as physical attacks, natural disasters, technical failures, and cyberattacks. These risks can cause disruptions, power outages, and other problems that can impact the reliability, security, and stability of the grid. Some prominent risks that can arise in each stage are presented below.

Aging infrastructure: Many power grids were built decades ago and are reaching the end of their useful life. As a result, components may begin to fail or become less efficient over time, leading to increased downtime and maintenance costs.

Cost risks: Legacy power systems can be expensive to operate and maintain, particularly if they rely on aging equipment or require significant investments in infrastructure to upgrade or modernize.

Environmental risks: Legacy power systems are often powered by fossil fuels, which can generate greenhouse gas emissions and contribute to climate change. In addition, the extraction and transport of fossil fuels can have negative environmental impacts.

Economic and political instability: Economic and political instability can also impact power grids. In some cases, governments may not have the resources to maintain or upgrade infrastructure, leading to increased downtime and outages. In other cases, political unrest could lead to deliberate attacks on power systems or disruptions to the supply of fuel or other resources needed to generate power.

Reliability risks: Legacy power systems often rely on a small number of large generators and transmission lines to supply power over long distances. This can make the system vulnerable to disruptions caused by equipment failures, natural disasters, or other events.

Grid stability risks: As renewable energy sources are increasingly integrated into legacy power systems, the variability of these sources can create challenges for grid stability and reliability. This requires careful planning and management to ensure that the grid can balance supply and demand in real time.

Interference with other systems: Legacy power systems can sometimes interfere with other systems, such as communication systems, due to electromagnetic interference or radio frequency interference. This can lead to communication disruptions, data loss, and equipment failures.

Limited visibility and control: Legacy power systems often lack the advanced monitoring and control systems found in the smart grid. This can make it difficult for grid operators to detect problems and respond quickly, which can lead to longer outages and higher costs.

Cyber security risks: As legacy power systems are increasingly connected to the Internet, they are exposed to various cyber security risks, which stem from a range of factors such as outdated hardware and software, insufficient encryption, inadequate access controls, limited incident response plans, and the absence of proper security assessments and audits.

As a result, the risks associated with legacy power systems include potential power outages, damage to critical infrastructure, and harm to the public due to physical risks such as equipment failure, natural disasters, and physical attacks. Cyber risks such as malware infections, ransomware attacks, and unauthorized access to critical systems can compromise the CIA triad of data and systems. The aging infrastructure and outdated technology of legacy power systems rising these risks by potentially making them more vulnerable to disruptions and cyberattacks.

1.7 Cyber physical asset in power system and part of vulnerabilities

The smart grid assets are the various interconnected devices and systems in a smart grid that combine physical and cyber components to manage and optimize the generation, distribution, and consumption of energy. These assets include a range of devices and systems such as sensors, smart meters, communication networks, power generation systems, and control systems.

These devices and systems are integrated into the smart grid to collect, analyze and use real-time data to make more informed decisions about energy generation and distribution. For example, smart meters installed in homes and businesses provide real-time data about energy consumption that can be used to optimize the distribution of energy across the grid. Energy management systems can also use data from sensors to optimize energy generation based on demand, weather, and other factors.

The integration of cyber-physical assets in the smart grid has significantly increased the efficiency and reliability of energy delivery. However, it has also introduced new vulnerabilities that can be exploited by cyber attackers. Cyber physical assets in power systems are susceptible to a range of security vulnerabilities, stemming from the convergence of cyber and physical systems.

Therefore, it is essential to secure these assets with appropriate cyber security measures to prevent cyberattacks that could disrupt the grid's operations and potentially cause significant damage. Some common vulnerabilities include inadequate access controls, lack of encryption, outdated software and hardware, insufficient incident response plans, and the absence of proper security assessments and audits. These vulnerabilities can be exploited by cyber attackers to disrupt power systems, causing physical damage, power outages, and even harm to the public. Moreover, as power systems increasingly rely on Internet-connected devices and new digital technologies, the attack surface for cyber threats continues to expand, making it more challenging to secure cyber-physical assets in power systems.

The assets are often spread across different geographic locations and operate on various communication protocols, which can make it difficult to implement consistent and comprehensive security measures. So, specific cyber security requirements must be met according to the characteristics of each asset. Here are prominent and generic key security considerations for the smart grid about asset vulnerabilities [13,15,25–29]:

Secure communication: The smart grid assets rely on communication networks to transmit data and control signals. Implementing secure communication protocols, such as encryption and authentication, can help protect against eavesdropping and unauthorized access to the network.

Encryption: Data encryption is an essential security measure that helps protect against data interception and manipulation. Lightweight encryption should be implemented for all sensitive data that are transmitted or stored within the assets.

Intrusion detection and prevention: Intrusion detection and prevention systems (IDS/IPS) are a security technology that can be used to identify potential vulnerabilities in assets within the smart grid. They are designed to monitor the network traffic and analyze it for potential security breaches, including attempts to exploit known vulnerabilities in smart grid assets.

Access control: Smart grid assets should be protected by strong access controls to ensure that only authorized personnel can access and modify the system. This includes both physical and digital access controls, such as access cards and passwords.

Incident response plan: Developing an incident response plan can help detect and respond to security incidents affecting smart grid assets, including procedures for reporting incidents, mitigating threats, and recovering from disruptions.

Security monitoring: Continuously monitoring the smart grid system for unusual activity and security events can help identify potential security threats and vulnerabilities.

Regular security assessments: Regular security assessments and audits can help identify and address vulnerabilities in the smart grid system.

Supplier risk management: Third-party vendors and contractors involved in the smart grid operations should have adequate security measures in place to prevent security incidents that could impact the smart grid assets.

Patch management: The smart grid relies on software and firmware to function. Regular patch management helps ensure that the latest security updates and fixes are applied to all assets.

Training and awareness: Regular training of employees on security best practices and the importance of maintaining the confidentiality, integrity, and availability of smart grid assets can help prevent human error and improve overall system security. Finally, personnel working with the smart grid must receive regular training and awareness programs to understand the latest cyber security threats and best practices for mitigating them.

In conclusion, securing cyber physical assets in the smart grid require a multilayered and collaborative approach that involves the identification and mitigation of potential vulnerabilities, the implementation of appropriate security controls, and the use of intrusion detection systems. By addressing these vulnerabilities and enhancing the security of cyber physical assets, the smart grid can become more resilient and better equipped to withstand potential cyberattacks.

1.8 Path forward and latest development

Smart grid security requires a holistic approach that considers both technical and nontechnical factors. State-of-the-art interoperability and cyber security solutions are essential for the success of the smart grid. As technology advances, it is important to stay up-to-date with the latest standards and best practices to ensure the continued success of the smart grid. Smart grid cyber security is a critical concern as the smart grid infrastructure becomes increasingly interconnected and reliant on technology.

The path forward involves the implementation of robust security measures to protect the smart grid from cyber threats. To enhance the cyber security of the smart grid, a path forward must involve a multilayered approach that addresses various aspects of the system. The following are some prominent key steps for the path forward:

Risk assessment: Conduct a comprehensive risk assessment to identify potential vulnerabilities, threats, and consequences associated with cyberattacks on smart grid assets. This includes evaluating the security posture of existing cyber physical assets, as well as the potential impact of emerging technologies.

Security policies and standards: A comprehensive cyber security policy framework should be established that sets standards and guidelines for the protection of cyber physical assets in the smart grid. The policies should be regularly reviewed and updated to keep pace with evolving threats. National Institute of Standards and Technology (NIST) and the North American Electric Reliability Corporation (NERC) have developed security standards for the smart grid.

Regular security assessments: Regular security assessments, such as penetration testing and vulnerability scans, should be conducted to identify and mitigate potential vulnerabilities in the system. This can help to identify and prioritize security measures that need to be implemented.

Appropriate security controls: Appropriate security controls, such as access controls, encryption, and intrusion detection systems, should be implemented to protect cyber physical assets in the smart grid. These controls should be regularly reviewed and updated to ensure their effectiveness.

Workforce development: Developing a skilled cyber security workforce is essential for maintaining the security of the smart grid. Investment in education and training programs can help to build a pipeline of cyber security professionals who can help secure the smart grid.

Awareness training: Training and education programs should be developed for personnel who operate and maintain the smart grid. This can help to raise awareness about potential cyber security threats and improve the overall security culture within the organization.

Collaborating and stakeholders: Collaboration with stakeholders, such as energy companies, government agencies, and cyber security experts, is critical to developing a comprehensive cyber security strategy for the smart grid. Stakeholders should work together to share information, best practices, and resources to enhance the overall security of the system.

Developing and deploying emerging technologies: Emerging technologies, such as AI, ML, blockchain can be leveraged to enhance the cyber security of the smart grid. These technologies can help to automate security processes and improve the resilience of the system against potential cyberattacks. As smart grid technologies are developed and deployed, it is essential to ensure that cyber security is built-in from the beginning. This involves designing systems and devices with security in mind, testing for vulnerabilities and weaknesses, and continuously monitoring and updating security measures.

By following these steps, a path forward can be established to enhance the cyber security of the smart grid and protect the cyber physical assets that are essential to the proper functioning of the system. Securing the smart grid against cyber threats requires a multifaceted approach that includes enhanced security standards, collaboration and information sharing, emerging technologies, workforce development, and ongoing monitoring and updating of security measures. By implementing these strategies, the smart grid can become more resilient and better equipped to withstand potential cyberattacks.

One of the latest developments in the smart grid and cyber security is the increasing use of IoT devices to monitor and control various aspects of the grid. This has led to new security concerns, as many IoT devices are not designed with security in mind and can be vulnerable to cyberattacks. As a result, there is a growing focus on developing new security protocols and standards for IoT devices in the smart grid.

Another development is the use of advanced analytics and AI to detect and respond to cyber threats in real time. ML algorithms can be trained to recognize patterns in data that may indicate a cyberattack and can help to quickly identify and contain the threat before it causes damage.

In addition, there is a growing trend toward using blockchain technology to secure data and transactions in the smart grid. Blockchain provides a decentralized and tamper-proof means that can help to prevent unauthorized access, tampering, and other malicious activities that can compromise the integrity and confidentiality of data.

Furthermore, there is an increasing focus on supply chain security, as the smart grid relies on a complex network of vendors, suppliers, and contractors. Ensuring that all components and systems are secure and resilient is essential to maintaining the overall security of the smart grid.

Lastly, there is a growing emphasis on collaboration and information sharing between utilities, vendors, and government agencies to improve the overall cyber security of the smart grid. This includes sharing threat intelligence, best practices, and lessons learned to enhance the resilience and security of the system.

1.9 Conclusion

Cyber security plays a pivotal role in safeguarding the integrity and reliability of power systems, renewable-integrated energy grids, and smart grids. Its core objectives are to maintain the integrity, availability, and confidentiality of data and control systems, ensuring the continued dependable operation of these critical infrastructures. The rising number of reported cyberattacks on a global scale serves as a stark reminder of the escalating threat landscape in these sectors, with risks spanning service disruptions to potential national security concerns. Safeguarding the cyber-physical assets within power systems, including those in smart grids, is paramount to prevent physical damage resulting from cyberattacks. As humankind reliance on these interconnected and digitally dependent systems continues to grow, fortifying cyber security measures is essential for maintaining stability and resilience in evolving energy landscape.

References

[1] Gündüz, M.Z., Daş, R., 2018. Internet of things (IoT): evolution, components and applications fields. Pamukkale University Journal of Engineering Sciences 24 (2), 327–335.

[2] Das, R., Gündüz, M.Z., 2019. Analysis of cyber-attacks in IoT-based critical infrastructures. International Journal of Information Security Science 8 (4), 122–133.

[3] Dileep, G., Feb. 2020. A survey on smart grid technologies and applications. Renewable Energy 146, 2589–2625. https://doi.org/10.1016/j.renene.2019.08.092.

[4] Khoei, T.T., Slimane, H.O., Kaabouch, N., Nov. 2022. Cyber-security of smart grids: attacks, detection, countermeasure techniques, and future directions. Communications and Network 14 (4). https://doi.org/10.4236/cn.2022.144009. Art. no. 4.

[5] Gündüz, M.Z., Das, R., 2020. Cyber-security on smart grid: threats and potential solutions. Computer Networks 169, 107094. https://doi.org/10.1016/j.comnet.2019.107094.

[6] Karthika, D., Jan. 2023. A study on artificial intelligence for monitoring smart environments. Materials Today: Proceedings 80, 2009–2013. https://doi.org/10.1016/j.matpr.2021.06.046.

[7] Zhao, N., Zhang, H., Yang, X., Yan, J., You, F., Feb. 2023. Emerging information and communication technologies for smart energy systems and renewable transition. Advances in Applied Energy 9, 100125. https://doi.org/10.1016/j.adapen.2023.100125.

[8] Gopstein, A., Nguyen, C., O'Fallon, C., Hastings, N., Wollman, D., 2021. NIST Framework and Roadmap for Smart Grid Interoperability Standards, Release 4.0. National Institute of Standards and Technology, Gaithersburg, MD. https://doi.org/10.6028/NIST.SP.1108r4. NIST SP 1108r4.

[9] Gündüz, M.Z., Das, R., 2021. Cyber-attack process and defense strategies in smart grid applications. Journal of Network Communications and Emerging Technologies 11 (11).

[10] Ghiasi, M., Niknam, T., Wang, Z., Mehrandezh, M., Dehghani, M., Ghadimi, N., Feb. 2023. A comprehensive review of cyber-attacks and defense mechanisms for improving security in smart grid energy systems: past, present and future. Electric Power Systems Research 215, 108975. https://doi.org/10.1016/j.epsr.2022.108975.

[11] Hasan, M.K., Habib, A.A., Shukur, Z., Ibrahim, F., Islam, S., Razzaque, M.A., Jan. 2023. Review on cyber-physical and cyber-security system in smart grid: standards, protocols, constraints, and recommendations. Journal of Network and Computer Applications 209, 103540. https://doi.org/10.1016/j.jnca.2022.103540.

[12] Singh, J., Wazid, M., Das, A.K., Chamola, V., Guizani, M., Aug. 2022. Machine learning security attacks and defense approaches for emerging cyber physical applications: a comprehensive survey. Computer Communications 192, 316–331. https://doi.org/10.1016/j.comcom.2022.06.012.

[13] Sun, C.-C., Hahn, A., Liu, C.-C., Jul. 2018. Cyber security of a power grid: state-of-the-art. International Journal of Electrical Power and Energy Systems 99, 45–56. https://doi.org/10.1016/j.ijepes.2017.12.020.

[14] Raj Kumar, G.V.S., Sai Prasad, A.V.H., Padma, B., Raj Koti, B., 2022. Smart Grid communication and information technologies for cyber security, data privacy, and policy issues. In: Deebak, B.D., Al-Turjman, F. (Eds.), Sustainable Networks in Smart Grid. Academic Press, pp. 1—29. https://doi.org/10.1016/B978-0-323-85626-3.00008-9.

[15] Habib, A.A., Hasan, M.K., Alkhayyat, A., Islam, S., Sharma, R., Alkwai, L.M., Apr. 2023. False data injection attack in smart grid cyber physical system: issues, challenges, and future direction. Computers and Electrical Engineering 107, 108638. https://doi.org/10.1016/j.compeleceng.2023.108638.

[16] Ahmetoglu, H., Das, R., 2022. A comprehensive review on detection of cyber-attacks: data sets, methods, challenges, and future research directions. Internet of Things 20, 100615. https://doi.org/10.1016/j.iot.2022.100615.

[17] Ma, C., Nov. 2021. Smart city and cyber-security; technologies used, leading challenges and future recommendations. Energy Reports 7, 7999—8012. https://doi.org/10.1016/j.egyr.2021.08.124.

[18] Shakarian, P., Shakarian, J., Ruef, A., 2013. Attacking Iranian nuclear facilities: Stuxnet. In: Shakarian, P., Shakarian, J., Ruef, A. (Eds.), Introduction to Cyber-Warfare. Syngress, Boston, pp. 223—239. https://doi.org/10.1016/B978-0-12-407814-7.00013-0.

[19] Sánchez, H.S., Rotondo, D., Escobet, T., Puig, V., Quevedo, J., Jan. 2019. Bibliographical review on cyber attacks from a control oriented perspective. Annual Reviews in Control 48, 103—128. https://doi.org/10.1016/j.arcontrol.2019.08.002.

[20] Daricili, A.B., 2018. Cyber Attacks Against Turkey Right After Shotting down Russian Su-24 Aircraft. Uludağ Journal of Economy & Society 37 (2), 1—10.

[21] Sullivan, J.E., Kamensky, D., Apr. 2017. How cyber-attacks in Ukraine show the vulnerability of the U.S. power grid. The Electricity Journal 30 (3), 30—35. https://doi.org/10.1016/j.tej.2017.02.006.

[22] Tam, T., Rao, A., Hall, J., Oct. 2021. The good, the bad and the missing: a narrative review of cyber-security implications for Australian small businesses. Computers and Security 109, 102385. https://doi.org/10.1016/j.cose.2021.102385.

[23] Tang, D., Fang, Y.-P., Zio, E., Jul. 2023. Vulnerability analysis of demand-response with renewable energy integration in smart grids to cyber attacks and online detection methods. Reliability Engineering and System Safety 235, 109212. https://doi.org/10.1016/j.ress.2023.109212.

[24] Pour, M.M., Anzalchi, A., Sarwat, A., 2017. A review on cyber security issues and mitigation methods in smart grid systems. In: Southeast Con 2017, pp. 1—4. https://doi.org/10.1109/SECON.2017.7925278.

[25] Berghout, T., Benbouzid, M., Muyeen, S.M., Sep. 2022. Machine learning for cybersecurity in smart grids: a comprehensive review-based study on methods, solutions, and prospects. International Journal of Critical Infrastructure Protection 38, 100547. https://doi.org/10.1016/j.ijcip.2022.100547.

[26] Demirol, D., Das, R., Hanbay, D., 2023. A key review on security and privacy of big data: issues, challenges, and future research directions. Signal, Image and Video Processing 17 (4), 1335—1343. https://doi.org/10.1007/s11760-022-02341-w.

[27] Singh, N.K., Majeed, M.A., Mahajan, V., Dec. 2022. Statistical machine learning defensive mechanism against cyber intrusion in smart grid cyber-physical network. Computers and Security 123, 102941. https://doi.org/10.1016/j.cose.2022.102941.

[28] Wang, W., Lu, Z., Apr. 2013. Cyber security in the smart grid: survey and challenges. Computer Networks 57 (5), 1344—1371. https://doi.org/10.1016/j.comnet.2012.12.017.

[29] Zhuang, P., Zamir, T., Liang, H., Jan. 2021. Blockchain for cybersecurity in smart grid: a comprehensive survey. IEEE Transactions on Industrial Informatics 17 (1), 3—19. https://doi.org/10.1109/TII.2020.2998479.

CHAPTER 2

Smart grids: Sources of security intimidations, challenges, and opportunities

Tapsi Nagpal and Prateema Sharma
Lingaya's Vidyapeeth, Faridabad, Haryana, India

2.1 Introduction

The electrical grid plays a crucial role in facilitating economic and social advancement by serving as a vital infrastructure for generating, transmitting, and distributing electricity. The main purpose of this is to disperse electricity throughout a certain area [1–4]. Despite being greatly relied upon by consumers and acknowledged as one of the greatest technical accomplishments of the 20th century, current methods of energy delivery lack flexibility.

Electricity is distributed to users using contemporary delivery systems, comprised of several transmission and distribution networks [5]. Nevertheless, this extensive and intricate system struggles to meet the increasing need for flexible, modifiable, and instantaneous capabilities. The lack of inherent self-correcting features, which are crucial in modern dynamic environments, implies that its ongoing functionality mostly relies on human intervention [6]. Despite continuous improvements, a comprehensive overhaul of the infrastructure is necessary to accommodate the increasing need for intelligent systems.

The implementation of the smart grid (SG) represents a fundamental shift in the method of electricity distribution. The SG incorporates advanced sensing and communication technology to enhance power distribution procedures [7]. The SG provides a bidirectional architecture that facilitates the simultaneous movement of information and electricity, in contrast to the traditional unidirectional grid [8]. This evolution ensures that everyone has access to information, enhances automation, promotes

Cyber Security Solutions for Protecting and Building the Future Smart Grid
ISBN 978-0-443-14066-2
https://doi.org/10.1016/B978-0-443-14066-2.00012-8

customer engagement, and fosters more cooperation in the functioning of the grid [7].

The use of state-of-the-art sensing and monitoring technologies by the SG diminishes the need for human intervention. The grid evolves to possess enhanced demand response, autonomous repair capabilities, and effortless integration of renewable sources. These advancements not only expand the choices available to providers and consumers, but they also enable the use of technologies that are now incompatible with our existing grid infrastructure [9]. Given the pressing need to tackle climate change, the Secretary-General proposes a feasible strategy to decrease carbon emissions in the power sector and seamlessly integrate renewable energy sources. The existing infrastructure may become insufficient to accommodate the growing demand, particularly in light of the projected 70% rise in energy output by 2040. The rise in demand emphasizes the pressing need to provide a dependable and sustainable electrical infrastructure capable of managing future challenges [10].

The objective of this chapter is to provide a comprehensive overview of the latest SG research, along with clear evaluations of different study directions. Our objectives are to gather the most up-to-date contributions, showcase the advancements in the area, and provide valuable perspectives on the future trajectory of the SG. The subsequent questions in the survey are as follows: Section 2.2 examines pertinent surveys in the field. Section 2.3 offers a comprehensive account of SG's historical background. Section 2.4 provides comprehensive coverage of security problems, threat categorizations, and recommended treatments. Section 2.6 examines the incorporation of renewable energy, whereas Section 2.5 discusses the prerequisites and challenges of interoperability. Section 2.8 concludes the topic, while Section 2.4 provides insights on the usage and prospects of the SG industry.

2.2 Introduction to smart grid

2.2.1 Conventional systems

The current system of physical infrastructure used to distribute energy is referred to as the electric power grid [11]. The electric grid consists of three main systems: power distribution, transmission, and generation. The integrated network comprises power plants for electricity production, substations for voltage transformation, and distribution facilities for energy distribution from substations to customers. Its purpose is to provide electricity to consumers. With the emergence of future technologies such as better metering and remote monitoring, the existing physical infrastructure will become more unsustainable. This is because it lacks automation and heavily relies on personnel services for maintenance and repairs. The current electric system poses several dangers due to outdated infrastructure and little resilience to disruptions, despite its significant dependence. Furthermore, it exhibits nonlinearity and a high level of intricacy [12].

The first and crucial action to address the increasing demand for environmentally sustainable policies is the incorporation of renewable energy sources into the power system. Regrettably, this step will provide extra challenges for controllers at all levels of the electrical grid, hence significantly complicating the already intricate system [12]. This clearly indicates that there will soon be a need for either a substantial increase in the electric infrastructure or a more environmentally friendly alternative.

2.2.2 Smart systems

In order to enhance the level of automation and effectiveness, intelligent systems incorporate advanced technology into existing processes and systems. They assess circumstances and make flexible and anticipatory decisions by leveraging combined functions such as sensing, actuating, and controls. Intelligent systems have the ability to acquire knowledge, engage in logical thinking, control their behavior, and perceive both their environment and their own state [13].

They possess the ability to interact with one other and with other components of the system [13]. The systems' capacity to consistently observe their environment allows them to update their internal information, which is essential for making educated judgments [13]. Intelligent systems could engage in reasoning, enabling them to adapt to dynamic situations and environmental objectives [14].

Smart systems provide several benefits and serve various purposes, including enhanced functionality, interconnectedness, and security. The integration of smart technology into the current electrical infrastructure will enable future modernization and the solutions of many developing issues.

2.2.3 Smart grid

The SG, which integrates intelligent gadgets with the current electrical grid, represents a novel and modernized grid system. By using a bidirectional flow of power and information, it establishes an automated delivery

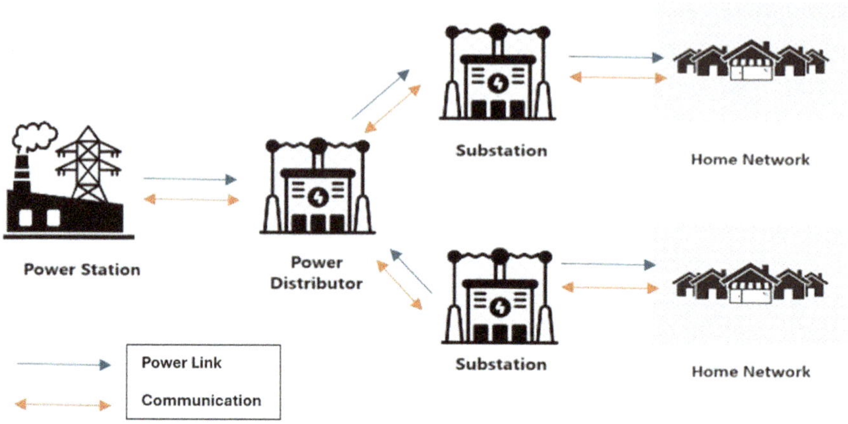

Figure 2.1 Basic SG [19].

method [15]. The objectives of this initiative are to improve demand-side management and energy efficiency, as well as to enhance grid safety via the use of self-healing systems [8]. The system incorporates many bidirectional intelligent devices, including meters, actuators, and sensors, to provide precise and real-time monitoring, balancing, and control always [13,16].

The SG has the capability to generate, transmit, and store options, whereas the conventional power grid is limited to the distribution of electrical energy. This enables the SG to improve the existing power infrastructure and facilitate collaborative, responsive, and natural operations [17]. The system integrates many advanced technological components together with enhanced monitoring, allowing a clearer and more comprehensive understanding of the actions and interactions of suppliers and customers [18]. The basic architecture of an SG is seen in Fig. 2.1, illustrating both the sequential power distribution and bidirectional information exchange [19].

2.2.3.1 Regulating and observing

ICT is used to operate, supervise, and control the grid. By using ICT, energy suppliers may effectively manage power demand, leading to a more efficient and reliable power supply [20]. The SG utilizes bidirectional links between electric power providers and customers to strategically plan and execute efficient operations using data acquired from consumers [21]. SG utilizes distributed computer agents to provide self-healing power system networks, central/grid control protection systems, grid computing, and extensive diagnostic monitoring of transmission equipment. Therefore, SG can be considered intelligent [22]. These technologies may be enhanced by using a Supervisory Control and Data Acquisition (SCADA) system [22]. SCADA, is a specialized control system that is used for the remote monitoring, operation, and management of critical equipment and processes. SCADA systems collect, transmit, and analyze data from sensors and equipment in the field. By providing up-to-date information on the condition of processes and equipment, they enable operators to make educated decisions, initiate control actions, and respond to warnings and malfunctions.

2.2.3.2 Smart meters

Smart meters are another unique component of the SG which enables advanced metering infrastructure (AMI). This allows for more data to be collected and in a much more effective manner. Smart meters collect data every minute, whereas old metering data was recorded hourly or monthly. To improve metered grid data (i.e., voltage and current phasors) accuracy at more frequent sampling intervals, phasor measurement units (PMUs) can be applied. The PMU's implementation into SG has been extensively researched (see Ref. [23]), significantly contributing to the SG's advancements in the control, estimation, and security of power grids. As a comparison, modern PMUs collect up to 60 data points per second, whereas the current SCADA systems collect a single data point every 1−2 s [12]. Applying the AMI in the power distribution system and the PMU in the power transmission grid will provide the power grid a much more in-depth look at grid performance compared to the data available from SCADA technology [12]. Incorporating the use of additional smart technology in consumer's homes will expand control and monitoring of multiple devices connected to the power grid.

2.2.3.3 Intended benefits

The integration of SGs is expected to provide several advantages, as first outlined by the National Institute of Standards and Technology (NIST) [24], and later redefined by Fang et al. [8]. The advantages are classified and explained as follows.

- Enhancing the dependability and excellence of the power grid
    1. Enhancing the caliber and dependability of electricity transmission
    2. Enhancing the ability to withstand and recover from disturbances in the grid, such as power outages and data interruptions.
    3. Introducing possibilities for enhancing grid security

- Enhancing the efficiency and optimizing the performance of the grid
  1. Maximizing the efficiency of facility use, thereby preventing the need for further power plant development.
  2. Improving the capability and effectiveness of current power grids
  3. Implementing automation for maintenance and varied activities.
- Grid modernization and adaptation
  1. Facilitating autonomous repair processes and proactive maintenance in response to disruptions.
  2. Accelerating the implementation of renewable energy sources
  3. Provision for accommodating distributed power sources
  4. Implementing uniform standards for emerging energy storage options and facilitating the shift toward plug-in electric cars.
- Environmental sustainability and efficiency
  1. Mitigating the release of greenhouse gases by promoting the widespread use of electric vehicles and harnessing contemporary energy sources.
  2. Reducing the production of energy in an inefficient manner at times of high demand, and thus decreasing the use of oil.
- Improvements related to the market and consumers.
  1. Enhancing the range of options available to consumers.
  2. Facilitating the implementation of new markets, services, and products

The advantages, as outlined by NIST in Ref. [24], provide a multitude of benefits to the people, nations, and corporations who embrace SGs. By integrating environmental, security, and operational advantages into the power grid, customers may anticipate a greater range of options when it comes to managing their energy usage. Additionally, they can look forward to the emergence of new markets, services, and products. Implementing Sustainable Goals (SGs) is crucial for decreasing reliance on foreign oil and promoting renewable energy production in the United States (US) and other nations. This will result in job creation and the development of new commodities, manufacturing techniques, and other advancements stemming from research in this field [24]. The first four primary items on the aforementioned list are intimately correlated with the latter three segments of this poll, namely cybersecurity, interoperability, and the integration of renewable energy in smart grids. Directing research endeavors toward incorporating these elements into the electrical grid will persistently provide the specified advantages, which are crucial for promoting sustainable living and securing our future. In 2008, the

Department of Energy (DOE) noted that a 5% increase in efficiency in existing power networks would result in energy savings equivalent to reducing the emissions of 53 million autos.

2.3 Challenges (security)

With the advancement of electrical networks, their susceptibility to various threats also grows [25]. There are several risks and threats associated with the integration of digital communications. These include cybersecurity and data privacy concerns, as well as issues related to dependability and technical failures. Additionally, changes in the way power companies and their customers interact can also pose risks and dangers [12].

The SG facilitates extensive engagement at every level of the electricity distribution, consumption, generating, and transmission systems. This indicates a significant amount of coordination and communication among the many tiers of the power grid system. This tight relationship creates several additional chances and a higher likelihood of cyberattacks and cascade failures, which have the potential to impact all the previously stated systems [12]. This has the potential to result in severe and disastrous outcomes, such as instability in the energy market.

Infrastructure failures, power outages, risks to human well-being, and impaired consumer devices [12]. There is a risk of experiencing more frequent but less severe repercussions, such as small-scale outages [12]. Given the importance of the electric grid, it is imperative to develop and deploy defensive systems that can effectively monitor large volumes of data, assess the state of the system, detect malfunctions, anticipate threats, and provide appropriate remedies [12].

According to Gunduz and Das [26], SG is regarded as one of the most significant implementations of the Internet of Things (IoT). An IoT network allows objects to connect either directly with one other or through an internet gateway. IoT-based SG systems are extensive and essential infrastructures that have intricate architectures and are susceptible to assaults. Using IoT applications is convenient for SGs, but it also presents many risks. Given that the monitoring and control activities are conducted using internet-based protocols, the SG may be attractive to hostile groups as vital infrastructure [26]. Considering this appeal, it is important to scrutinize vulnerabilities in the security systems of an SG component, as well as any cyber security risks inside its infrastructure [26,27].

2.3.1 Classifying types of attacks

Given the vastness of the system, including a wide geographical area and intricate infrastructure, it would be exceedingly challenging to detect and categorize every potential assault on an individual basis. Instead, three primary security objectives are established and must be integrated into SG. The three security objectives, namely confidentiality, integrity, and availability, are commonly referred to as the CIA triad. This may be shown in Fig. 2.2. The CIA trinity is crucial for effectively managing, operating, and safeguarding system and communication infrastructures [28].

Cyberattacks in SGs aim to target at least one of the objectives of the CIA triad, either to exploit information for the benefit of the attackers or to do harm to others [26]. Certain assaults are strategically planned to take advantage of many components of the SG and execute simultaneous attacks. A coordinated assault is the most difficult to fight against since it may overcome typical defenses. To counter such attacks, many layers of security solutions with strong methods are required [26], as discussed in the analysis conducted in Refs. [29—31]. These attacks specifically aim at compromising all the elements, prerequisites, and security goals within an SG [26].

These cyberattacks can happen at many communication tiers inside SG [32]. The network/transport layer, the MAC layer, and the physical layer are each impacted in distinct ways by different types of assaults [26]. Being able to recognize different categories of cyber security risks allows for the implementation of effective and suitable countermeasures [33]. Effective security solutions for the SG may be established by examining the cybersecurity needs based on different attack types and the impacted communication levels [26].

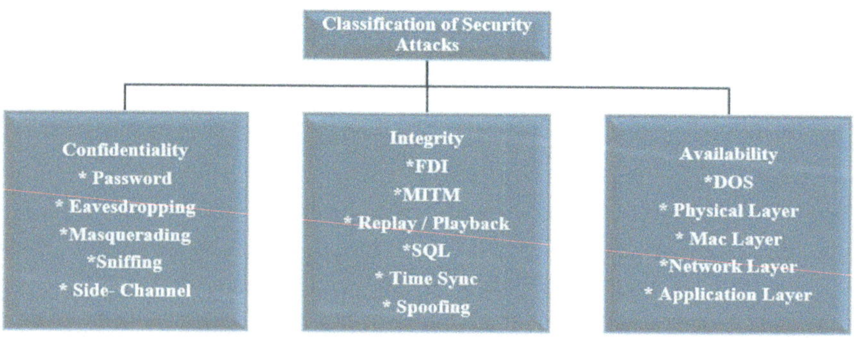

Figure 2.2 Classification of security attacks.

2.3.2 Confidentiality attacks

Confidentiality attacks aim to preserve the integrity of transferred information across power networks. However, these assaults are employed with the aim of gaining illegal access to information [34]. This data comprises the customer's account particulars and electricity use. In the event that hostile entities acquire consumer account information, they might potentially engage in activities such as identity theft, privacy breaches, phishing, and the sale of client information. By acquiring power use data, individuals might potentially engage in fraudulent billing and theft of electricity.

Confidentiality attacks in SG often have negligible impact on communication functionality. Nevertheless, the significance of safeguarding client privacy has garnered more attention with the advancement of technology. The following table, Table 2.1, provides a description of the techniques discussed These attacks include password assaults, masquerade attacks, and side-channel attacks.

2.3.3 Integrity attacks

Integrity-targeting attacks seek to alter or disrupt the transmission of data inside SG [32], potentially resulting in safety concerns for both equipment and individuals [37]. Integrity attacks specifically target the substance of the data that was initially transmitted, such as billing data, account data, control directives, sensor and voltage values, and functioning status of devices. These attacks involve intercepting the data and then modifying, rearranging, or delaying it [26,38,39]. Data integrity checks are frequently

Table 2.1 Types and description of confidentiality attacks.

Type of attack	Description
Password attack	Unauthorized attempts to obtain access by utilizing the password of an authorized individual.
Eavesdropping attack	Intercept wireless communication on a local area network (LAN), capture IP packets in SG networks or secretly listen in on messages exchanged between nodes.
Masquerade attack	Attackers fake authority to gain access to privileges.
Side channel attack	Implemented to obtain cryptographic keys.
Sniffing attack	Conducted in order to get TCP/IP packets or PMU messages that are sent via a network.

employed to counteract these assaults as this data is both personal to customers and important to organizations. Additionally, there exist approaches that are resistant to data integrity assaults, as outlined in Refs. [40,41]. While attacks targeting data integrity are sometimes perceived as less "brute-force" compared to other approaches, they are more complex than assaults targeting other categories of the CIA trinity [32]. The following table, Table 2.2, provides a description of the attacks discussed in this study that specifically aim to compromise integrity.

Table 2.2 Types and description of integrity attacks.

Type of attack	Description
False Data Injection (FDI) attack	Erroneous data is entered into NAN measurements or smart meters. They compromise the reliability of monitoring and measuring subsystems, leading to an ensuing response of impaired decision-making throughout the SG network.
MITM attack	The action is carried out to disrupt the communication between the smart meter and data concentrator equipment.
SQL injection	Manipulates databases by injecting script commands or malicious queries into the database, leading to control over the system, alteration or deletion of data, and the insertion of altered data.
Replay/playback attack	This process involves retransmitting or delaying communications that have been obtained through masquerade assaults. Initiated to enable attackers to redirect energy to various spots and induce physical harm to the system.
Time sync attack	Focuses on collecting timing data in Singapore, mostly from PMUs and WAPMCs. The TSA has the potential to produce inaccurate location inaccuracies, which can lead to a false alert suggesting a problem. This false alarm can then induce disruptions in communication lines, leading in a series of failures.
Spoofing attack	A perpetrator can assume the identity of other devices by taking advantage of the accessibility of the address fields on the MAC layer, so gaining the capability to transmit deceptive information.

2.3.4 Availability attacks

Denial-of-service (DoS) attacks, also known as attacks targeting availability, have the objective of disrupting, obstructing, or delaying communications in SGs [32,45]. A DoS attack has the potential to hinder the functioning of electronic devices and significantly reduce the communication performance of a power system, as it is intended to be continually accessible [32]. Adversaries inundate the network's transmission channels with substantial amounts of traffic, leading to the loss of valid data packets and their subsequent failure to be processed [42]. The SG is susceptible to DoS attacks because to its utilization of the TCP/IP stack and IP protocols, which inherit vulnerabilities [46]. In order to ensure the availability of SG, it is necessary to implement advanced and effective countermeasures to protect against these specific assaults [26]. The following table, Table 2.3, provides a description of the methods discussed in this study that specifically aim to compromise availability.

The DoS attacks can happen at several communication tiers inside the SG, each with different impacts and levels of severity [32], as explained in Table 2.3.

Table 2.3 Types and description of availability attacks.

Type of attack	Description
DOS attacks	Executed with the intention of sabotaging, obstructing, or impeding SG communications.
Physical layer attacks	The purpose is to inundate wireless communication channels with disruptive signals, so impeding the connection between users and smart meters.
MAC layer attacks	Conducted with the intention of altering the MAC characteristics of a device to intentionally reduce the performance of other devices that share the same communication channel.
Network and transport layer attacks	These assaults are carried out with the purpose of initiating a decline in the overall performance of end-to-end communication.
Application layer attacks	It specifically refers to the deliberate attempt to deplete computer resources, such as the CPU or I/O bandwidth.

2.3.5 Attack security methods

Although there are established procedures to protect the SG against well-studied confidentiality, integrity, and availability threats, there has been a rise in research on emerging assaults, such FDI attacks, and more advanced protection strategies. The research studies considered in this area are summarized in Table 2.4, shown below.

2.3.5.1 Standard integrity attack guarding mechanisms

To carry out integrity or confidentiality assaults, the attacker must have gained authorized access to a user's sensitive information and communication networks [26,32]. To prevent different forms of integrity attacks, it is essential to implement end-to-end encryption and authentication mechanisms [26]. Sanjab et al. [44] propose the use of sequence numbers and timestamps as an effective countermeasure against replay attacks in SG applications. Sha sough et al. [47] describes cryptographic algorithms as techniques used to safeguard against data integrity assaults. Rawat and Bajracharya [34] propose many methods to enhance security in smart grid monitoring. These methods include enhanced use of the PMU, implementation of volt-VAR control-based schemes, utilization of power fingerprinting techniques, and avoidance of integrity attacks on data through trusted network connection-based ways.

By implementing the steps proposed by Bedi et al. in their study [43], the occurrence of SQL injection attacks may be greatly reduced in SG networks. The SQL protections listed include conducting penetration testing, doing static code checking, implementing positive pattern matching, validating input types, avoiding the usage of dynamic SQL, restricting remote user access to the database, and filtering semicolons during type verification [43].

To mitigate Man-in-the-Middle (MITM) attacks, security gateways can be employed to encrypt network traffic [26]. Security gateways provide the capability to establish VPN tunnels to facilitate network connectivity. They also provide data encryption at the source and decryption at the target [26]. Encryption operations usually take place within hardware solutions. In addition, it is crucial to authenticate both the target and source of information in order to prevent and halt MITM attacks, as demonstrated in Ref. [48]. Metke and Ekl [49] also explored the incorporation of reliable computing techniques and public key infrastructure (PKI), which links a user's identity to a public key through a digital certificate. The authors

Table 2.4 Summary of attack defense papers.

Year	Title	Focused attack type	Method used
2013	An analysis of smart grid attacks and countermeasures	Confidentiality	Data encryption
2015	Smart grid security issues	Confidentiality	Authentication process
2018	Review of IoT in electric power and energy systems	Confidentiality	Constructing discrete infrastructure for communication
2016	Smart grid security: Threats, challenges, and solutions	Integrity, replay	Timestamps and sequence numbers
2015	Smart grid cyber security: Challenges and solutions	Integrity	Cryptography
2015	Cyber security for smart grid systems: Status, challenges, and perspectives, in: Southeast	Integrity	Increased secure monitoring with PMU, volt-VAR control scheme, power fingerprinting, and reliable network-based methods
2018	Cyber-security in smart grid: Survey and challenges	Integrity, MITM	Authenticating source and target of information Including PKI and trusted computing methods
2010	Security technology for smart grid networks	Integrity	Authenticating source and target of information Including PKI and trusted computing methods

Continued

Table 2.4 Summary of attack defense papers.—cont'd

Year	Title	Focused attack type	Method used
2017	Real-time detection of false data injection attacks in smart grid: A deep learning-based intelligent mechanism	Integrity, FDI	Real-time detection using deep learning-based methods
2019	Cognitive dynamic system for control and cyber-attack detection in smart grid	Integrity, FDI	Entropic state
2019	Cognitive risk control for mitigating cyber-attack in smart grid	Integrity, FDI	Entropic state
2021	Detection of false data injection attacks in smart grid based on a new dimensionality-reduction method	Integrity, FDI	Statistical FDI attack Detection approach based on a novel dimensionality

argued that the use of PKI technologies is the optimal method for ensuring security in the SG, encompassing the establishment, administration, updating, and revocation of certificates, in addition to software and hardware enhancements [49].

2.3.5.2 Confidentiality attack guarding mechanisms

Methods such as data encryption [35], authentication processes [36], maximizing communication channel capacity, identifying saturated channels, and building separate infrastructure for power grid device communication [26] are effective in protecting against confidentiality attacks like eavesdropping, masquerade, spoofing, and side-channel attacks, as explained by Bedi et al. [43].

Data encryption and authentication mechanisms are often employed to prevent confidentiality threats in almost all online interactions. Gao et al. investigate a hybrid approach to data encryption in smart grids, where data is compressed and encrypted simultaneously in a single phase [50]. The EncryCS technique was shown to enhance both security and transmission efficiency [50]. In contrast, Syed et al. focused on utilizing homomorphic encryption to ensure the secure training of deep learning networks, specifically for the purpose of fault diagnosis and localization in smart grids [51]. Saxena et al. present a combined authentication/authorization scheme for confidentiality attacks that necessitates simultaneous verification of these processes [52]. Additionally, Chim et al. propose their Privacy-preserving Authentication Scheme for Smart grid networks (known as PASS), which is based on tamper-resistant devices [53].

2.3.5.3 Availability attack guarding mechanisms

Effective strategies for mitigating availability attacks include implementing techniques such as huge pipes, traffic filtering, utilizing air-gapped networks, and employing anomaly detection methods, as mentioned in Ref. [43]. Given that DoS assaults pose a significant risk to IoT-based smart grid systems, employing software solutions at the network layer might serve as an effective approach to preventing such attacks, as suggested in Ref. [26].

Both options serve as both security measures and anomaly detection solutions. Preventing availability attacks with air-gapped networks entails ensuring that there are no automated procedures or physical connections between systems. Scientists in Ref. [54] employ an air-gapped technique to identify and resolve such incursions without the need for controller contact.

Another technique, known as traffic filtering, is used to identify and mitigate availability threats by analyzing network traffic and removing any misleading or malicious signals. This approach was demonstrated in Ref. [55]. Jokar and Leung [56] did research on using machine learning approaches to avoid availability attacks on different tiers. They suggested a detection and prevention strategy based on ZigBee.

2.3.5.4 FDI attack guarding mechanisms

A comprehensive investigation has been carried out to counteract a wide range of different SG assaults. FDI assaults are a recent and highly perilous threat to SG, making them a hot subject of investigation.

He, Mendis, and Wei [57] introduced a real-time detection system that utilizes deep learning-based techniques. At the time of publishing, the authors assert that, based on their understanding, they were the pioneers in suggesting real-time procedures for detecting FDI attacks. By utilizing deep learning approaches, the researchers are able to identify the behavioral patterns of assaults by analyzing historical measurement data. They then apply these identified characteristics to detect real-time FDI attacks, resulting in effective implementation [57]. The simulation findings demonstrate the robustness of their detection technique against different levels of manipulated data.

The detection thresholds of the state vector estimator (SVE) and the intensity of ambient noise are closely related [57]. The findings also indicate that the utilized technique can achieve a high level of detection accuracy, even when occasional operation errors occur [57].

The concept of the entropic state was initially introduced for application in smart grids in Ref. [42], and then elaborated upon by the same authors in Ref. [58]. Within a cognitive dynamic system (CDS), the primary objective is to effectively manage the system's state while simultaneously reducing the quantity of unfamiliar input in the preceptor. The latter task is accomplished by reducing the level of disorder in the system, which is achieved through the dynamic optimization of the estimating process [42]. The authors propose the use of entropic states as a novel metric for the SG. This metric serves two main purposes: detecting FDI assaults and assessing the SG's health over successive cycles [42]. Oozeer and Haykin illustrated the capability of the entropic state to identify FDI threats and manipulate them into a manageable state through the implementation of cognitive risk management (CRC). By implementing task-switching management, the CDS successfully empowered a new executive with a distinct set of activities. This enhanced the system configuration and effectively reduced the risk to a reasonable level during an assault [42]. The entropic state is a measure of the level of uncertainty or information gap, and it may be used to identify an assault, as demonstrated in Ref. [58]. The simulations demonstrated the efficacy of their solution for future SG systems [58]. The CRC technique has been implemented in several systems, including vehicle-to-vehicle (V2V) communication systems [59]. Additionally, the CDS approach in control systems is now a prominent area of research.

Shi et al. [60] introduced a method for detecting statistical FDI attacks using a novel dimensionality reduction technique and a Gaussian mixture model. The authors suggested a method that consists of two stages: the first stage entails reducing the dimensionality, while the second stage employs a semi-supervised learning approach using the Gaussian mixture model [60]. If the output of the Gaussian mixture model surpasses a pre-established threshold, then FDI assaults are detected in addition to the regular pattern [60]. The simulation results from the testing show that the suggested strategy accurately recognized the FDI assaults with a high level of precision and achieved the intended discriminating performance [60].

2.4 Industry applications and future trends

2.4.1 Small scale implementation

2.4.1.1 Research/publications

Several universities worldwide have specialized laboratories focused on studying SG, including Canadian institutions such as York University [62], Toronto Metropolitan University [63], and the University of Wa-terloo [64], as well as the University of Melbourne in Australia [65]. These laboratories have undertaken extensive study and published several academic publications and surveys on various aspects of SG. With the growing desire for a more environmentally friendly way of supplying electricity, there is an increasing focus on research dedicated to SG.

2.4.1.2 Prototypes

In addition to doing research and producing academic publications, prototypes of several features of smart grids, such as demand side management [66] or remote monitoring [67], have been developed to design and evaluate the performance of smart grid components. Small-scale grids are optimal for study and experimentation with novel technologies, as their outcomes may be scaled up and their impacts on larger urban areas can be extrapolated. In recent times, a limited number of colleges and organizations have developed larger and more comprehensive prototypes.

The Microgrid Energy Management System (MG-EMS) prototype, which is shown in the Figure above, is a notable achievement in SG prototyping and has received an award. This prototype, situated at the Laboratory for Clean Energy Research (LaCER) in the School of Electrical and Electric Engineering at Nanyang Technological University in Singapore, utilizes software programs to handle sensed data and execute generation and load management algorithms [68]. The microgrid is expanded to a smart grid prototype by utilizing NI LabVIEW software, CompactRIO hardware with FPGA technology, and NI DAQ hardware [68]. This approach was used to establish a maximum power point tracking (MPPT) module for the photovoltaic (PV) system. Additionally, a control scheme for the battery energy storage system (BESS) was developed to combine energy management systems for buildings (HEMS, BEMS), solar PV technology, and energy storage with the existing microgrid prototype [68].

The objective of Green Empowerment's SGs for Small Grids initiative is to provide engineers and technicians in distant areas with intelligent and open-source technologies. The organization collaborates with local partners to construct sustainable electricity micro-grids in isolated indigenous villages in South Asia [69]. Green Empowerment engages in the development and construction of prototypes for laboratory testing using real-world appliances, in conjunction with simulated challenges. They collaborate closely with their partners to evaluate the financial viability. During the years 2022 and 2023, they were able to conduct trials of their prototype appliance controllers and monitoring devices in Malaysia and the Philippines [69].

2.4.2 Global implementation

Several governments worldwide have also commenced efforts on the SG, either by undertaking experimental programs or initiating testing and research. Countries including the United States, Australia, Britain, China, Japan, and South Korea have also regarded SGs as a means of diminishing carbon emissions and enhancing energy security [70].

NIST is actively engaged in researching and promoting collaboration across many nations to establish worldwide standards for the SG [32]. Canada, Mexico, Brazil, Japan, South Korea, Australia, India, and China are among the nations that have either previously investigated or are planning to investigate the significant SG infrastructure [71]. This investigation aims to establish a foundation of energy standards that other countries will soon adopt.

2.4.2.1 Australia

In 2009, investments proposal, the Australian government expressed its interest in making a $100 million investment. In addition, they expressed interest in increasing consumer knowledge regarding energy use and implementing generation management and distributed demand systems [70]. Five locations in New South Wales were selected for the establishment of SG facilities. Energy Australia collaborated with IBM, GE Energy, and Grid Net to work on this project. The objective was to construct a SG based on Worldwide Interoperability for Microwave Access (WIMAX) technology, with the potential to automatically manage substations, support up to 50,000 smart meter connections, and handle electric cars [70].

In 2021, the Australian Energy Security Board put out a proposal to reorganize the energy sector by 2025 [72]. They proposed a variety of measures to make the transition to the most decentralized power system in the world, focusing on the use of the SG and renewable energy sources [72]. By implementing and enforcing changes, Australia's electrical system will become more consumer-oriented, resulting in increased flexibility, dependability, and affordability [72].

2.4.2.2 India

The Indian government has launched countrywide smart grid programs, aiming to identify and address the obstacles and problems related to the implementation, such as customer acceptability. The Puducherry Electricity Department in Puducherry successfully launched a minor SG initiative, since customers were eager to accept SG. This initiative aimed to include a diverse group of 34,000 users with varying monthly budgets and power usage. Each consumer would be equipped with modern metering technology [73]. A multitude of initiatives have been suggested and executed throughout several towns in India, including Ajmer, Tripura, Haryana, and Manesar [73]. Currently, there are a total of 12 SG pilot projects listed in the power distribution sector. These projects involve the implementation of various functionalities, such as advanced metering technologies, load management, substation automation (including SCADA), distribution transformer monitoring, distributed generation, outage management, power quality measuring, micro grid, electric vehicle charging infrastructure, home energy management center, and cyber security and training infrastructure [73].

2.4.2.3 United Kingdom

Since 1990, the UK has been actively reducing its greenhouse gas and carbon dioxide emissions. Presently, the United Kingdom is during implementing the SG across the whole country. This is being done gradually, under the guidance of the industry regulator Ofgem and the Department of Energy and Climate Change (DECC) [74]. In the year 2021, a total of 8 million smart meters were installed and linked to the Data Communications Company (DCC). The DCC claims that their total of 17 million units at that time contribute to a reduction of 500,000 tons of carbon dioxide annually [74]. Their objective is to deploy 53 million intelligent meters by 2025. As of 2022, they have released a paper suggesting changes to the energy market, digitalizing the energy system, and embracing innovative smart technologies [74]. Ofgem's proposed reforms seek to empower customers with greater influence over energy usage while also boosting the reduction of net-zero greenhouse gas emissions.

2.4.2.4 China

In September 2020, the Chinese government set a goal to reach the highest level of carbon dioxide emissions by 2030 and achieve carbon neutrality by 2060 [75]. Subsequently, their efforts have been focused on constructing and enhancing its energy infrastructure [75]. China is projected to take the lead in the adoption of advanced metering infrastructure, since they have initiated the replacement of their initial generation of smart meters with more sophisticated systems [76].

2.4.2.5 Canada

SG initiatives are taking place in several provinces and regions across Canada. The SG initiative, a program under Natural Resource Canada, allocates $100 million over a span of 4 years (2019—23) to advance the demonstration of SG technologies and the implementation of SG integrated systems [77]. Over the course of the 4-year program, grantees provided updates on the initial implementation of the grid and its impact throughout their project, even up to 5 years after their first investment in the project [77]. The data provided will be utilized by Natural Resource

Canada to assess the effects on the power system and shape future initiatives and policies. Presently, the SG program has provided funding for a total of 21 projects around Canada.

Furthermore, Canada's Energy Innovation Program and Smart Grid Demonstration offers financial support to initiatives that showcase advancements in smart grid systems or technology [78]. The project, which has recently concluded its request for proposals, aims to expedite the modernization of the grid, enhance consumer accessibility, fill market gaps, and promote diversity, inclusivity, equity, and accessibility in this sector [78].

2.4.2.6 United States

The United States DOE received a funding of $4.5 billion from the Recovery Act specifically allocated for the purpose of modernizing the electrical system [79,80]. The two most significant programs are the SG Demonstration Program (SGDP) [81] and the SG Investment Grant (SGIG) [79,80]. The SGDP focuses on developing cutting-edge SG and energy storage devices, while also assessing their performance for potential future uses. The whole project has a total of 32 initiatives, out of which 16 are specifically focused on regional demonstrations to establish their validity, assess expenses, and evaluate the feasibility of scalable business models [81]. The remaining 16 initiatives are focused on the development of energy storage systems, similar to the ones mentioned before. The SGIG aims to expedite the use of current SG methods, tools, and technologies to enhance the performance of the contemporary grid. It encompasses almost $8 billion from the Act. The initiative encompasses 99 projects, which entail power supply firms enhancing their systems to conduct trials on distribution and transmission networks [79]. In addition, there are several additional SG programs available, including those focused on job training, interoperability, and cybersecurity, as well as renewable and distributed systems [80].

2.4.2.7 Japan

In 2015, the Japanese government highlighted six crucial segments necessary for expediting the advancement and implementation of Smart Grid technology in the country. The crucial components encompassed smart meters as the foremost essential, in addition to communication technologies, solar PV, battery energy storage systems, energy management systems, and the liberalization of the electrical market [76]. The countries' endeavors toward sustainable growth were expedited during the staging of the 2020 Olympic games, with the objective of guaranteeing a reliable energy provision. The implementation of the nations' Sustainable Development Goals (SG) has been postponed due to the interruptions caused by COVID-19. As a result, their efforts are expected to extend until 2030 [76]. An Ida et al. book provides a comprehensive overview of the present economic situation of SGs in Japan, including their progress and the outcomes of four field trials carried out in various locations [82]. The field studies conducted in Kitakyushu city, Keihanna Science City, Toyota city, and Yokohama city were intended as randomized control trials on actual electricity systems in residential areas [82]. The studies' validity is supported by the inclusion of this random element, as well as by discoveries related to behavioral economics and the utilization of extensive data sets [82].

2.4.3 Future research

A significant amount of study has been devoted to the direct or indirect advancement of SGs. The quest for enhancing the dependability and effectiveness of electricity networks will persist as long as engineering remains a discipline.

Fields also do so. Continuous advancements are being made in the domains of renewable resources and cyber security, leading to the development of increasingly potent and precise methodologies. For instance, there is ongoing research being conducted on identifying attacks and developing protection mechanisms, as well as finding more efficient ways to generate and store power. These research efforts have implications outside the purview of SG. The literature on strategies to enhance existing techniques is extensive, encompassing areas such as the development of faster computers and the optimization of power production turbine blades. The progress in interoperability and grid designs is a result of the expansion of cybersecurity and the integration of renewable energy, along with the development of society in general. The issue of integrating new technologies with the electrical grid or optimizing its operation will persist as advancements continue to be made. As people see the alterations in the environment and population, the way energy is generated or distributed from one place to another is an equally important factor to consider.

While the current sections of the SGs have been well researched and developed, it is important to also examine the incorporation of fresh research or addressing contemporary issues within the same framework. Considering the present condition of climate change and the prevalence of the digital era, the following areas are expected to get increased attention in the future: cyber security, particularly in relation to FDI assaults; enhancing the precision of forecasts; integrating micro-grids; and developing energy and demand management systems [70]. Based on recent research in fields like security and forecasting [61,83], it may be inferred that the use of deep learning and massive datasets will be crucial in addressing future difficulties.

2.4.3.1 Constraints

The cost is a primary limitation that hinders the progress and application of the SG [84]. The distribution and transmission networks incur substantial expenses, together with the accompanying technology and equipment, like as smart meters. Implementing the SG will be a particularly arduous limitation for underdeveloped nations. In 2017, Young conducted a research that examined the potential of SG in various nations. The analysis involved determining the projected cost of SG and the ratio of each nation's GDP to the cost of development. The researchers conducted Kepner-Tregoe analyses [84] to evaluate and compare several aspects such as the availability of resources, proximity to metropolitan areas, accessibility of education and training, political stability, and development cost. This analysis revealed a wide range of ratios, ranging from a minimum of 2 to a maximum of 47. This demonstrates the significant financial difficulties associated with implementing SGs and highlights the selection of five developing nations that are best suitable for investment in SGs [84].

2.4.4 Future implementation expectations

The ultimate objective is to universally adopt the SG, establishing it as the novel means of delivering and distributing electrical power. The SG will enhance the efficiency of the power system by improving its resiliency, flexibility, and dependability. It will achieve this by adapting a self-healing grid architecture, strengthening power management, and optimizing the usage of existing electrical assets [77]. Adopting global standards would lead to enhanced productivity for firms and will promote competitiveness among suppliers [85]. Implementing this will result in cost reduction, which will be advantageous for both utilities and customers [77]. The implementation of the SG will also result in an increase in job opportunities within the energy industry. The SG is much more eco-friendly compared to the alternative conventional power infrastructure. It offers a novel method to facilitate the greater integration of renewable energy sources and decrease greenhouse gas emissions.

2.5 Conclusion

This study provides a comprehensive overview and analysis of the key components and challenges associated with SGs. Highlighted are major measures taken by several countries/continents toward the implementation of Sustainable Development Goals. These actions include suggested future redesigns including the SG, as well as significant investments of millions of dollars into testing, developing, and implementing them.

Implementing SG technology and analyzing and addressing real obstacles and issues related to SG deployment. Finally, remarks are provided about the present condition of the SG. The importance of SGs was also highlighted, as it relates to a more collaborative and environment friendly approach to transmitting and distributing power. The poll primarily focused on the topics of cyber security, interoperability, and the integration of renewable energy.

The literature review revealed that the issues encountered in the regions studied were diverse and need more investigation and inquiry. FDI assaults offer a significant and emerging threat to the cyber security of SG. As a result, it has been a hot study issue because to its recent and hazardous nature. Interoperability is a key characteristic of SG, enabling the seamless flow of information. It is crucial to provide excellent communication among all components in SG in order to achieve this exchange. The integration of renewable energy is becoming increasingly important for SG, as the demand for additional renewable and sustainable means of energy generation rises. Incorporating these resources into the SG will be imperative to address the energy requirements and environmental obstacles of the 21st century.

The most potential avenues for future study vary depending on the many components of SG. FDI assaults represent a significant and conspicuous threat in the field of cyber security. Although several methods have been suggested to mitigate and defend against FDI assaults, further investigation, experimentation, and simulations are necessary to verify that the offered strategies, when implemented in SG, would offer a highly accurate detection system and strong resistance against attacks. Considering the studies on the effectiveness and achievements of data-driven and deep learning techniques in identifying and rectifying attacks, it is logical to propose that cybersecurity solutions utilizing these approaches should be further researched, including the previously mentioned CDS framework.

Enhancing interoperability testing is crucial in addition to establishing standards and techniques of interoperability. These tests are crucial for verifying that components can effectively work together. It is crucial to develop techniques for managing the adverse effects of renewable energy integration (REI) in the power grid. However, equal attention should also be devoted to exploring energy storage solutions. Energy storage is essential for ensuring continuous supply of electricity, particularly considering the intermittent and unpredictable characteristics of renewable energy sources. Further exploration of new and creative storage technologies, particularly the growing V2G and V2X (vehicle-to-everything) approaches, is warranted due to the increasing popularity of electric cars.

References

[1] Office of Electric Transmission, Distribution, Grid 20230 - A National Vision for Electricity's Second 100 Years, Tech. Rep, United States Department of Energy, Washington, D.C, 2003.

[2] Energy Sector Control Systems Working Group, Roadmap to Achieve Energy Delivery Systems Cyber- Security, Tech. Rep, United States Department of Energy, Washington, D.C, 2011.

[3] Department of Energy, Climate Change, the uk Renewable Energy Strategy, Tech. Rep, The HM Gov- ernment, United Kingdom, 2009.

[4] Electricity Advisory Committee, Smart Grid: Enabler of the New Energy Economy, Tech. Rep, United States Department of Energy, Washington, D.C, 2008.

[5] L. Gelazanskas, K. Gamage, Demand side management in smart grid: a review and proposals for future direction, Sustainable Cities and Society 11 (2013), https://doi.org/10.1016/j.scs.2013.11.001.

[6] Y. Cunjiang, Z. Huaxun, Z. Lei, Architecture design for smart grid, Energy Procedia 17 (2012) 1524–1528, https://doi.org/10.1016/j.egypro.2012.02.276, 2012 in- Ternational Conference on Future Electrical Power and Energy System.

[7] S. Blumsack, A. Fernandez, Ready or not, here comes the smart grid!, Energy, 7th Biennial Inter- national Workshop, Advances in Energy Studies 37 (1) (2012) 61–68, https://doi.org/10.1016/j.energy.2011.07.054.

[8] X. Fang, S. Misra, G. Xue, D. Yang, Smart grid — the new and improved power grid: a survey, IEEE Communications Surveys & Tutorials 14 (4) (2012) 944–980, https://doi.org/10.1109/SURV.2011.101911.00087.

[9] R. Hledik, How green is the smart grid? The Electricity Journal 22 (3) (2009) 29–41.

[10] World Energy Outlook, Energy and Air Pollution, Tech. Rep, International Energy Agency (IEA0), Paris, France, 2016.

[11] M. Moretti, S.N. Djomo, H. Azadi, K. May, K. De Vos, S. Van Passel, N. Witters, A systematic review of environmental and economic impacts of smart grids, Renewable and Sustainable Energy Reviews 68 (2017) 888–898, https://doi.org/10.1016/j.rser.2016.03.039.

[12] H. Khurana, M. Hadley, N. Lu, D. Frincke, Smart-grid security issues, security and privacy, IEEE 8 (2010) 81–85, https://doi.org/10.1109/MSP.2010.49.
[13] M. Romero, W. Guedria, H. Panetto, B. Barafort, Towards a characterisation of smart systems: a systematic literature review, Computers in Industry 120 (2020) 103224, https://doi.org/10.1016/j.compind.2020.103224.
[14] M. Romero Aquino, A Smart Assessment of Business Processes for Enterprises Decision Support, University of Lorraine, 2021. Ph.D. thesis.
[15] M. Hentea, Building an Effective Security Program for Distributed Energy Resources and Systems, John Wiley & Sons, 2021.
[16] E.Y. Song, G.J. FitzPatrick, K.B. Lee, Smart sensors and standard-based interoperability in smart grids, IEEE Sensors Journal 17 (23) (2017) 7723–7730, https://doi.org/10.1109/JSEN.2017.2729893.
[17] A. Kitsios, K. Bousakas, T. Salame, B. Bogno, P. Papageorgas, G. Vokas, F. Mauffay, P. Petit, M. Ail- lerie, J.-P. Charles, Renewable energy sources, the internet of things and the third industrial revolution: smart grid and contemporary information and communication technologies, in: Technologies and Materials for Renewable Energy, Environment and Sustainability vol 1814, 2017, p. 020070, https://doi.org/10.1063/1.4976289.
[18] G. Venayagamoorthy, Potentials and promises of computational intelligence for smart grids, in: 2009 IEEE Power and Energy Society General Meeting, PES '09, 2009, pp. 1–6, https://doi.org/10.1109/PES.2009.5275224.
[19] T.-T. Tran, O.-S. Shin, J.-H. Lee, Detection of replay attacks in smart grid systems, in: 2013 International Conference on Computing, Management and Telecommunications (ComManTel), 2013, pp. 298–302, https://doi.org/10.1109/ComManTel.2013.6482409.
[20] A. Rajfallovska, Green Telecommunications in Smart Grid, International Atomic Energy Agency, 2016.
[21] E.F. Camacho, T. Samad, M. Garcia-Sanz, I. Hiskens, Control for renewable energy and smart grids, the impact of control technology, Control Systems Society 4 (8) (2011) 69–88.
[22] S. Paul, M. Rabbani, R. Kundu, S. Zaman, A review of smart technology (smart grid) and its features, in: Proceedings of 2014 1st International Conference on Non Conventional Energy (ICONCE 2014), 2014, pp. 200–203, https://doi.org/10.1109/ICONCE.2014.6808719.
[23] G. Paramo, A. Bretas, S. Meyn, Research trends and applications of pmus, Energies 15 (15) (2022), https://doi.org/10.3390/en15155329.
[24] NIST, Nist Framework and Roadmap for Smart Grid Interoperability Standards, Release 1.0, 2010.
[25] A. Berl, M. Niedermeier, H. Meer, Smart Grid Considerations: Energy Efficiency vs. Security, vol 88, Elsevier, 2013, pp. 159–198, https://doi.org/10.1016/B978-0-12-407725-6.00004-6. Ch. 4.
[26] M. Gunduz, R. Das, Cyber-security on smart grid: threats and potential solutions, Computer Networks 169 (C) (2020), https://doi.org/10.1016/j.comnet.2019.107094.
[27] M. Gündüz, R. Das, Analysis of cyber-attacks on smart grid applications, in: 2018 International Conference on Artificial Intelligence and Data Processing (IDAP), 2018, pp. 1–5, https://doi.org/10.1109/IDAP.2018.8620728.
[28] C.A. Ardagna, V. Bellandi, E. Damiani, M. Bezzi, C. Hebert, Big data analytics-as-a-service: bridging the gap between security experts and data scientists, Computers & Electrical Engineering 93 (2021) 107215, https://doi.org/10.1016/j.compeleceng.2021.107215.

[29] J. Chen, M.A. Mohamed, U. Dampage, M. Rezaei, S.H. Salmen, S.A. Obaid, A. Annuk, A multi- layer security scheme for mitigating smart grid vulnerability against faults and cyber-attacks, Applied Sciences 11 (21) (2021), https://doi.org/10.3390/app11219972.

[30] M. Tian, M. Cui, Z. Dong, X. Wang, S. Yin, L. Zhao, Multilevel programming-based coordinated cyber physical attacks and countermeasures in smart grid, IEEE Access 7 (2019) 9836–9847, https://doi.org/10.1109/ACCESS.2018.2890604.

[31] Z. Zhang, S. Huang, Y. Chen, B. Li, S. Mei, Cyber-physical coordinated risk mitigation in smart grids based on attack-defense game, IEEE Transactions on Power Systems 37 (1) (2022) 530–542, https://doi.org/10.1109/TPWRS.2021.3091616.

[32] W. Wang, Z. Lu, Cyber security in the smart grid: survey and challenges, Computer Networks 57 (2013) 1344–1371, https://doi.org/10.1016/j.comnet.2012.12.017.

[33] M.Z. Gündüz, R. Das, A comparison of cyber-security oriented testbeds for iot-based smart grids, in: 2018 6th International Symposium on Digital Forensic and Security (ISDFS), 2018, pp. 1–6, https://doi.org/10.1109/ISDFS.2018.8355329.

[34] D.B. Rawat, C. Bajracharya, Cyber security for smart grid systems: status, challenges, and perspectives, in: SoutheastCon 2015, 2015, pp. 1–6, https://doi.org/10.1109/SECON.2015.7132891.

[35] Z.A. Baig, A.-R. Amoudi, An analysis of smart grid attacks and countermeasures, Journal of Communication 8 (8) (2013) 473–479, https://doi.org/10.12720/jcm.8.8.473-479.

[36] V. Delgado-Gomes, J.F. Martins, C. Lima, P.N. Borza, Smart grid security issues, in: 2015 9th International Conference on Compatibility and Power Electronics (CPE), 2015, pp. 534–538, https://doi.org/10.1109/CPE.2015.7231132.

[37] H.S.Y. Yan, Y. Qian, D. Tipper, A survey on cyber security for smart grid communications, IEEE Communications Surveys and Tutorials 14 (4) (2012) 998–1010, https://doi.org/10.1109/SURV.2012.010912.00035.

[38] B. Khelifa, S. Abla, Security concerns in smart grids: threats, vulnerabilities and countermeasures, in: 2015 3rd International Renewable and Sustainable Energy Conference (IRSEC), 2015, pp. 1–6, https://doi.org/10.1109/IRSEC.2015.7454963.

[39] S. Tan, D. De, W.-Z. Song, J. Yang, S.K. Das, Survey of security advances in smart grid: a data driven approach, IEEE Communications Surveys & Tutorials 19 (1) (2017) 397–422, https://doi.org/10.1109/COMST.2016.2616442.

[40] S. Jadidi, H. Badihi, Y. Zhang, Active fault-tolerant and attack-resilient control for a renewable microgrid against power-loss faults and data integrity attacks, IEEE Transactions on Cybernetics (2023) 1–16, https://doi.org/10.1109/TCYB.2023.3305240.

[41] H. Liu, T. Gu, Y. Liu, J. Song, Z. Zeng, Fault-tolerant privacy-preserving data aggregation for smart grid, Wireless Communications and Mobile Computing 2020 (September 2020), https://doi.org/10.1155/2020/8810393.

[42] M.I. Oozeer, S. Haykin, Cognitive dynamic system for control and cyber-attack detection in smart grid, IEEE Access 7 (2019) 78320–78335, https://doi.org/10.1109/ACCESS.2019.2922410.

[43] G. Bedi, G.K. Venayagamoorthy, R. Singh, R.R. Brooks, K.-C. Wang, Review of internet of things (iot) in electric power and energy systems, IEEE Internet of Things Journal 5 (2) (2018) 847–870, https://doi.org/10.1109/JIOT.2018.2802704.

[44] A. Sanjab, W. Saad, I. Guvenc, A. Sarwat, S. Biswas, Smart Grid Security: Threats, Challenges, and Solutions, 2016, p. 06992, arXiv:1606.

[45] R.K. Pandey, M. Misra, Cyber security threats — smart grid infrastructure, in: 2016 National Power Systems Conference (NPSC), 2016, pp. 1–6, https://doi.org/10.1109/NPSC.2016.7858950.

[46] F. Aloul, A. Al-Ali, R. Al-Dalky, M. Al-Mardini, W. El-Hajj, Smart grid security: threats, vulnerabilities and solutions, International Journal of Smart Grid and Clean Energy 1 (1) (2012) 1–6.

[47] S. Shapsough, F. Qatan, R. Aburukba, F. Aloul, A.R. Al Ali, Smart grid cyber security: challenges and solutions, in: 2015 International Conference on Smart Grid and Clean Energy Technologies (IC- SGCE), 2015, pp. 170–175, https://doi.org/10.1109/ICSGCE.2015.7454291.

[48] Z.E. Mrabet, N. Kaabouch, H.E. Ghazi, H.E. Ghazi, Cyber-security in smart grid: survey and challenges, Computers & Electrical Engineering 67 (2018) 469–482, https://doi.org/10.1016/j.compeleceng.2018.01.015.

[49] A.R. Metke, R.L. Ekl, Security technology for smart grid networks, IEEE Transactions on Smart Grid 1 (1) (2010) 99–107, https://doi.org/10.1109/TSG.2010.2046347.

[50] J. Gao, X. Zhang, H. Liang, X.S. Shen, Joint encryption and compressed sensing in smart grid data transmission, in: 2014 IEEE Global Communications Conference, 2014, pp. 662–667, https://doi.org/10.1109/GLOCOM.2014.7036883.

[51] D. Syed, S.S. Refaat, O. Bouhali, Privacy preservation of data-driven models in smart grids using homomorphic encryption, Information 11 (7) (2020), https://doi.org/10.3390/info11070357.

[52] N. Saxena, B.J. Choi, R. Lu, Authentication and authorization scheme for various user roles and devices in smart grid, IEEE Transactions on Information Forensics and Security 11 (5) (2016) 907–921, https://doi.org/10.1109/TIFS.2015.2512525.

[53] T. Chim, S. Yiu, L.C. Hui, V.O. Li, Pass: privacy-preserving authentication scheme for smart grid network, in: 2011 IEEE International Conference on Smart Grid Communications (SmartGridComm), 2011, pp. 196–201, https://doi.org/10.1109/SmartGridComm.2011.6102316.

[54] S. Etigowni, M. Cintuglu, M. Kazerooni, S. Hossain, P. Sun, K. Davis, O. Mohammed, S. Zonouz, Cyber-air-gapped detection of controller attacks through physical interdependencies, in: 2017 IEEE International Conference on Smart Grid Communications (SmartGridComm), 2017, pp. 102–107, https://doi.org/10.1109/SmartGridComm.2017.8340685.

[55] T. Liu, Y. Sun, Y. Liu, Y. Gui, Y. Zhao, D. Wang, C. Shen, Abnormal traffic-indexed state estimation: a cyber–physical fusion approach for smart grid attack detection, Future Generation Computer Systems 49 (2015) 94–103, https://doi.org/10.1016/j.future.2014.10.002.

[56] P. Jokar, V.C.M. Leung, Intrusion detection and prevention for zigbee-based home area networks in smart grids, IEEE Transactions on Smart Grid 9 (3) (2018) 1800–1811, https://doi.org/10.1109/TSG.2016.2600585.

[57] Y. He, G.J. Mendis, J. Wei, Real-time detection of false data injection attacks in smart grid: a deep learning-based intelligent mechanism, IEEE Transactions on Smart Grid 8 (5) (2017) 2505–2516, https://doi.org/10.1109/TSG.2017.2703842.

[58] M.I. Oozeer, S. Haykin, Cognitive risk control for mitigating cyber-attack in smart grid, IEEE Access 7 (2019) 125806–125826, https://doi.org/10.1109/ACCESS.2019.2939089.

[59] S. Feng, S. Haykin, Cognitive risk control for anti-jamming v2v communications in autonomous vehicle networks, IEEE Transactions on Vehicular Technology 68 (10) (2019) 9920–9934, https://doi.org/10.1109/TVT.2019.2935999.

[60] H. Shi, L. Xie, L. Peng, Detection of false data injection attacks in smart grid based on a new dimensionality-reduction method, Computers & Electrical Engineering 91 (2021) 107058, https://doi.org/10.1016/j.compeleceng.2021.107058.

[61] J. Wang, X. Niu, L. Zhang, Z. Liu, X. Huang, A wind speed forecasting system for the construction of a smart grid with two-stage data processing based on improved elm and deep learning strategies, Expert Systems with Applications 241 (2024) 122487, https://doi.org/10.1016/j.eswa.2023.122487.
[62] Department of Electrical Engineering and Computer Science, York University, Smart Grid Research Lab, 2024. https://smartgrid.eecs.yorku.ca/.
[63] Centre for Urban Energy, Schneider Electric Smart Grid Laboratory, 2024. https://www.torontomu.ca/cue/testing/schneider-electric-smart-grid-lab/.
[64] High Voltage Engineering Laboratory HVEL, Smart Grid, 2024. https://uwaterloo.ca/high-voltage-engineering-laboratory/research/smart-grid.
[65] Department of Electrical and Electronic Engineering, The University of Melbourne, Smart Grid Lab, 2024. https://electrical.eng.unimelb.edu.au/power-energy/smart-grid-lab.
[66] S.P.T. Nambiar, C.R.A. Tyagi, V.D.M. Nagadharni, A laboratory prototype of a smart grid based demand side management, in: ISGT2011-India, 2011, pp. 28–31, https://doi.org/10.1109/ISET-India.2011. 6145343.
[67] K.P. Natarajan, S. Bhagavath Singh, Fpga based remote monitoring system in smart grids, Indian Journal of Science and Technology 10 (2017) 1–5, https://doi.org/10.17485/ijst/2017/v10i5/108829.
[68] P.H. Cheah, Developing a Functional Smart Grid Prototype Using Ni Labview, Ni Compactrio, and Ni Daq, Nanyang Technological University, 2024. Available, https://www.ni.com/en-ca/innovations/case-studies/19/developing-a-functional-smart-grid-prototype.html.
[69] G. Empowerment, Smart Grids for Small Grids, Green Empowerment, 2024. Available, https://greenempowerment.org/technical-resources/smart-grid-for-small-grids/.
[70] O. Majeed Butt, M. Zulqarnain, T. Majeed Butt, Recent advancement in smart grid technology: future prospects in the electrical power network, Ain Shams Engineering Journal 12 (1) (2021) 687–695, https://doi.org/10.1016/j.asej.2020.05.004.
[71] NIST, Smart Grid International Coordination, 2018. https://www.nist.gov/programs-projects/smart-grid-national-coordination/international-coordination.
[72] E.S. Board, Post 2025 Electricity Market Design, Energy Security Board, 2023. Available: https://esb-post2025-market-design.aemc.gov.au/.
[73] National Smart Grid Mission, Sg Projects, Ministry of Power, Government of India, 2024. Available, https://www.nsgm.gov.in/en/sg-pilot.
[74] A. Cameron, The uk Smart Grid: How it Started & How It's Going, Fortra, 2022. Available, https://www.tripwire.com/state-of-security/uk-smart-grid-how-it-started-how-its-going.
[75] Globe Newswire, Global and china Smart Meters Market Report 2022-2027: Growing Number of Households Spurs Demand, Research and Markets, 2022. Available, https://www.globenewswire.com/en/news-release/2022/08/02/2490054/28124/en/Global-and-China-Smart-Meters-Market-Report-2022-2027-_Growing-Number-of-Households-Spurs-Demand.html.
[76] N. Nhede, Smart Grid's Role in Energy Transition and the Top Five Market Leaders, Smart Energy International, 2021. URL, https://www.smart-energy.com/industry-sectors/smart-meters/smart-grids-role-in-energy-transition-and-the-top-five-market-leaders/.
[77] Government of Canada, Smart Grid Program, Green Infrastructure Programs, 2023. URL, https://natural-resources.canada.ca/climate-change/green-infrastructure-programs/smart-grids/19793.

[78] Government of Canada, Energy Innovation Program Smart Grid Demonstration Call for Proposals, 2022. https://natural-resources.canada.ca/science-and-data/funding-partnerships/opportunities/grants-incentives/energy-innovation-program/energy-innovation-program-smart-grid-demonstration-call-for-proposals/energy-innovation-program-smart.
[79] United States Department of Energy, Recovery Act: Smart Grid Investment Grant Program, 2024. https://www.smartgrid.gov/archive/recovery_act/overview/smart_grid_investment_grant_program.
[80] United States Department of Energy, Recovery Act Smart Grid Program, SmartGrid.gov, 2024. URL, https://www.smartgrid.gov/recovery_act/index.html.
[81] United States Department of Energy, Recovery Act: Smart Grid Demonstration Program, 2024. https://www.smartgrid.gov/archive/recovery_act/overview/smart_grid_demonstration_program.html.
[82] T. Ida, M. Tanaka, K. Ito, Smart Grid Economics: A Field Experimental Approach to Demand Response, vol 32, Springer Nature, 2024.
[83] J. Ruan, G. Liang, J. Zhao, H. Zhao, J. Qiu, F. Wen, Z.Y. Dong, Deep learning for cybersecurity in smart grids: review and perspectives, Energy Conversion and Economics 4 (4) (2023) 233−251, https://doi.org/10.1049/enc2.12091.
[84] J.R. Young, Smart Grid Technology in the Developing World, Seattle Pacidic Library, 2017. URL, https://digitalcommons.spu.edu/honorsprojects/68.
[85] R. Kappagantu, S.A. Daniel, Challenges and issues of smart grid implementation: a case of indian scenario, Journal of Electrical Systems and Information Technology 5 (3) (2018) 453−467, https://doi.org/10.1016/j.jesit.2018.01.002.

Further readings

[1] C.M. Colson, et al., Towards Real-Time Power Management of Microgrids for Power System Integration: A Decentralized Multi-Agent Based Approach, Ph.D. thesis, Montana State University-Bozeman, College of Engineering, 2012.
[2] H. Farhangi, The path of the smart grid, Power and Energy Magazine, IEEE 8 (2010) 18−28, https://doi.org/10.1109/MPE.2009.934876.
[3] I. Colak, S. Sagiroglu, G. Fulli, M. Yesilbudak, C.-F. Covrig, A survey on the critical issues in smart grid technologies, Renewable and Sustainable Energy Reviews 54 (C) (2016) 396−405, https://doi.org/10.1016/j.rser.2015.10.03.
[4] G. Dileep, A survey on smart grid technologies and applications, Renewable Energy 146 (2020) 2589−2625, https://doi.org/10.1016/j.renene.2019.08.092.
[5] M.A. Judge, A. Khan, A. Manzoor, H.A. Khattak, Overview of smart grid implementation: frame- works, impact, performance and challenges, Journal of Energy Storage 49 (2022), https://doi.org/10.1016/j.est.2022.104056.
[6] I. Alotaibi, M.A. Abido, M. Khalid, A.V. Savkin, A comprehensive review of recent advances in smart grids: a sustainable future with renewable energy resources, Energies 13 (23) (2020), https://doi.org/10.3390/en13236269.
[7] A.I. Kawoosa, D. Prashar, A review of cyber securities in smart grid technology, in: 2021 2nd International Conference on Computation, Automation and Knowledge Management (ICCAKM), 2021, pp. 151−156, https://doi.org/10.1109/ICCAKM50778.2021.9357698.
[8] Electricity Advisory Committee, Smart Grid: An Introduction, Tech. Rep, United States Department of Energy, Washington, D.C, 2008.
[9] M.Z. Gunduz, R. Das, Internet of things (iot): evolution, components and applications fields, Pamukkale Eniversity Journal of Engineering Sciences 24 (2) (2018) 327−335.

[10] Y. Yang, T. Littler, S. Sezer, K. McLaughlin, H.F. Wang, Impact of cyber-security issues on smart grid, in: 2011 2nd IEEE PES International Conference and Exhibition on Innovative Smart Grid Technologies, 2011, pp. 1–7, https://doi.org/10.1109/ISGTEurope.2011.6162722.

[11] C. Peng, H. Sun, M. Yang, Y.-L. Wang, A survey on security communication and control for smart grids under malicious cyber attacks, IEEE Transactions on Systems, Man, and Cybernetics: Systems 49 (8) (2019) 1554–1569, https://doi.org/10.1109/TSMC.2018.2884952.

[12] T.A. Rizzetti, P. Wessel, A.S. Rodrigues, B. Menezes Da Silva, R. Milbradt, L.N. Canha, Cyber security and communications network on scada systems in the context of smart grids, in: 2015 50th International Universities Power Engineering Conference (UPEC), 2015, pp. 1–6, https://doi.org/10.1109/UPEC.2015.7339762.

[13] Y. Liu, P. Ning, M.K. Reiter, False data injection attacks against state estimation in electric power grids, ACM Transactions on Information and System Security 14 (1) (2011), https://doi.org/10.1145/1952982.1952995.

[14] Y. Yuan, Z. Li, K. Ren, Modeling load redistribution attacks in power systems, IEEE Transactions on Smart Grid 2 (2) (2011) 382–390, https://doi.org/10.1109/TSG.2011.2123925.

[15] L. Ma, Z. Chu, C. Yang, G. Wang, W. Dai, Recursive watermarking-based transient covert attack detection for the industrial cps, IEEE Transactions on Information Forensics and Security 18 (2023) 1709–1719, https://doi.org/10.1109/TIFS.2023.3251857.

[16] A. Abdelwahab, W. Lucia, A. Youssef, Set-theoretic control for active detection of replay attacks with applications to smart grid, in: 2020 IEEE Conference on Control Technology and Applications (CCTA), 2020, pp. 1004–1009, https://doi.org/10.1109/CCTA41146.2020.9206373.

[17] L. Pavithra, D. Rekha, Prevention of replay attack for isolated smart grid, in: Next Generation Infor- Mation Processing System, Springer, Singapore, 2021, pp. 251–258.

[18] J.A. Larcom, H. Liu, Modeling and characterization of gps spoofing, in: 2013 IEEE International Conference on Technologies for Homeland Security (HST), 2013, pp. 729–734, https://doi.org/10.1109/THS.2013.6699094.

[19] Z. Zhang, S. Gong, A.D. Dimitrovski, H. Li, Time synchronization attack in smart grid: impact and analysis, IEEE Transactions on Smart Grid 4 (1) (2013) 87–98, https://doi.org/10.1109/TSG.2012.2227342.

[20] U.K. Premaratne, J. Samarabandu, T.S. Sidhu, R. Beresh, J.-C. Tan, An intrusion detection system for iec61850 automated substations, IEEE Transactions on Power Delivery 25 (4) (2010) 2376–2383, https://doi.org/10.1109/TPWRD.2010.2050076.

[21] M. Strasser, C. Popper, S. Capkun, M. Cagalj, Jamming-resistant key establishment using uncoordinated frequency hopping, in: 2008 IEEE Symposium on Security and Privacy (Sp 2008), 2008, pp. 64–78, https://doi.org/10.1109/SP.2008.9.

[22] C. Popper, M. Strasser, S. Capkun, Jamming-resistant broadcast communication without shared keys, in: USENIX Security Symposium, 2009, pp. 231–248.

[23] Y. Liu, P. Ning, H. Dai, A. Liu, Randomized differential dsss: jamming-resistant wireless broadcast communication, in: 2010 Proceedings IEEE INFOCOM, 2010, pp. 1–9, https://doi.org/10.1109/INFCOM.2010.5462156.

[24] S. Mohagheghi, J. Stoupis, Z. Wang, Communication protocols and networks for power systems- current status and future trends, in: 2009 IEEE/PES Power Systems Conference and Exposition, 2009, pp. 1–9, https://doi.org/10.1109/PSCE.2009.4840174.

[25] T.-I. Choi, K.Y. Lee, D.R. Lee, J.K. Ahn, Communication system for distribution automation using cdma, IEEE Transactions on Power Delivery 23 (2) (2008) 650–656, https://doi.org/10.1109/TPWRD.2007.910991.

[26] H.J. Zhou, C.X. Guo, J. Qin, Efficient application of gprs and cdma networks in scada system, in: IEEE PES General Meeting, 2010, pp. 1–6, https://doi.org/10.1109/PES.2010.5588206.

[27] R. Leon, V. Vittal, G. Manimaran, Application of sensor network for secure electric energy infrastructure, IEEE Transactions on Power Delivery 22 (2) (2007) 1021–1028, https://doi.org/10.1109/TPWRD.2006.886797.

[28] D. Pendarakis, N. Shrivastava, Z. Liu, R. Ambrosio, Information aggregation and optimized actuation in sensor networks: enabling smart electrical grids, in: IEEE INFOCOM 2007 - 26th IEEE International Conference on Computer Communications, 2007, pp. 2386–2390, https://doi.org/10.1109/INFCOM.2007.286.

[29] Z. Lu, W. Wang, C. Wang, From jammer to gambler: modeling and detection of jamming attacks against time-critical traffic, in: 2011 Proceedings IEEE INFOCOM, 2011, pp. 1871–1879, https://doi.org/10.1109/INFCOM.2011.5934989.

[30] A. Stavrou, A.D. Keromytis, Countering dos attacks with stateless multipath overlays, in: Proceedings of the 12th ACM Conference on Computer and Communications Security, CCS '05, Association for Computing Machinery, New York, NY, USA, 2005, pp. 249–259, https://doi.org/10.1145/1102120.1102153.

[31] E.-K. Lee, M. Gerla, S.Y. Oh, Physical layer security in wireless smart grid, IEEE Communications Magazine 50 (8) (2012) 46–52, https://doi.org/10.1109/MCOM.2012.6257526.

[32] C. Schuba, I. Krsul, M. Kuhn, E. Spafford, A. Sundaram, D. Zamboni, Analysis of a denial of service attack on tcp, in: Proceedings. 1997 IEEE Symposium on Security and Privacy (Cat. No.97CB36097), 1997, pp. 208–223, https://doi.org/10.1109/SECPRI.1997.601338.

[33] A. Yaar, A. Perrig, D. Song, Pi: a path identification mechanism to defend against ddos attacks, in: 2003 Symposium on Security and Privacy, 2003, 2003, pp. 93–107, https://doi.org/10.1109/SECPRI.2003.1199330.

[34] J. Mirkovic, P. Reiher, A taxonomy of ddos attack and ddos defense mechanisms, SIGCOMM Comput, The Communication Review 34 (2) (2004) 39–53, https://doi.org/10.1145/997150.997156.

[35] D. Jin, D.M. Nicol, G. Yan, An event buffer flooding attack in dnp3 controlled scada systems, in: Proceedings of the 2011 Winter Simulation Conference (WSC), 2011, pp. 2614–2626, https://doi.org/10.1109/WSC.2011.6147969.

[36] S. Ranjan, R. Swaminathan, M. Uysal, A. Nucci, E. Knightly, Ddos-shield: ddos-resilient scheduling to counter application layer attacks, IEEE/ACM Transactions on Networking 17 (1) (2009) 26–39, https://doi.org/10.1109/TNET.2008.926503.

[37] F. Ayadi, I. Colak, R. Bayindir, Interoperability in smart grid, in: 2019 7th International Conference on Smart Grid (icSmartGrid), 2019, pp. 165–169, https://doi.org/10.1109/icSmartGrid48354.2019.8990680.

[38] H.J. Kim, C.M. Jeong, J.-M. Sohn, J.-Y. Joo, V. Donde, Y. Ko, Y.T. Yoon, A comprehensive review of practical issues for interoperability using the common information model in smart grids, Energies 13 (6) (2020), https://doi.org/10.3390/en13061435.

[39] R. Ma, H.-H. Chen, Y.-R. Huang, W. Meng, Smart grid communication: its challenges and opportunities, IEEE Transactions on Smart Grid 4 (1) (2013) 36–46, https://doi.org/10.1109/TSG.2012.2225851.

[40] T. King, Sensing & measurement research activities, in: Department of Energy, Grid Moderization Laboratory Consortium, 2018, pp. 1–23.

[41] A. Goldstein, Advanced Electrical Power System Sensors Workshop Report, 2017-11-13 2017, https://doi.org/10.6028/NIST.SP. 1500-11.

[42] S.T. Mak, Sensor data output requirements for smart grid applications, in: IEEE PES General Meeting, 2010, pp. 1–3, https://doi.org/10.1109/PES.2010.5589580.

[43] S.Y. Yurish, Sensors: smart vs. intelligent, Sensors & Transducers 114 (3) (2010), 6–I,II,III,IV,V,VI.
[44] M. Farrokhabadi, B.V. Solanki, C.A. Canizares, K. Bhattacharya, S. Koenig, P.S. Sauter, T. Leibfried, S. Hohmann, Energy storage in microgrids: compensating for generation and demand fluctuations while providing ancillary services, IEEE Power and Energy Magazine 15 (5) (2017) 81–91, https://doi.org/10.1109/MPE.2017.2708863.
[45] A. Bari, J. Jiang, W. Saad, A. Jaekel, Challenges in the smart grid applications: an overview, International Journal of Distributed Sensor Networks 10 (2) (2014) 974682, https://doi.org/10.1155/2014/974682.
[46] M. Saadatmand, b. ramezy, B. Mozafari, Review of communication technologies for smart grid ap- plications, in: The National Conference on New Approaches in Power Industry, 2017, pp. 1–10.
[47] V.C. Gungor, B. Lu, G.P. Hancke, Opportunities and challenges of wireless sensor networks in smart grid, IEEE Transactions on Industrial Electronics 57 (10) (2010) 3557–3564, https://doi.org/10.1109/TIE.2009.2039455.
[48] M. di Bisceglie, C. Galdi, A. Vaccaro, D. Villacci, Cooperative sensor networks for voltage quality monitoring in smart grids, in: 2009 IEEE Bucharest PowerTech, 2009, pp. 1–6, https://doi.org/10.1109/PTC. 2009.5282012.
[49] D.-K. Kim, A. Alaerjan, L. Lu, H. Yang, H. Jang, Toward interoperability of smart grids, IEEE Communications Magazine 55 (8) (2017) 204–210, https://doi.org/10.1109/MCOM.2017.1600392.
[50] E.Y. Song, G.J. FitzPatrick, K.B. Lee, A.M. Gopstein, P.A. Boynton, Interoperability testbed for smart sensors in smart grids, in: 2018 IEEE Power & Energy Society Innovative Smart Grid Technologies Conference (ISGT), 2018, pp. 1–5, https://doi.org/10.1109/ISGT.2018.8403332.
[51] H. Li, S. Gong, L. Lai, Z. Han, R.C. Qiu, D. Yang, Efficient and secure wireless communications for advanced metering infrastructure in smart grids, IEEE Transactions on Smart Grid 3 (3) (2012) 1540–1551, https://doi.org/10.1109/TSG.2012.2203156.
[52] D.S. Markovic, D. Zivkovic, I. Branovic, R. Popovic, D. Cvetkovic, Smart power grid and cloud computing, Renewable and Sustainable Energy Reviews 24 (2013) 566–577, https://doi.org/10.1016/j.rser. 2013.03.068.
[53] Z. Feng, Z. Yuexia, Study on smart grid communications system based on new generation wire- less technology, in: 2011 International Conference on Electronics, Communications and Control (ICECC), 2011, pp. 1673–1678, https://doi.org/10.1109/ICECC.2011.6066343.
[54] H. Shahinzadeh, A.-s. Mirhedayati, M. Shaneh, H. Nafisi, G.B. Gharehpetian, J. Moradi, Role of joint 5g-iot framework for smart grid interoperability enhancement, in: 2020 15th International Conference on Protection and Automation of Power Systems (IPAPS), 2020, pp. 12–18, https://doi.org/10.1109/IPAPS52181.2020.9375539.
[55] A.S. Alaerjan, Model-driven interoperability layer for normalized connectivity across smart grid domains, IEEE Access 9 (2021) 98639–98653, https://doi.org/10.1109/ACCESS.2021.3096043.
[56] A. Alaerjan, D.-K. Kim, H. Ming, K. Malik, Using dds based on unified data model to improve inter- operability of smart grids, in: 2018 IEEE International Conference on Smart Energy Grid Engineering (SEGE), 2018, pp. 110–114, https://doi.org/10.1109/SEGE.2018.8499513.
[57] S. Cavalieri, Semantic interoperability between iec 61850 and onem2m for iot-enabled smart grids, Sensors 21 (7) (2021), https://doi.org/10.3390/s21072571.

[58] H. Bergmann, C. Mosiman, A. Saha, S. Haile, W. Livingood, S. Bushby, G. Fierro, J. Bender, M. Poplawski, J. Granderson, M. Pritoni, Semantic Interoperability to Enable Smart, Grid-Interactive Efficient Buildings, Summer Study on Energy Efficiency in Buildings, 2020, https://doi.org/10.20357/B7S304.
[59] I.S. Association, Ieee standard for electric power systems communications-distributed network pro- tocol (dnp3), in: IEEE Std 1815-2012 (Revision of IEEE Std 1815-2010), 2012, pp. 1–821, https://doi.org/10.1109/IEEESTD.2012.6327578.
[60] M. Ginocchi, A. Ahmadifar, F. Ponci, A. Monti, Application of a smart grid interoperability test- ing methodology in a real-time hardware-in-the-loop testing environment, Energies 13 (7) (2020), https://doi.org/10.3390/en13071648.
[61] J. Durand, Conformance Interoperability and Portability Testing: Proposed Procedures and Practices, 2012.
[62] I. Papaioannou, et al., Smart Grid Interoperability Testing Methodology, JRC Publications Repository, 2018, https://doi.org/10.2760/08049.
[63] E.Y. Song, K.B. Lee, G.J. FitzPatrick, Y. Zhang, Interoperability test for iec 61850-9-2 standard- based merging units, in: 2017 IEEE Power & Energy Society Innovative Smart Grid Technologies Conference, ISGT), 2017, pp. 1–6, https://doi.org/10.1109/ISGT.2017.8086084.
[64] N. Chen, Passive Interoperability Testing for Communication Protocols, Ph.D. thesis, Universite´ de Rennes, 2013. URL, https://theses.hal.science/tel-00869819.
[65] E.Y. Song, G.J. FitzPatrick, K.B. Lee, E. Griffor, A methodology for modeling interoperability of smart sensors in smart grids, IEEE Transactions on Smart Grid 13 (1) (2022) 555–563, https://doi.org/10.1109/TSG.2021.3124490.
[66] M.A. Mohamed, A.M. Eltamaly, H.M. Farh, A.I. Alolah, Energy management and renewable energy integration in smart grid system, in: 2015 IEEE International Conference on Smart Energy Grid Engineering (SEGE), 2015, pp. 1–6, https://doi.org/10.1109/SEGE.2015.7324621.
[67] L.F. Ochoa, G.P. Harrison, Minimizing energy losses: optimal accommodation and smart operation of renewable distributed generation, IEEE Transactions on Power Systems 26 (1) (2011) 198–205, https://doi.org/10.1109/TPWRS.2010.2049036.
[68] Z. Ming, H. Lixin, Y. Fan, J. Danwei, Retracted article: research of the problems of renewable energy orderly combined to the grid in smart grid, in: 2010 Asia-Pacific Power and Energy Engineering Conference, 2010, pp. 1–4, https://doi.org/10.1109/APPEEC.2010.5448334.
[69] M. Liserre, T. Sauter, J.Y. Hung, Future energy systems: integrating renewable energy sources into the smart power grid through industrial electronics, IEEE Industrial Electronics Magazine 4 (1) (2010) 18–37, https://doi.org/10.1109/MIE.2010.935861.
[70] G.M. Shafiullah, A.M.T. Oo, D. Jarvis, A.B.M.S. Ali, P. Wolfs, Potential challenges: integrating renewable energy with the smart grid, in: 2010 20th Australasian Universities Power Engineering Conference, 2010, pp. 1–6.
[71] M. Fadaeenejad, A. Saberian, M. Fadaee, M. Radzi, H. Hizam, M. AbKadir, The present and future of smart power grid in developing countries, Renewable and Sustainable Energy Reviews 29 (2014) 828–834, https://doi.org/10.1016/j.rser.2013.08.072.
[72] A. Gaviano, K. Weber, C. Dirmeier, Challenges and integration of pv and wind energy facilities from a smart grid point of view, Energy Procedia 25 (2012) 118–125, https://doi.org/10.1016/j.egypro.2012. 07.016.
[73] F. Ayadi, I. Colak, I. Garip, H.I. Bulbul, Impacts of renewable energy resources in smart grid, in: 2020 8th International Conference on Smart Grid (icSmartGrid), 2020, pp. 183–188, https://doi.org/10.1109/icSmartGrid49881.2020.9144695.

[74] I. Worighi, A. Maach, A. Hafid, O. Hegazy, J. Van Mierlo, Integrating renewable energy in smart grid system: architecture, virtualization and analysis, Sustainable Energy, Grids and Networks 18 (2019) 100226, https://doi.org/10.1016/j.segan.2019.100226.
[75] F. Akhtar, M.H. Rehmani, Energy replenishment using renewable and traditional energy resources for sustainable wireless sensor networks: a review, Renewable and Sustainable Energy Reviews 45 (2015) 769–784, https://doi.org/10.1016/j.rser.2015.02.021.
[76] E. Bitar, P. Khargonekar, K. Poolla, Systems and control opportunities in the integration of renewable energy into the smart grid, IFAC Proceedings Volumes 44 (1) (2011) 4927–4932, https://doi.org/10.3182/20110828-6-IT-1002.01244.
[77] J. Moura, Accommodating High Levels of Variable Generation, the Canadian Wind Energy Association's 2009 Wind Matters Conference, 2009.
[78] D. Eltigani, S. Masri, Challenges of integrating renewable energy sources to smart grids: a review, Renewable and Sustainable Energy Reviews 52 (2015) 770–780, https://doi.org/10.1016/j.rser.2015.07. 140.
[79] L. Xie, P.M.S. Carvalho, L.A.F.M. Ferreira, J. Liu, B.H. Krogh, N. Popli, M.D. Ilić, Wind integration in power systems: operational challenges and possible solutions, Proceedings of the IEEE 99 (1) (2011) 214–232, https://doi.org/10.1109/JPROC.2010.2070051.
[80] I.E. Commission, et al., Grid Integration of Large-Capacity Renewable Energy Sources and Use of Large- Capacity Electrical Energy Storage, International Electrotechnical Commission, 2012.
[81] I. Wasiak, Z. Hanzelka, Integration of distributed energy sources with electrical power grid, Bulletin of the Polish Academy of Sciences, Technical Sciences 57 (No 4) (2009) 297–309, https://doi.org/10.2478/v10175-010-0132-1.
[82] Energy Great Lakes Regional Wind, Eastern Wind Integration and Transmission Study, Technical Re- Port, National Renewable Energy Laboratory (NREL), 2010.
[83] G. Energy, et al., Western Wind and Solar Integration Study, Tech. Rep, National Renewable Energy Laboratory, 2010.
[84] J. DeCesaro, K. Porter, M. Milligan, Wind energy and power system operations: a review of wind integration studies to date, The Electricity Journal 22 (10) (2009) 34–43, https://doi.org/10.1016/j.tej.2009.10.010.
[85] J. Marcos, L. Marroyo, E. Lorenzo, D. Alvira, E. Izco, Power output fluctuations in large scale pv plants: one year observations with one second resolution and a derived analytic model, Progress in Photovoltaics: Research and Applications 19 (2) (2011) 218–227, https://doi.org/10.1002/pip.1016.
[86] Caribbean Renewable Energy Development Programme CREDP, Wind-Grid Integration Brief, Solar Water Heating Market, 2004.
[87] M. Ourahou, W. Ayrir, B. EL Hassouni, A. Haddi, Review on smart grid control and reliability in presence of renewable energies: challenges and prospects, Mathematics and Computers in Simulation 167 (2020) 19–31, https://doi.org/10.1016/j.matcom.2018.11.009.
[88] P. Hewage, M. Trovati, E. Pereira, A. Behera, Deep learning-based effective fine-grained weather forecasting model, Pattern Analysis & Applications 24 (1) (2021) 343–366, https://doi.org/10.1007/s10044-020-00898-1.
[89] P. Cartwright, Connecting renewables: the challenge of integrating large offshore wind farms, Refocus 7 (1) (2006) 24–26, https://doi.org/10.1016/S1471-0846(06)70513-3.
[90] N.T. Linh, Power quality investigation of grid connected wind turbines, in: 2009 4th IEEE Conference on Industrial Electronics and Applications, 2009, pp. 2218–2222, https://doi.org/10.1109/ICIEA.2009. 5138593.

[91] H.H. El-Tamaly, M.A. Wahab, A.H. Kasem, Simulation of directly grid-connected wind turbines for voltage fluctuation evaluation, International Journal of Applied Engineering Research 2 (1) (2007) 15–30.

[92] E. Bossanyi, Z. Saad-Saoud, N. Jenkins, Prediction of flicker produced by wind turbines, Wind Energy 1 (1) (1998) 35–51, https://doi.org/10.1002/(SICI)1099-1824(199809)1:1<35::AID-WE11>3.0. CO;2-J.

[93] S.K. Khadem, M. Basu, M. Conlon, Power quality in grid connected renewable energy systems: role of custom power devices, in: International Conference on Renewable Energies and Power Quality, Granada, 2010.

[94] F.H. Gandoman, A. Ahmadi, A.M. Sharaf, P. Siano, J. Pou, B. Hredzak, V.G. Agelidis, Review of facts technologies and applications for power quality in smart grids with renewable energy systems, Renewable and Sustainable Energy Reviews 82 (2018) 502–514, https://doi.org/10.1016/j.rser.2017.09. 062.

[95] O. Ceaki, G. Seritan, R. Vatu, M. Mancasi, Analysis of power quality improvement in smart grids, in: 2017 10th International Symposium on Advanced Topics in Electrical Engineering (ATEE), 2017, pp. 797–801, https://doi.org/10.1109/ATEE.2017. 7905104.

[96] P. Gope, B. Sikdar, Lightweight and privacy-friendly spatial data aggregation for secure power supply and demand management in smart grids, IEEE Transactions on Information Forensics and Security 14 (6) (2019) 1554–1566, https://doi.org/10.1109/TIFS.2018.2881730.

[97] A. Razzak, M.T. Islam, P. Roy, M.A. Razzaque, M.R. Hassan, M.M. Hassan, Leveraging deep q-learning to maximize consumer quality of experience in smart grid, Energy 290 (2024) 130165, https://doi.org/10.1016/j.energy.2023.130165.

[98] F. Liu, T. Dong, Q. Liu, Y. Liu, S. Li, Combining fuzzy clustering and improved long short-term memory neural networks for short-term load forecasting, Electric Power Systems Research 226 (2024) 109967, https://doi.org/10.1016/j.epsr.2023.109967.

[99] N.A. Kumar, R. Daniel, P.K. Pasam, A Novel Electrical Load Forecasting Model Using a Deep Learning Approach, the Internet of Energy: A Pragmatic Approach towards Sustainable Development, 2024, p. 67.

[100] S. Kishore, L.V. Snyder, Control mechanisms for residential electricity demand in smartgrids, in: 2010 First IEEE International Conference on Smart Grid Communications, 2010, pp. 443–448, https://doi.org/10.1109/SMARTGRID.2010. 5622084.

[101] A.Y.A. Bani Ahmad, P. William, D. Uike, A. Murgai, K.K. Bajaj, A. Deepak, A. Shrivastava, Frame- work for sustainable energy management using smart grid panels integrated with machine learning and iot based approach, International Journal of Intelligent Systems and Applications in Engineering 12 (2s) (2023) 581–590.

[102] F. Benzi, N. Anglani, E. Bassi, L. Frosini, Electricity smart meters interfacing the households, IEEE Transactions on Industrial Electronics 58 (10) (2011) 4487–4494, https://doi.org/10.1109/TIE.2011.2107713.

[103] S.R. Salkuti, Challenges, issues and opportunities for the development of smart grid, International Journal of Electrical and Computer Engineering 10 (2) (2020) 1179–1186, https://doi.org/10.11591/ijece.v10i2.pp1179-1186.

[104] O. Hegazy, Advanced Power Electronics Interface and Optimization for Fuel Cell Hybrid Electric Vehicles Applications, Ph.D. thesis, Vrije Universiteit Brussel, 2012.

[105] Department of Energy, Distribution automation: results from the smart grid investment grant program, Technical report, United States Department of Energy (2016).

[106] K.M. Tan, T.S. Babu, V.K. Ramachandaramurthy, P. Kasinathan, S.G. Solanki, S.K. Raveendran, Empowering smart grid: a comprehensive review of energy storage technology and application with renewable energy integration, Journal of Energy Storage 39 (2021) 102591, https://doi.org/10.1016/j.est. 2021.102591.

[107] C. Pang, P. Dutta, M. Kezunovic, Bevs/phevs as dispersed energy storage for v2b uses in the smart grid, IEEE Transactions on Smart Grid 3 (1) (2012) 473–482, https://doi.org/10.1109/TSG.2011.2172228.
[108] R. Arghandeh, M. Pipattanasomporn, S. Rahman, Flywheel energy storage systems for ride-through applications in a facility microgrid, IEEE Transactions on Smart Grid 3 (4) (2012) 1955–1962, https://doi.org/10.1109/TSG.2012.2212468.
[109] S. Samineni, B. Johnson, H. Hess, J. Law, Modeling and analysis of a flywheel energy storage system for voltage sag correction, IEEE Transactions on Industry Applications 42 (1) (2006) 42–52, https://doi.org/10.1109/TIA.2005.861366.
[110] H. Chamandoust, A. Hashemi, S. Bahramara, Energy management of a smart autonomous electrical grid with a hydrogen storage system, International Journal of Hydrogen Energy 46 (34) (2021) 17608–17626, https://doi.org/10.1016/j.ijhydene.2021.02.174.
[111] D. Enescu, G. Chicco, R. Porumb, G. Seritan, Thermal energy storage for grid applications: current status and emerging trends, Energies 13 (2) (2020), https://doi.org/10.3390/en13020340.
[112] M. Abujubbeh, F. Al-Turjman, M. Fahrioglu, Software-defined wireless sensor networks in smart grids: an overview, Sustainable Cities and Society 51 (2019) 101754, https://doi.org/10.1016/j.scs.2019. 101754.
[113] D.B. Avancini, J.J. Rodrigues, S.G. Martins, R.A. Rabêlo, J. Al-Muhtadi, P. Solic, Energy meters evolution in smart grids: a review, Journal of Cleaner Production 217 (2019) 702–715, https://doi.org/10.1016/j.jclepro.2019.01.229.
[114] D.K. Panda, S. Das, Smart grid architecture model for control, optimization and data analytics of fu- ture power networks with more renewable energy, Journal of Cleaner Production 301 (2021) 126877, https://doi.org/10.1016/j.jclepro.2021.126877.
[115] M. Uzunoglu, M. Alam, Dynamic modeling, design, and simulation of a combined pem fuel cell and ultracapacitor system for stand-alone residential applications, IEEE Transactions on Energy Conversion 21 (3) (2006) 767–775, https://doi.org/10.1109/TEC.2006.875468.
[116] L. Shi, M.L. Crow, Comparison of ultracapacitor electric circuit models, in: 2008 IEEE Power and Energy Society General Meeting - Conversion and Delivery of Electrical Energy in the 21st Century, 2008, pp. 1–6, https://doi.org/10.1109/PES.2008.4596576.
[117] Y. Kim, V. Raghunathan, A. Raghunathan, Design and management of battery-supercapacitor hybrid electrical energy storage systems for regulation services, IEEE Transactions on Multi-Scale Computing Systems 3 (1) (2017) 12–24, https://doi.org/10.1109/TMSCS.2016.2627543.
[118] U. Akram, M. Khalid, A coordinated frequency regulation framework based on hybrid battery- ultracapacitor energy storage technologies, IEEE Access 6 (2018) 7310–7320, https://doi.org/10.1109/ACCESS.2017.2786283.
[119] A.A. Aquino-Lugo, T. Overbye, Agent technologies for control applications in the power grid, in: 2010 43rd Hawaii International Conference on System Sciences, 2010, pp. 1–10, https://doi.org/10.1109/HICSS.2010.43.
[120] Y.M. Atwa, E.F. El-Saadany, M.M.A. Salama, R. Seethapathy, Optimal renewable resources mix for distribution system energy loss minimization, IEEE Transactions on Power Systems 25 (1) (2010) 360–370, https://doi.org/10.1109/TPWRS.2009.2030276.

CHAPTER 3

Emerging smart grid technologies and their impact on grid operations

Divya Asija, R.K. Viral, Tushar Tomar, Manoj Kumar and Aman Kumar
Department of Electrical and Electronics Engineering, Amity School of Engineering and Technology, Amity University Uttar Pradesh, Noida, India

3.1 Introduction

Smart grid technologies utilize recent cyber advancements to enhance control and monitoring functions throughout the electric power grid. The smart grid incorporates various individual technical initiatives such as advanced metering infrastructure (AMI), demand response (DR), WAMPAC systems based on phasor measurement units (PMUs), large scale renewable integration in the form of wind and solar generation, and plug-in hybrid electric vehicles (PHEVs) [1]. Of these initiatives, AMI and WAMPAC rely heavily on cyber infrastructure and its data transmitted via multiple communication protocols to utility control centers and consumers. Cybersecurity issues within the communications and computing infrastructure can allow attackers to manipulate either power applications or the physical system. Cyber-attacks can take many forms depending on their target. System resources can also be made unavailable through Denial of Service (DoS) attacks by flooding the network or system with unnecessary data. Even secure cyber systems can be attacked by insider threats, where a trusted individual can use system privileges to steal data or affect system operations. Weaknesses in communication protocols also allow attackers to steal or manipulate data in transit.

AMI is based on the deployment of smart meters on the consumer and utility side [2]. This gives the utility the ability to provide real-time pricing

Cyber Security Solutions for Protecting and Building the Future Smart Grid
ISBN 978-0-443-14066-2
https://doi.org/10.1016/B978-0-443-14066-2.00009-8

data to consumers, gather information on current usage, and perform more advanced fault analysis in the distribution system. Because AMI is connected to the distribution system, it is usually necessary to compromise a huge volume of consumer meters to create a substantial impact on the reliability of the bulk energy system. This is in stark contrast to the impact that a coordinated cyber-attack on WAMPAC would have on the reliability of the bulk power system. Therefore, the focus of this chapter is to study the related cyber physical security issues of WAMPAC [3]. However, it is important to note that there are several cybersecurity and privacy issues related to AMI and are beyond the scope of this chapter.

3.2 Digital communication system

A smart grid (SG) is a digital, self-healing electrical energy system. It has the features of observability, controllability, integrity to ensure reliable power and reduced carbon emissions. In this paper, communication technologies employed in smart grid such as Wireless Area network (WAN), Neighborhood Area Network (NAN) and Home (local) area network (HAN/LAN) are elaborated [4] and shown in Fig. 3.1.

3.2.1 Wide area network (WAN)

WAN is deployed and operated within vast area of tens of kilometers, and it consists of several NANs and LDCs. Moreover, the communication of all smart grid's components including operator control center, main and renewable energy generation, transmission, and distribution, is based on WAN [4]. The WAN has very high transmission data rate up to few Gbps. The WAN can be implemented by Ethernet networks, WiMAX, 3G/LTE, and microwave transmission.

3.2.2 Neighborhood area network (NAN)

NAN is deployed and operated within area of hundreds of meters which is few urban buildings. Several HANs can be connected to one NAN, and they transmit data of energy consumed by each house to the NAN network. The NAN network delivers this data to Local Data Centers for storage [5]. This data storage is important for charging the consumers and data analysis for energy generation-demand pattern recognition. The NAN has up to 2 Kbps transmission data rate. The NAN can be implemented by PLC, Wi-Fi, and cellular technologies.

3.2.3 HAN

HAN is deployed and operated within a small area (tens of meters), usually a house or a small office. The HAN has relatively low transmission data rate compared to other two networks, hundreds of bits per second (bps) [5]. In a typical implementation, a HAN consists of a broadband Internet connection that is shared between multiple users through a wired or wireless modem. It enables the communication and sharing of resources between computers, mobile and other devices over a network connection. In smart grid implementation, all smart home devices that consume energy and smart meters can be connected to HAN. The devices data is acquired and transmitted through HAN to the smart meters. HAN allows more efficient home energy management. HAN can be implemented by ZigBee or Ethernet technologies (Fig. 3.1).

3.3 Description of smart grid communication technologies

3.3.1 Zigbee

ZigBee is based on an IEEE 802.15 standard. ZigBee is used in applications that require a low data rate, long battery life, low cost, and secure networking [6]. Applications include wireless light switches, electrical meters with in-home-displays, traffic management systems, and other consumer and industrial equipment that requires short-range wireless transfer of data at relatively low rates. ZigBee allows connection of up to 60,000 devices to its network. ZigBee has a defined rate between 20 and 250 kbs, best suited for periodic or intermittent data or a single signal

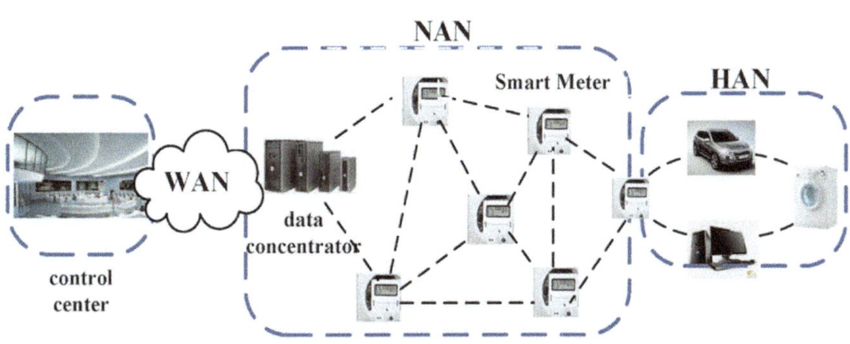

Figure 3.1 A typical smart grid communication in WAN, NAN, HAN.

transmission from a sensor or input device. The technology defined by the ZigBee specification is intended to be simpler and less expensive than other wireless personal area networks (WPANs), such as Bluetooth or Wi-Fi. ZigBee networks are secured by 128-bit symmetric encryption keys.

3.3.2 WLAN

A wireless local area network (WLAN) links two or more devices using spread-spectrum or Orthogonal Frequency Division Multiplexing (OFDM) [15] and usually providing a connection through an access point to the wider Internet. This gives users the ability to move around within a local coverage area and still be connected to the network. Most modern WLANs are based on IEEE 802.11 standards, marketed under the Wi-Fi brand name. WLANs have become popular in the home due to ease of installation, and in commercial complexes offering wireless access to their customers.

3.3.3 WiMAX

WiMAX is one of the hottest broadband wireless technologies around today. WiMAX systems are expected to deliver broadband access services to residential and enterprise customers in an economical way.

3.3.4 Bluetooth

Bluetooth technology allows devices to communicate with each other without cables or wires. Bluetooth relies on short range radio frequency, and any device that incorporates the technology can communicate as long as it within the required distance.

3.3.5 Cellular communication

The cellular communication technology is the best option for communication between far nodes, utility, and smart meters. Currently, 2G, 2.5G, 3G, 4G, GSM, LTE, and WiMAX are the available cellular communication technologies [7]. GSM/GPRS technologies are a good option for communicating between smart meters and the utility, which transfers data and control signals over long distances. Global System for Mobile (GSM) is considered among the most secure communication networks. General Packet Radio Service (GPRS) employs wireless packet based on GSM network. If the infrastructure exists, extra cost for building the communications infrastructure will not be needed.

3.4 Communication architecture of smart grids

3.4.1 ICT model of power system

With evolving Smart Grid Architecture, the Information and Communication Technology (ICT) becomes the backbone for establishing reliable control functionality for secure operation of large, interconnected Grid. The current research areas of Energy Storage and Smart Grids scores high in timeliness and appropriateness [9]. Advances in energy storage systems are crucial if renewable energy has to progress from being aside-player in the electricity arena to becoming a more central contributor. RE technologies based on solar and wind energy suffer from poor Plant Load Factor (PLF) because of intermittency in generation and the asynchronous relationship between demand and supply. Current energy storage technologies, especially in the electrochemical category, suffer from low energy density and therefore, are not cost-effective options above a certain scale. Therefore, there is a compelling need for cutting-edge research in energy storage options. The current proposal seeks to further research on energy storage technologies in a manner that addresses three important dimensions—Technology, Economics and Practical Implementation [8]. The potential of renewable energy technologies as decentralized solutions for underserved segments as well as important substitutes for fossil fuels can be realized only when they are integrated with the grid through appropriate technologies and a web of interconnected devices, sensors, and digital communication. The promise of smart grids is still a few years away from being realized to the full and therefore, the proposed research areas can play a critical role in advancing knowledge and accelerating action. It thus scores high in timelines sand appropriateness.

3.4.2 Substation automation system

Substation automation is the act of automatically controlling a substation using instrumentation and control devices. Substation automation refers to using data from Intelligent Electronic Devices (IED's), control and automation capabilities within the second, and control commands from remote users using Supervisory Control and Data Acquisition System (SCADA) to control power-system (switchyard) devices. A substation automation system is commonly used to control, protect and monitor a substation. However, over the years advances in electronics, information and communications technology have brought about sweeping changes in the way substations are operated. The advent of software-based substation automation systems connected by serial links rather than rigid parallel copper wiring gradually became the norm rather than the exception. Though successful and widely accepted, these systems were based on either the manufacturers' own proprietary communication solutions or the defined use of communication standards from other application domains [7].

Substation Automation (SA) is a system to enable an electric utility to remotely monitor, control and coordinate the distribution components installed in the substation.

There are three levels of Substation automation system:

1. Station level
2. Bay level
3. Process level

3.4.2.1 Station level

A redundant PC-based HMI enables local station control through the software package Micro SCADA Pro, which contains an extensive range of SCADA functions. The station level contains station-oriented functions that cannot be implemented at the field level, e.g., a list of alarms or events related to the entire substation, a gateway for communication with remote control centers. A dedicated master clock for the synchronization of the entire system shall be provided.

3.4.2.2 Bay level

A bay comprises of circuit breaker and associated isolators, earth switches and instrument transformers. At bay level, the IEDs provide all bay level functions such as control (command outputs), monitoring (status indications, measured values) and protection. The IEDs are directly connected to the switchgear without any need for additional interposing or transducers. Each bay control IED is independent of the others and its functioning is not affected by any fault occurring in any of the other bay control units of the station.

3.4.2.3 Process level

It consists of all switchgear that is hardwired using copper cables and uses fiber optic cables to connect shaft level IEDs used for control and protection (Fig. 3.2).

3.4.3 SCADA

SCADA is a remote monitoring and control system that uses coded signals across communication channels to operate (usually employing one

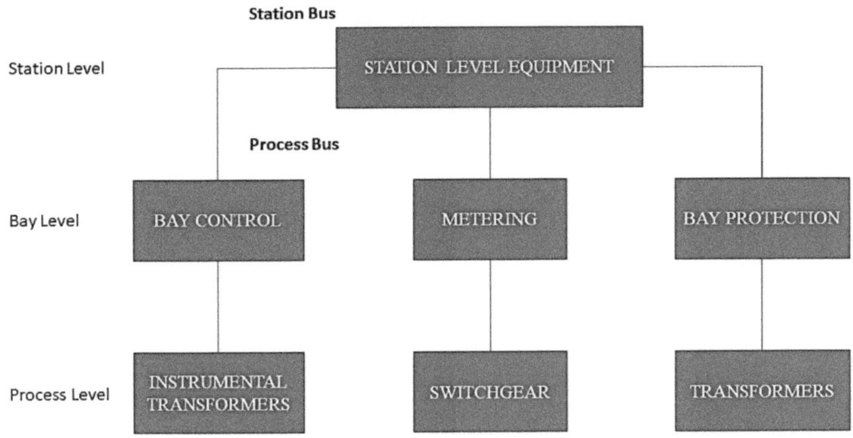

Figure 3.2 Substation architecture at process, bay, and station level.

"communication channel per remote station"). By adding the use of coded signals across communication channels to get information about the condition of the remote equipment for display or for recording purposes, the control system may be coupled with a data acquisition system. It's a particular kind of industrial control system (ICS). Computer-based systems known as industrial control systems monitor and regulate industrial operations that take place in the real world. SCADA systems have traditionally set themselves apart from other ICS systems by being large-scale operations that may span several substations and distances.

An SCADA system usually consists of the following subsystems:

i. A central host computer server or servers (sometimes called an SCADA Center, master station, or Master Terminal Unit (MTU).

ii. One or more field data interface devices, usually RTUs, or PLCs, which interface to field sensing devices and local control switchboxes and valve actuators

iii. A communications system used to transfer data between field data interface devices and control units and the computers in the SCADA central host. 4. A collection of standard and/or custom software [sometimes called Human Machine Interface (HMI) of MMI (Man-Machine Interface) software systems used to provide the SCADA central host and operator terminal application, support the communications system, and monitor and control remotely located field data interface devices.

3.4.4 IEDs (intelligent electronics devices)

Electric energy is a commodity that like any other is bought under a contract between a customer and utility. There are many aspects to electrical power management apart from the price and uniform delivery they are great importance. This is equipment downtime minimization, maintaining energy quality indices indoors limits set by the relevant electricity quality standards, integration of renewable energy sources. Every one of them aspects requires comprehensive data acquisition from grid. This must be done using IEDs placed usually on a substation level, capable of accurate and fast voltage and current measurement. The smart grid at the substation level is composed of IEDs connected to transducers and sensors, a communication structure and special software. The trend is to apply the IEC 61,850 standard for substation automation. The present IEC 61,850 standard suite involves communication and relevant data models, and the work is going on to standardize the methods and the devices needed for online monitoring. The standard allows power utilities to cost effectively integrate power system applications, including SCADA, EMS/DMS, protection system and field equipment. The IEDs in the grid have many functions—protection, power quality analysis, equipment diagnosis, energy metering, grid monitoring. The information needed by each of these functions has to be extracted from a single raw input data which are sequences of the samples of phase voltage and current signals. For some functions, e.g., Circuit Breaker diagnosis [1], the input signal samples sequences are transmitted to external processing devices. Other functions are implemented in IEDs and the processing is done by embedded processor. The basic architecture of advanced IED has been shown in Fig. 3.3.

3.4.5 Phasor measurement unit

A PMU is a device used to estimate the magnitude and phase angle of an electrical phasor quantity (such as voltage or current) in the electricity grid using a common time source for synchronization. Time synchronization is usually provided by GPS or IEEE 1588 Precision Time Protocol, which allows synchronized real-time measurements of multiple remote points on the grid. PMUs are capable of capturing samples from a waveform in quick succession and reconstructing the phasor quantity, made up of an angle measurement and a magnitude measurement. The resulting measurement is known as a synchro phasor. These time synchronized measurements are important because if the grid's supply and demand are not perfectly

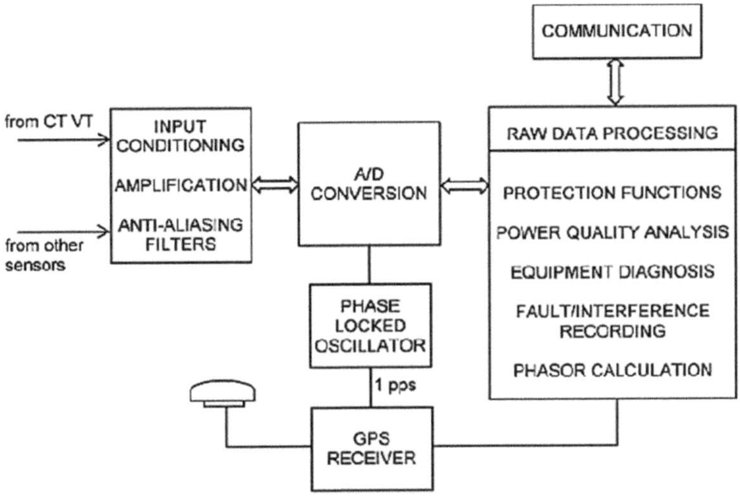

Figure 3.3 Design of advanced many function IED.

matched, frequency imbalances can cause stress on the grid, which is a potential cause for power outages. PMUs can also be used to measure the frequency in the power grid [2]. A typical commercial PMU can report measurements with very high temporal resolution, up to 120 measurements per second. This helps engineers in analyzing dynamic events in the grid which is not possible with traditional SCADA measurements that generate one measurement every 2 or 4 s.

A phasor is a mathematical representation of a sinusoidal waveform. The phase angle at a given frequency is determined with respect to a time reference. Synchro phasors are phasor values that represent power system sinusoidal waveforms referenced to the nominal power system frequency and coordinated universal (UTC) time. The phase angle of a synchro phasor is governed by the waveform, the system frequency, and the instant of measurement. Thus, with a universal precise time reference, power system phase angles can be accurately measured throughout a power system. The global positioning system (GPS) technology provides an economic option for the same. An important advantage of the GPS technology is that its receiver can automatically detect accurate synchronization. The device which provides synchronized phasor measurements is called a PMU. A number of widely distributed PMUs in the power system may be utilized for [3]:

- Real time monitoring and control
- State estimation
- Protection and control for distributed generation
- Network congestion management
- Angular and voltage stability monitoring
- Post-Mortem analysis of disturbances and faults

A PMU is an electronic device that uses a digital signal processing element to measure and convert an AC waveform into a phasor, depending on the system frequency, and to synchronize those measurements under the control of a GPS Reference source. The analog signal is sampled and processed by a recursive phasor algorithm to generate voltage and current phasors [4]. The various components of the PMU are illustrated by the block diagram in Fig. 3.4.

3.4.6 PDC (phasor data concentrator)

The storage space and cost of the smart grid dataset has grown exponentially due to the high data rates of various sensor readings from Advanced Metering Infrastructure (AMI) and PMU. This whitepaper focuses on phasor data concentrators (PDCs) that aggregate data from PMUs [5]. The PMU measures voltage, current and frequency parameters in real time on the power grid. A typical PDC can handle data from 10 to 40 PMUs. This paper exploits the need for adequate security and data compression challenges simultaneously. As a result, the optimal compression scheme ER1c for efficient storage of IREG and C37.118 timestamped PDC records is investigated. We estimate that our approach can significantly reduce the storage cost requirements of commercial PDCs (SEL3373, GE Multiline

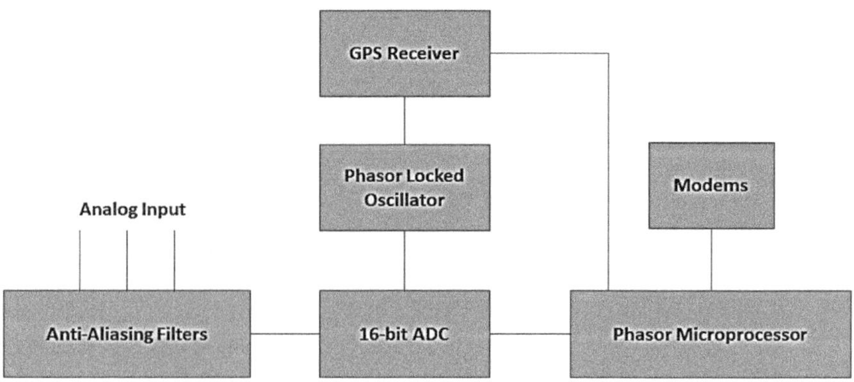

Figure 3.4 Components of a phasor measurement unit.

P30) by 80%. For example, 2 years of PDC data storage space can easily be replaced with just 10 days of storage. Additionally, the approach at combined with AES 256 encryption allows PDC data at larger scales to be protected according to the National Institute of Standards and Technology (NIST) standard [6].

The number of PMU deployments is growing exponentially in North America at, so the amount of data stored even for a short period of time is staggering [7]. PDC storage is limited (estimated at 100 GB) and varies by vendor. Therefore, data should be stored in alternative formats such as DVDs or hard drives. The main idea of this white paper is to optimize the storage of data sets such as with large data rates (30 samples/sec).

VCL-PDC is a high-performance platform for managing synchro phasor data and processing streaming data in real time. The VCL-PDC receives and synchronizes phasor data from multiple PMUs and produces a real-time time-stamped output data stream. PDC can store PMU data simultaneously and send to multiple locations. By using multiple Her PDCs, you can create multiple levels of concentration within the synchro phasor data system. Various utilities are included to streamline the initial setup, configuration, and testing process. VCL-PDC is a very versatile and easy-to-use phasor data concentrator.

This optimal reduction in file or storage sizes of PDC data can help in reducing storage cost, efficiently secure and organize SQL queries and retrieval with faster download-time. The parameters (compression rates, data retrieval time) have considered as a benchmark performance metrics at super PDC level.

The designed PDC (iPDC) follows the same standard [8] while combining received frames. The iPDC can receive (data and configuration) frames from either the PMU or the PDC. As frames are received, iPDC time aligns them into one frame. Those frames with the same SOC and FRACSEC are combined into one frame in the order of IDCODE of PMU/PDC. A list of PMU/iPDC IDCODEs is maintained. The IDCODEs in the list are in the order the iPDC receives configuration frames from the PMU or other iPDCs. Therefore, data frames that arrive out of order must be arranged in the order of IDCODEs listed. These combined frames are sent by other iPDCs upon request. When an iPDC receives data from another iPDC, it uses the internal data of this frame to combine with the main frame. This data includes multiple PMU data. Fig. 3.5 is the combined configuration and data frame respectively.

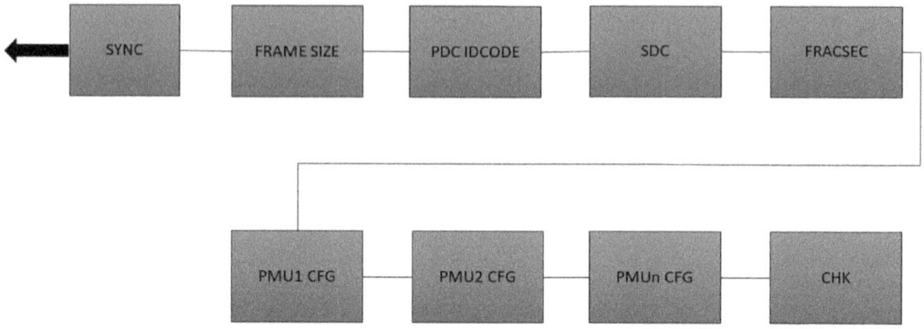

Figure 3.5 Combined configuration frame.

3.5 Distribution system

3.5.1 AMI

Smart grid concept has emerged from efforts which make electrical grids with developing technological possibilities more efficient, safer, more environmentally friendly and manageable covering all processes from production of energy to the distribution and consumption. Recently, have been worked on the subject, and studies have been started to be performed on many applications such as communication, information technologies, control systems, semi-conductor technologies [9,10].

Smart grid inspects and observes energy flow from first generation unit to end consumer. This situation is more sustainable for oncoming energy implementation because of existing electricity system and advanced metering infrastructure problems [4]. Smart grid is equipped with Advanced Metering Infrastructure (AMI). The most basic component of advanced measurement infrastructures are smart meters. According to Lord Kelvin, if you can measure what you are talking about and you can tell it with numbers, you know something about it. So as to know something about it which technology we exert, we need to measure and observe correctly [11].

According to the definition of electric power research institute (EPRI), which is based in the United States, established by electric service providers, AMI is a system consisting of modern electronic-digital hardware and software, which enables data measurement intermittently and remote communication continuously. These systems have an infrastructure that can make detailed measurements, collect time-based information constantly and share this information with the parties as needed [12].

Advanced Metering is a metering system that measures customer consumption hourly or more frequently. AMI is based on smart meters. A smart meter has two functions: measurement and communication. For quantitative measurement, smart meters must be able to accurately measure the amount of medium using a variety of physical principles, topologies, and methods. A smart meter consists of control and calibration devices and functions. Customers should be able to see meter information as this information is the basis for billing. Demand management on the customer side also needs an indication. Time synchronization is important to ensure data is sent to a central hub or other collection system for data analysis and billing. All necessary data and variables can be collected by measuring and storing them in different memory units. After analyzing this saved data, variables and momentary conditions are available on the screen of the display. All this prose is done by advanced measurement infrastructure.

Automatic smart reading (AMR) systems are separated from AMI. AMRs performs with less developed technologies such as turn off appliances manually, collecting data without real time (monthly) etc. but advanced metering infrastructures have real time operations, customer/pricing/utility options, remote control and management units etc. Thus transition to advanced metering infrastructures is crucial to build smart grid structures [13] (Fig. 3.6).

3.5.2 DERs (distributive energy resources)

Distributed Energy Resources (DER) early uses as a backup generation has been progressing toward permanent Distributed Generation (DG), along with the development and enhancement of new technologies over small-

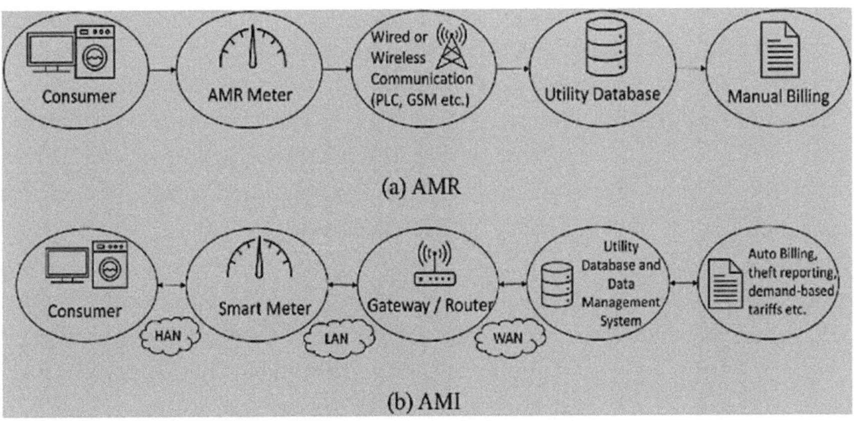

Figure 3.6 AMR and AMI structures.

scale generation. Over last few years, increasing penetration of renewables in the distribution networks at consumer level raises concerns on protection, control, stability and reliability. Considering the DG integration and wide variations in operating conditions of the microgrid, relays experience protection issues at fault current level violating important tripping decision rules.

DER generation covers multiple technologies, including those from renewable energy sources. The DER generation source with the most capacity integration to date has been solar photovoltaic (PV). Motivated by falling solar generation prices and evolving rate structures, there has been widespread adoption of PV. Other renewable DERs include those from wind and water, with small-scale run-of-river being well established as a technology, but potential wave and tidal DERs not yet having a clear technology forerunner or paradigm. Biomass and waste-to-energy are increasingly being coupled with agriculture by-products and municipal waste systems, respectively. Electrochemical DER generation includes fuel cells and electrolyzers, although these also have energy-storage and energy-consumption (for hydrogen production) properties that can give these technologies DER storage and demand-side characteristics. Non-renewable DERs include small-scale fossil-fired generation and cogeneration for heat and electricity production, such as combined heat and power (CHP) generation.

Many of the DERs described here have power electronic interfaces (PEIs) to the grid and are inverter-based. Legacy inverters were programmed to trip in response to a fault or disturbance on the grid with negligible effect. However, at higher penetrations of inverter-based DER, the tripping leads to a significant portion of the generation going offline, thus exacerbating the instability of the grid under fault. This was seen in the Canyon fire event [14] and the Blue Cut fire event [15]. Though most of the generation lost in the two fire events was connected to the bulk system, the ride-through capability is becoming increasingly important for the distribution-connected, inverter-based DERs. Output from some DERs, such as PV, can be highly variable when deployed without storage. These increase switching operations and wear on the load tap changers and voltage regulators that maintain voltage within operating limits.

3.5.3 Distribution automation

The most important application of the Advanced Distribution Automation is fault diagnosis by monitoring the faults in the grid, then identifying the root cause of the occurred fault and then restoring the system. Automated fault and root-cause identification can also be investigated using the state of art of technologies such as cause-effect network, Artificial Intelligence or Bayesian inference. The deployment of wireless sensors, distributed actuators, and the present day information and communication technologies would generate more accurate real-time data about the states, making automated fault diagnosis feasible and applicable in future intelligent smart grids.

Distribution Automation technologies and systems can achieve substantial grid impacts and benefits:

i. Improvement in location of fault, isolation, and service restoration capabilities that result in a smaller number of outages, less operational and maintenance costs, less chances of equipment failure or damage, and lesser inconvenience at the consumer end.

ii. Improved distribution system resilience to extreme weather events by automatically limiting the extent of major outages and improving operator ability to diagnose and repair damaged equipment.

iii. More effective equipment monitoring and preventative maintenance that reduces operating costs, enables more efficient use of capital assets, reduces the likelihood of equipment failures.

iv. More efficient use of repair crews and truck rolls that reduces operating costs, enables faster service restoration, and lowers environmental emissions.

v. Improved grid integration of selected DER such as thermal storage for commercial and municipal buildings (Fig. 3.7).

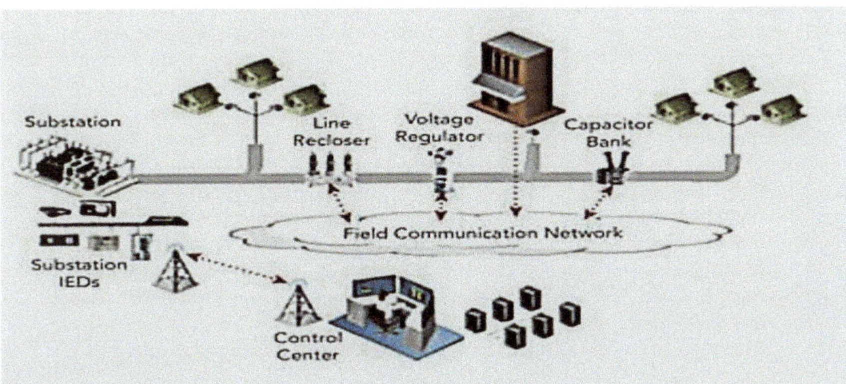

Figure 3.7 Illustration of distribution automation system distribution automation field devices includes.

3.6 Conclusion

In synopsis, the integration of SG technologies is ushering in a transformative phase in electricity distribution, yielding benefits for stakeholders and consumers. However, challenges, especially in distribution, persist. This chapter offers a comprehensive overview of strides in SG implementation, spanning technological advancements and foundational infrastructure. Central is the notion of transactive energy, a decentralized approach involving automated energy transactions. Leveraging SG intelligence and interconnectivity, it holds potential for diverse benefits. Integration of these transactions into local markets, coupled with existing SG-enabled infrastructure, marks a critical step toward local energy market emergence. In closure, this chapter underscores transactive energy's symbiotic role in SG realization. As the energy landscape evolves, the synergy between intelligent grid architectures and transactive energy not only addresses operational complexities and energy management but also nurtures participatory energy dynamics. The trajectory of this potential depends on technological, market, and regulatory confluence, shaping future energy distribution and consumption paradigms.

References

[1] W.W. Clark II, G. Cooke, in: Global Sustainable Communities Handbook, 2014.

[2] F. Lezama, S. Widergren, Local Electricity Markets, 2021.

[3] Z. Mhadhbi, S. Zairi, C. Gueguen, B. Zouari, Validation of a distributed energy management approach for smart grid based on a generic colored Petri NetsModel, Journal of Clean Energy Technologies 6 (1) (January 2018).

[4] V. Tamilmaran, D.P. Kothari, Smart grid: an overview, Smart Grid and Renewable Energy 2 (4) (November 2011).

[5] F. Hertrampf, S. Apel, S. Späthe, Requirements and architecture concept for a data processing system in local area smart grid, Journal of Electrical Engineering 6 (2018) 1–10.

[6] Roger N. Anderson, R. Ghafurian, H. Gharavi, Smart grid the future of the electric energy system, IEEE Explore (June 2020) 1–11. https://www.researchgate.net/publication/326426286_Smart_Grid-_The_Future_of_the_Electric_Energy_System.

[7] Y.E.M. Hamouda, S.J. Dwedar, Optimally Automated Home Management for Smart Grid System Using Sensors, Springer Nature Singaporepte Ltd, August 2020.

[8] S. Kumar Suman, M. Aqib, S.K. Singh, A security apptoach for smart grid, APTIKOM Journal on Computer Science and Information Technologies 2 (1) (2017).

[9] J. Kamto, L. Qian, J. Fuller, J. Attia, Light-Weight Key Distribution and Management for Advanced Meteringinfrastructure, December 2011.

[10] C. Efthymiou, G. Kalogridis, Smart Grid Privacy via Anonymization of Smart Metering Data, Toshiba Research Europe Ltd., November 2010.

[11] A. Nowakowski, A. Lisowiec, Z. Koáodziejczyk, MV bay controller with integrated protection, power quality analysis and extended self-testing functions, Proceedings of the 11th. Spanish-Portuguese Conference on Electrical Engineering (11Chlie).
[12] S. Horowitz, A. Phadke, B. Renz, The future of power transmission, IEEE Power and Energy Magazine 8 (2) (2010) 34–40.
[13] M. Hurtgen, J.-C. Maun, Advantages of power system state estimation using phasor measurement units, in: Proc. 16th PSCC, Glasgow, Scotland, July 2008.
[14] Bose, Smart transmission grid applications and their supporting infrastructure, IEEE Transactions on Power Systems 1 (2010) 1949–3053.
[15] M.F. Wen, V.O.K. Li, Optimal phasor data compression unit installation for wide-area measurement systems—an integer linear programming approach, IEEE Transactions on Smart Grid (2015) 1949–3053, https://doi.org/10.1109/tsg.2015.2503425.

Further readings

[1] K.E. Martin, J.R. Carrol, Phasing in the technology, IEEE Power and Energy Magazine 6 (5) (2008) 24–33.
[2] A.R. Khatib, R.F. Nuqui, M.R. Ingram, A.G. Phadke, Real-time estimation of security from voltage collapse using synchronized phasor measurements, IEEE Power Engineering Society General Meeting 1 (2004) 582–588.
[3] K. Martin, IEEE standard for synchrophasors for power systems, in: IEEE C37.118 -2005 (Revision of IEEE 1344-1995), 2006.
[4] I.K. Yang, N.J. Jung, Y.I. Kim, Status of advanced metering infrastructure development in Korea, in: Transmission & Distribution Conference & Exposition: Asia and Pacific, Seoul South Korea, 26-30 Oct, IEEE, 2009, pp. 1–3.
[5] F.M. Cleveland, Cyber security issues for advanced metering infrastructure (AMI), in: Power and Energy Society General Meeting-Conversion and Delivery of Electrical Energy in the 21st Century, Pittsburgh, PA, USA, 20-24 Jul, IEEE, 2008, pp. 1–5.
[6] US. Energy Information Administration Website, 2016 (online), Available: http://www.iea.org/topics/electricity/subtopics/smartgrids.
[7] Lord Kelvin Quotations Website, 2017 (online), Available: http://zapatopi.net/kelvin/quotes/.
[8] A.N. Akpolat, E. Dursun, From microgrid to smart grid: a review of campus projects, in: International Engineering, Science and Education Conference, Diyarbakir, Turkey, 1-3 Dec, 2016, pp. 379–387.
[9] S. Nimbargi, S. Mhaisne, S. Nangare, M. Sinha, Review on AMI technology for smart meter, in: Advances in Electronics, Communication and Computer Technology (ICAECCT), 2016 IEEE International Conference on, Pune, India, 2-3 Dec, IEEE, 2016, pp. 21–27.
[10] IEEE, in: IEEE Standard for Interconnection and Interoperability of Distributed Energy Resources with Associated Electric Power Systems Interfaces. Piscataway, NJ, 2018.
[11] NERC, 900 MW Fault Induced Solar Photovoltaic Resource Interruption Disturbance Report, 2018. https://www.nerc.com/pa/rrm/ea/October%209%202017%20Canyon%202%20Fire%20Disturbance%20Report/900%20MW%20Solar%20Photovoltaic%20Resource%20Interruption%20Disturbance%20Report.pdf.

CHAPTER 4

Multi-sensor data fusion perspective for smart grid analytics

Resul Daş[1] and Gurkan Tuna[2]
[1]Software Engineering, Technology Faculty, Firat University, Elazığ, Türkiye; [2]Computer Technologies, Trakya University, Edirne, Türkiye

4.1 Introduction

The Internet of Things (IoT) can be described as a system of physical things embedded with one or more sensor(s), electronics, software and connectivity to allow it to perform better by exchanging information with other connected devices or the operator. An increasing number of IoT devices are created for consumer market and examples of such devices are found in various applications including home automation, connected car, entertainment, mobile health and domestic.

The number of Internet users has been increasing day by day [1,2]. In the last couple of years, as well as traditional devices such as computers, tablets, smart phones, a diverse set of objects have been a part of the Internet community in parallel with the advancements in data communications, network devices and embedded systems. In parallel with this a more inclusive term, the Internet of Everything (IoE), has been taking the place of IoT. On the other hand, this revolutionary transformation causes researchers, engineers, developers and users to encounter many challenges. In this massive and crowded field, there are various types of devices, architectures, communication channels, link layer technologies, protocols, middleware and services [2–4] and will possibly be new ones. The scope of IoT systems expands day by day [5,6] and accordingly IoT systems are used in an increasing number of existing and novel application scenarios including smart houses [7,8], smart cities [4], smart healthcare applications [1], intelligent agriculture [9], defense systems [10], environmental observation [11,12], human-computer interaction applications [13], training

Cyber Security Solutions for Protecting and Building the Future Smart Grid
ISBN 978-0-443-14066-2
https://doi.org/10.1016/B978-0-443-14066-2.00006-2

applications [14]. Obtaining meaningful data from complex IoT architectures that consist of multiple sensors and based on a heterogeneous structure is highly critical. Therefore, in the last couple of years, multi-sensor data fusion became one of the most investigated research topics in IoT and IoE domains.

Data is the most basic item of IoT systems. While individual data itself is not very beneficial, volumes of it can be used to identify patterns and trends [15,16]. Transferring data in an error-free, complete and consistent manner between the entities (from online systems and devices to a storage media or database) is very important. Besides, the accuracy, consistency and reliability of the data should be ensured and preprocessing phase of the data should be error-free. Data fusion is a general context in data science and includes many processes according to different applications. Data fusion context includes multi-sensor data fusion [17,18], multi-sensor fusion [19], sensor fusion, sensor data fusion, information fusion [20–22] data integration, the combination of evidence [23], synthesis of observation [24,25]. Moreover, it covers several concepts such as merging [26], combination [27], synergy, integration and aggregation [28,29], sensor management, sensor coordination, sensor planning and control, too [3,30].

This study surveys the literature over the period 1990–2020 on data fusion as they apply to IoT applications. With this objective in mind, we first carried out an extensive survey and created a list of abbreviations and taxonomy for the reviewed papers as listed in Tables 4.1 and 4.2, respectively. Second, we reported the fusion methods in the literature and listed them in Table 4.3. Then, we presented the opportunities and challenges encountered in data fusion applications in IoT environments. The remainder of this paper is divided into the following sections. Section 4.2 focuses on the background of data fusion and presents the common methodology used in data fusion processes. Section 4.3 reports models and approaches used in data fusion as well as providing taxonomies for the reviewed papers and data fusion methods. Section 4.4 highlights the opportunities and challenges of multi-sensor data fusion in IoT systems. Section 4.5 outlines future research directions explored in this study. Finally, Section 4.6 concludes this paper.

4.2 Data fusion

Raw data changes over time and transforms into the knowledge. In a typical multi-sensor data fusion process, sensors, raw data, multi-sensor data, information and knowledge are involved. Raw data acquired by a group of sensors are first preprocessed to make it ready for the fusion process. Then,

Table 4.1 Table of abbreviations.

Abbreviation	Definition
NN	Neural network
ANN	Artificial Neural network
AI	Artificial intelligence
k-NN	*k* Nearest Neighbors
DT	Decision Tree
KF	Kalman filter
EKF	Extended Kalman filter
PF	Particle filter
SVM	Support vector Machines
WSN	Wireless sensor network
BSN	Body sensor network
IoT	Internet of things
IoE	Internet of Everything
DF	Data fusion
SF	Sensor fusion
MSDF	Multi-sensor data fusion
JDL	Joint Directors of Laboratories
EHR	Electronic health record
RSSI	Received signal strength Indicator
DSS	Decision support system
NLP	Natural language processing
LKF	Linear Kalman filter
IMU	Inertial measurement unit
RF	Random Forest
HTTP	Hyper text Transfer protocol
PCA	Principal component analysis
NB	Naive Bayes
PSO	Particle Swarm Optimization
WUM	Web usage mining
HCI	Human computer interaction
IF	Information fusion
DMFS	Distributed memory file system
HDFS	Hadoop distributed file system
CWA	Cognitive work analysis
CTA	Cognitive task analysis
DCP	Distributed cognitive process
GCR	General Controlled review

it is processed by the fusion system. With the help of the fusion system and complementary techniques pattern discovery is realized. Finally, the information obtained in the previous step is analyzed to extract knowledge.

Table 4.2 Taxonomy of data fusion in the literature.

References number	Year	Type	Objectives and topics	Content	Fusion Type(s)	Fusion architecture Model(s)	Focused Application(s)
[31]	1992	Overview	Development of a taxonomy of the multi-sensor data fusion	Formulation of sensing for data fusion	MSDF	Sequential data fusion	Manufacturing
[24]	1997	Overview	Introduction of data fusion applications for defense systems	Processing architectures and model	MSDF	JDL	Military
[32]	2002	Comprehensive review	Overview of the basic concepts of sensor fusion.	Principles of sensor fusion, including types, limitations, categorizations, architectures, methods and applications	SF	JDL, Waterfall, Boyd, LAAS, and Omnibus	GCR
[33]	2004	Comprehensive Review and Analysis	Analysis and revision of the fusion levels of JDL model as well as partitioning of data fusion functions	Data fusion process and resource management Levels and concepts	DF	JDL	GCR
[25]	2007	Overview	Implementation for a data fusion process	Process models for data fusion	DF	JDL, Waterfall, Boyd and Omnibus	GCR
[34]	2009	Comprehensive review	Analysis of data fusion process to obtain complete and consistent data, and presentation of the challenges of data fusion.	Data integration process and transformation and data fusion techniques	DF	N/A	Information systems
[35]	2009	Overview	Presentation of advances in multi-sensor satellite fusion.	Algorithms for sensor fusion, remote sensing	MSDF	N/A	Image fusion
[36]	2010	Review	Description of the modeling and simulation of the verification, validation and accreditation in simulation projects.	Modeling and Simulation issues, IF and Models	IF	JDL-U	Virtual manufacturing

[27]	2010	Overview	Exploration of the processes identified for data fusion process model	Data analysis for situational awareness and The security analysis.	DF	JDL	Cybersecurity
[37]	2012	Overview	Overview of the clustering methods for multi-sensor data fusion	Clustering	MSDF	N/A	Active and passive Radar initiation
[38]	2012	Survey	Review of information fusion in data privacy	Data privacy and protection	IF	N/A	Data privacy
[21]	2012	Case study	Analysis of the interaction between users and information fusion process.	CWA, CTA, DCP, fusion process	IF	JDL, OODA loop and DFIG 2004 model	HCI
[10]	2012	Overview	Overview of JDL data fusion model in cyber defense.	Fusion process	DF	JDL and Omnibus model	Cyber defense
[39]	2012	Review	Review of data fusion models	Collaborative framework	DF	JDL	Military and non-military
[40]	2013	Re-examination	Assessment of the JDL model for information exploitation.	Information fusion and its management, System-level management of context	IF	JDL	information exploitation, information management And data mining
[41]	2013	Review	Exploration of the different context information fusion techniques	Classification of context information fusion techniques	IF	JDL, DFD, Omnibus and object-Oriented model	Context information Applications
[20]	2013	Review	Classification of algorithms and methods in data fusion field.	Data fusion and association techniques	DF	JDL	GCR
[42]	2013	Comprehensive review	Critical review and analysis of data fusion conceptualizations	Challenging problems of multi-sensor data Fusion	MSDF, DF	JDL	GCR
[43]	2015	Overview	Comparison of data fusion architecture models	Data fusion architecture models	DF	JDL and Bowman DF&RM model, the Luo and Kay, the Pau, LAAS	GCR

Continued

Table 4.2 Taxonomy of data fusion in the literature.—cont'd

References number	Year	Type	Objectives and topics	Content	Fusion Type(s)	Fusion architecture Model(s)	Focused Application(s)
[44]	2015	Comprehensive review	A comprehensive review of data fusion architectures	Fusion architectures and their comparison	MSDF	JDL and Bowman DF&RM model, the Luo and Kay, the Pau, LAAS	GCR
[45]	2015	Overview	Discussion of data processing techniques designed for WSNs	Data types in WSNs and data fusion	DF	N/A	Big data in WSN
[46]	2015	Survey	Review of sensor fusion methods for wearable robots	Unimodal and multi-modal sensor fusion And algorithms	Sensor fusion	N/A	Wearable robotics
[17]	2015	Overview	Determination of a feasible methodology for the fusion of the data obtained from a mobile device.	MSDF techniques	DF	N/A	Identification of activities of daily living
[47]	2015	Survey	Analysis of the contextual information	IF process, context sources types	Context-based IF	JDL	Transformation, discrimination, access to context functions
[48]	2016	Survey	Review of information fusion techniques in Opinion mining context and proposal of a simple framework for that purpose	Opinion mining techniques and its most fundamental aspects	IF	JDL	Opinion mining
[49]	2016	Survey	Analysis of data fusion in intelligent transportation systems	Data fusion architectures in intelligent transportation systems	DF, IF	JDL	Intelligent transportation system

[50]	2016	Comprehensive review	Review of information fusion, taxonomy of the comprehension and management of connected smart cars in Vehicular ad-hoc networks.	Vehicular ad-hoc networks, situation Awareness in Internet of cars and Architecture	Low and high-level IF	JDL	Connected smart vehicles
[51]	2016	Comprehensive review	Review of sensor fusion in sports and biomedical applications	Use of smart sensors and sensor fusion In sports and biomedical applications.	MSDF	N/A	Sports and biomedical applications
[28]	2016	Comprehensive review	Analysis and comparison of current data fusion techniques.	Sensor classification, data acquisition, processing and imputation	MSDF	All methods	Mobile devices
[18]	2016	Comparative study	Comparison of hybrid architectures for multi-sensor fusion systems	Back propagation neural network and taxonomy of MSDF	MSDF	Hybrid	Water pollution monitoring System
[3]	2017	Survey	Comprehensive review and analysis of the DF mathematical methods for specific IoT environment.	Opportunities and challenges for each DF Method.	DF	All methods	IoT environment
[19]	2017	Overview	Review of the sensor fusion types and techniques encountered in sensor fusion applications.	sensor fusion methodologies, categories, Topologies	MSDF	JDL and Waterfall	GCR
[52]	2017	Survey	Comprehensive review and analysis of the MSDF strategies in a wearable BSN	BSN, sensor fusion strategies, fusion levels	MSDF	Includes different fusion levels	Human activity Recognition in Wearable technologies

Continued

Table 4.2 Taxonomy of data fusion in the literature.—cont'd

References number	Year	Type	Objectives and topics	Content	Fusion Type(s)	Fusion architecture Model(s)	Focused Application(s)
[30]	2017	Review	Overview of data fusion in healthcare monitoring	Data fusion techniques and algorithms for healthcare applications	DF	Centralized hierarchical data fusion model	Wearable health monitoring.
[53]	2017	Short paper	Mathematical analysis of data fusion problems	Analysis of the fusion problem and conservative fusion rules	DF	N/A	Intersection of the inverse Covariance ellipsoids For predict problems.
[22]	2017	Survey	Analysis of the biometric IF approach applied to web usage mining	Web mining, biometric information fusion	Text-based IF	JDL	Web user access logs.

Table 4.3 Fusion methods used in IOT applications.

References number	Fusion approach	Supported methods	Objectives and topics	Outcome	Used software tools	Devices and sensors	Networking	Collected data	Application	Results
[54]	Characterized Support vector Hyperspheres (CSVH)	SVM, LDSVM, TWSVM, MVSVM, NN	A novel approach For Atrial hypertrophy disease Diagnosis based on CSVH	Time complexity occurs When CSVH is compared With TWSVM, MVSVM, SVM and NN methods	MATLAB R2011b, Windows 10	Multiple medical sensors	Not specified	Two real Electrocardiogram Datasets acquired From medical sensors (250 Patient instances And 40 healthy people Instances)	Atrial Hypertropy Disease Diagnosis	The proposed method achieves the Highest diagnostic accuracy
[55]	KF-based sensor fusion	SVM, NN, NLP, KF, DSS	Track to track comparison, Orientation comparison, accelerometer Measures comparison	The main limitation of the proposed System is the Trade-off between Coverage and accuracy Of the sensors	Not specified	Health Bracelets and Belts, WSN, and a set of sensors	IEEE 802.15.1, Bluetooth v4.0, IEEE 802.15.4	RSSI datasets, Electronic health Record, NWKR [56]	Elderly people care	The performance of the proposed system was good even when not all sensors Were available.
[57]	Fuzzy logic fusion	Fuzzy Classifier	Monitoring system for the elderly	Further tests are needed in In order to have an effective Evaluation.	Labwindows CVI, C++	Contact sensor, infrared sensor, Hearth monitoring sensor, a set of other sensors and equipment.	RF, ZigBee, HTTP, TCP/IP	Elderly physiological And behavioral data, environmental datad	Healthcare Monitoring of the Elderly people In smart homes	The proposed monitoring System ensures Health monitoring for the elderly.
[58]	LKF algorithm	NN	A novel sensor fusion approach To estimate roll Angle of a vehicle.	The NN + LKF-based estimator was validated.	TruckSim simulation	IMU, Mercedes Sprinter, Steering angle sensor, Dual antenna data logger, GPS antennas, Potentiometers	GPS	Experimental data Obtained from the Installed system	Vehicle Systems	The proposed method provides better Results in 100% of the analyzed Experimental cases.
[59]	EKF-based Sensor fusion	Complementary filter	A sensor fusion approach for Vertical channel stabilization By fusing IMU measurements. In addition, a whitening filter was developed.	Since the reliability Of the linear acceleration Measurements were overestimated, The proposed approach was not successful.	Visual basic, MATLAB	IMU sensors, barometric altimeter, smart phone, PC, DC motor	WSN, Bluetooth, WiFi	Baro-IMU sensor data	Vertical Channel Stabilization by Fusing measurements from IMU and a Barometric Altimeter	Successful validation was achieved In different experimental conditions.

Continued

Table 4.3 Fusion methods used in IOT applications.—cont'd

References number	Fusion approach	Supported methods	Objectives and topics	Outcome	Used software tools	Devices and sensors	Networking	Collected data	Application	Results
[11]	Dynamic Bayesian Network	Not specified	An adaptive, context-aware and self-optimizing System For data fusion	The use of all of the provided sensors together cannot be deemed as a good Strategy when energy Efficiency and computational Burden is considered.	Not specified	Heterogenous sensors	WSN	Raw data (obtained From environment by Heterogeneous sensors)	Smart home environment	The proposed contextaware Multi-sensor fusion outperformed Static approaches in terms of energy Savings.
[60]	PSO	Not specified	The measurement Precision of multi-sensor data Fusion was improved by the improved PSO Method	Multi-dimensional distribution Problem was handled by KNN and a probabilistic data association algorithm	Embedded Workbench Software	Emulator, ZigBee, RFID Modules, RFID Nodes, RFID nodes, RFID tag, active tags, etc.	ZigBee and RFID	Raw data was obtained From environment By heterogeneous Sensors.	Smart home environment	Experimental results demonstrate the efficiency of the improved PSO Method for multi-sensor data fusion.
[61]	PNN Classifier	Signal processing, feature Extraction	A multi-sensor data Fusion technology based smart Home system and its intelligent Monitoring interface.	Remote Control, indoor positioning, intelligent energy Management, and fire Prevention was achieved.	Not specified	Wearable inertial sensor module, RF modules, Microcontroller	RF	Multiple sensors data	Smart home environment	The experimental results have successfully Validated the effectiveness Of the proposed system and its interface.
[62]	Unsupervised learning method	Not specified	Development of data processing Modules for computing the Space utilization	Characterization of the noise signature to identify false positives	Not specified	PIR sensors, analog sound sensors, environmental Monitoring sensor Nodes	Renewable WSN	sensor data obtained By PIR and sound sensor	Public space Utilization	Fusing of the data from two sensors was successful.

[63]	Statistical Correlation and Covariance Analysis, Multivariable data analysis, PCA, multiple Linear Regression Analysis	SVM, RF, kNN, NB	On-body physiological Signals was fused with environmental sensory Data.	Decision fusion of emotion Recognition based on the two Modalities resulted in an increase In the performance.	EnvBodySense, Microsoft Wristband	Mobile sensors, wearable body Sensors, mobile devices	WSN, GPS	Data from participants Walking around Nottingham city by Using mobiles sensors	On-body Environmental Sensing	Multivariate linear regression models For on-body sensor data were Developed.
[64]	Activity Recognition, feature fusion and decision fusion models	kNN, NB, DT	Categorization of IMU displacement conditions.	Multisensor Models Suffer from the effects of Displacement	Context Recognition Network Toolbox	Wearable inertial sensors, IMU, Laptop, video Camera	Bluetooth, WSN, BSN	Open access benchmark Dataset, REALDISP	Wearable Technologies And BSN	Accuracy rates Above 90% and 95% were achieved.
[65]	DMFS, HMFS	Jaccard coefficient, Cosine similarity, Hamming Distance, Hashing Methods, HDFS	A new system was developed for Storing multi-sensor object tracking data on HDFS	In the same cluster, object tracking related data from a group of sensors are merged into the same Large files.	IndoorSTG Simulation tool	RFID sensors, servers, virtual RFID and Bluetooth Readers	RFID and sensor Networks	HDFS	Multi-sensor Object-tracking In indoor Spaces	The experimental results show The efficiency of proposed Methods.
[66]	Joint error Optimization Approach	Modified Geometric link Reconstruction, Modified Linear Algebraic Reconstruction, Non-linear least Squares	A combined Multi-sensor attenuation model For the object and mobile user localization	Development of the combined Attenuation model from Multi-sensor data.	A gateway device And Lab-View	IRIS board, sensor boards, mote interface, and light and sound Sources	A gateway device, WSN nodes, IEEE 802.15.4 protocol,	Multi-sensor data	Multi-sensor data	Experimental results on node localization Indicate reasonable And feasible improvements.
[7]	Not defined explicitly	Energy Saving and shifting Benefits were explored.	Energy management	Content analysis To analyze data about smart Home products for energy savings and demand Shifting.	Mobile applications	Load monitor, in home display, Smart thermostat, smart light, Smart plug/switch, smart appliance, hub.	WSN, ZigBee, ZWave	Raw data obtained from Individual HEM Technologies	Smart home technology	The limitation of the study was the use of Consumer-facing technologies.

In a typical IoT application designed with extended periods of functioning time in mind, there are some inherent limitations due to its components. For instance, most of the devices have limited processing capacity and most of the sensors have low sampling rates so that the hardware be simple and power requirements be low [67]. In view of the fact that there are a large number of sensors, if not a critical application will be deployed, commercially available, low cost sensors are preferred. For example, high precision IMUs are typically expensive for an ordinary IoT application scenario. On the other hand, measurements from low cost sensor arrays involving a gyroscope, an accelerometer, and a magnetometer can be fused to obtain high precision data. Sensor fusion is a process by which data obtained from a group of, typically different types, sensors are fused to compute something more than could be determined by any one sensor alone. It is an inevitable process in most state-of-the-art sensor systems. For instance, a magnetic and inertial measurement unit (MIMU) involve a 3-axis gyroscope, accelerometer and magnetometer. It is widely used in many applications such as human motion tracking, mobile navigation, and unmanned aerial vehicles. In a MIMU, the angular rate of a moving object is measured by the gyroscope, the acceleration of a certain object is measured by the accelerometer, and the magnetic field is measured by the magnetometer [68]. Nevertheless, low-cost sensors used in most MIMUs have one or more drawbacks, such as temperature drift, random walk and nonlinearity. MIMU sensor measurements are fused together using complementary filters [67,69] or Kalman filter-based approaches [70] to obtain reliable attitude data.

The outputs of sensors found in low cost MIMUs are characterized by time-varying additive biases and high noise levels [69]. In MIMUs, the heading is obtained using the data from the gyroscope and the magnetometer and the change in rotation is supplied by the integration of the gyroscope from a known initial orientation. On the other hand, because of the inherent noise and bias, the gyroscope has a long-term drift; therefore, these errors must be corrected. This is where the magnetometer plays a role in the minimization of the drift in the horizontal orientation. Fig. 4.1 shows a set of raw measurements belonging to the same time domain from a gyroscope, an accelerometer and a magnetometer. Fig. 4.2 shows the result of the fusion process when Madgwick approach [67] was used and Fig. 4.3 shows the result of the fusion process when Mahony approach [69] was preferred. When compared with the ground truth, it was seen that except for a few seconds and minor differences, the Euler angles obtained

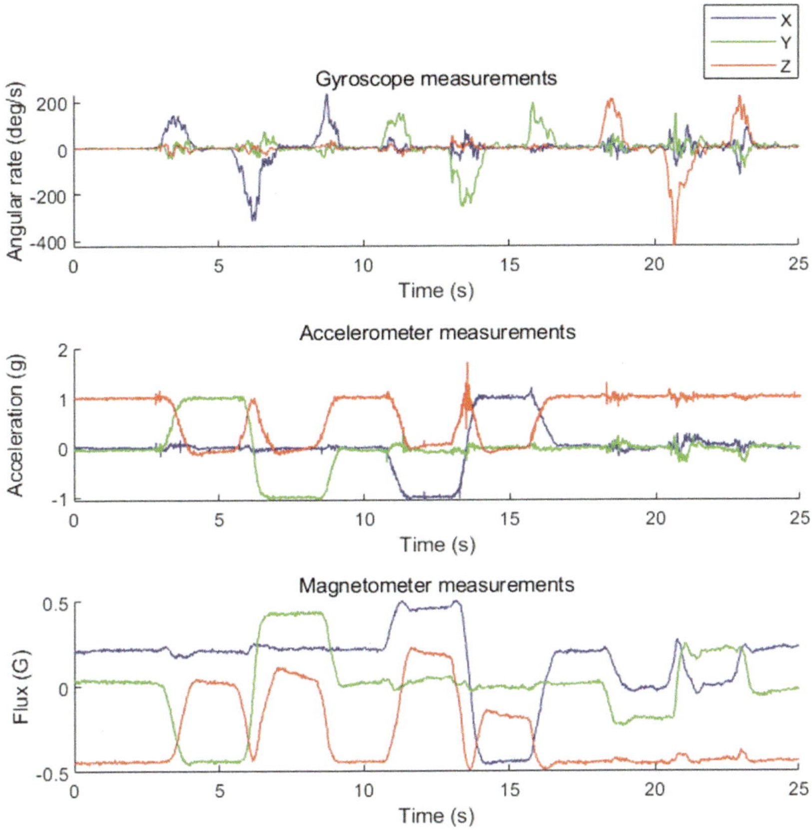

Figure 4.1 Sample raw measurements from a gyroscope, an accelerometer and a magnetometer.

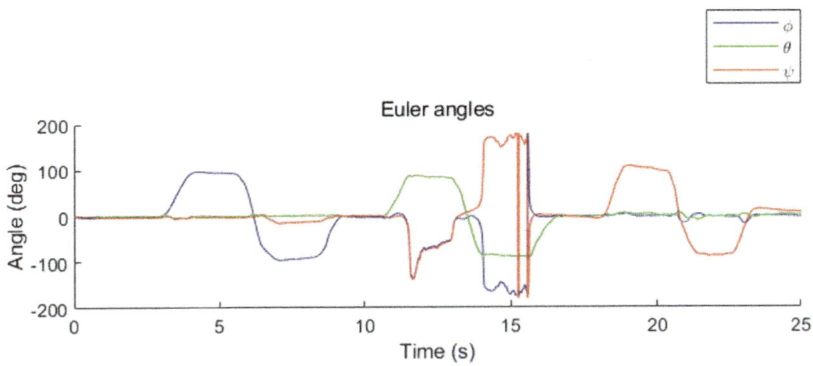

Figure 4.2 The result of the Madgwick approach-based fusion as Euler angles.

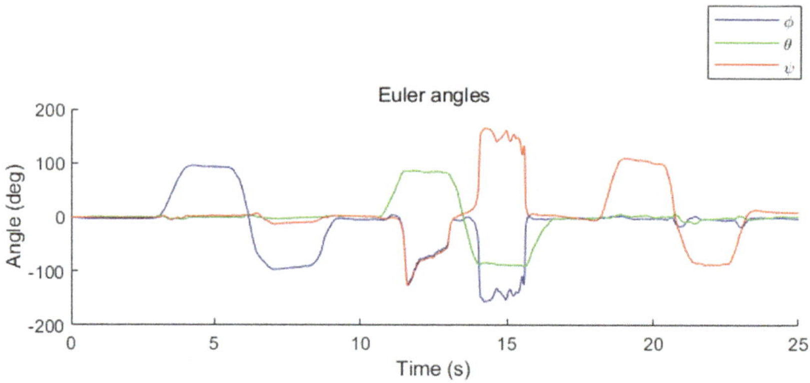

Figure 4.3 The result of the Mahony approach-based fusion as Euler angles.

Madgwick and Mahony approaches were quite reliable. However, considering the limited processing capability of some IoT devices the processing load and memory requirements of filtering techniques should be considered. It was shown that compared to Madgwick, Mahony approach generally takes less time to compute [68].

Sensor fusion combines devices responding to a stimulus. Information fusion combines data to create knowledge, as shown in Fig. 4.4. Display fusion presents data simultaneously to support humanitarian operations. Cognitive fusion integrates information in user's mental model [24,71].

Due to the lack of a general and regular taxonomy of methods, especially the names of mathematical methods are categorized incorrectly or incompletely. Moreover, although it is the most commonly used model in information fusion processes, JDL is used differently than the original.

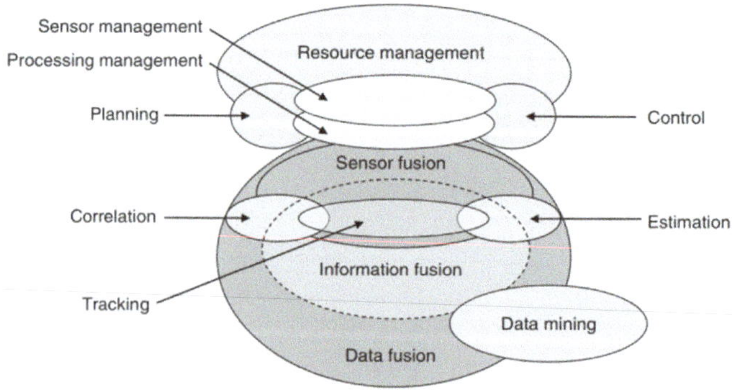

Figure 4.4 Fusion-related processes.

4.3 Related work

In recent years, the focus of multi-sensor data fusion has shifted to IoT and IoE applications, including wearable technologies [13,64] smart homes [55,57,61,72] smart cities [62], smart vehicles [56,73]. We note also that approaches that could be classified as data fusion are present in a plethora of other applications, such as web usage mining [16,74] context mining [11], text mining [2,16] where server data is used instead of sensor data. From the architecture point of view, fusion models can be categorized into three classes as abstract models, generic architectures and rigid architectures. Abstract models typically serve as a way to explain an aspect of a fusion system but they do not guide the engineers, designers and researchers in their implementations [75]. As a result, the fusion system may involve references to two or more abstract models. Examples of abstract models are Waterfall model [76] and Boyd control loop [77]. Generic architectures provide a generic design but do not specify several design decisions regarding hardware, operating system, communications system, or database management system [75]. Examples of generic architectures are JDL and Omnibus models [78]. On the other hand, different from generic architectures, rigid architectures specify these aspects by providing existing hardware designs, tools, and source code at the cost of flexibility due to the fact that some design decisions have already been made [75]. Rigid architectures guide the engineers and designers well in their implementations and can be quickly realized. Examples of rigid architectures are DFuse [79], Mr. Fusion [80], LAAS architecture [81] and Time-Triggered Sensor Fusion Model [82].

In this study, we endeavor to provide a more systematic means of approaching the relevant literature surrounding the multi-sensor fusion of, mainly, sensor data in the IoT and IoE context. We provide a taxonomy which includes a survey of published survey papers as well as a survey of recently produced research results in the area of data fusion with strong ties to IoT/IoE. However, first we need to provide the existing context for describing sensor fusion, and to this end we introduce a few key models that serve as the framework to organize the remainder of this paper.

4.3.1 JDL data fusion model

In the mid-1980, the Joint Directors of Laboratories formed the Data Fusion Group. This group endeavored to systematize the terminology related to data fusion. The objective was colored by the intention to develop a common terminology and understanding for the purposes of the US Department of Defense and its contractors. Hence, the viewpoint adopted by JDL is of seeing various objects as "targets" with an emphasis on image and remote— sensing data, even though it is, in principle, applicable to a wider variety of settings. The developed data fusion model is known as the *JDL model* and is composed of various processes and functions. In particular, the model defines at least five levels, as shown in Fig. 4.5, plus components to express data inputs, human-computer interaction, source preprocessing, database management, etc. [24,33,71,83,84].

In the remainder of this subsection, we briefly review the JDL Model components and remark on their correspondence to today's sensor systems in IoT settings.

- *Sources*: They are any kinds of data originator, such as raw sensor data, live data, a file, a database, etc. Because there are many types of data sources, they follow different formats for storage and communication. Because of those differences, they are processed in different ways [36,74]. The similarity to IoT sensing should be evident at this stage.
- *Level 0 — Signal refinement*: It is a preprocessing step, aimed usually to combine pixel or signal level data to obtain initial information about a target's characteristics. At this point, the signals are still treated at "face value", i.e., without any attempt to attach characteristics dependent on inferences about the objects involved or to associate any semantics. There are many operations that fall under this level, for example data cleaning techniques to de-noise and normalize the signals, data alignment to a consistent temporal or spatial reference, extraction of low-level features, elementary clustering of signals, etc. It is fair to state that most of the IoT literature contributes to some degree to the definition of Level 0 processes within the context of a specific application.

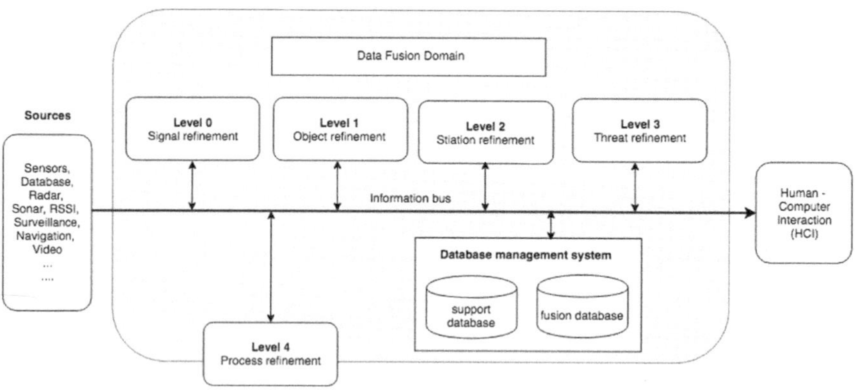

Figure 4.5 The original JDL model for data fusion [24,71].

- *Level 1 — Object refinement*: It is performed by combining sensor data so that the identity of an object can be determined and its position, velocity, and other attributes can be predicted most reliably and accurately. What one observes at this level is that characteristics are attached to the observations. Identification and disambiguation of the identity of the entities involved is also introduced at this level. An example from IoT is the identification of one or more individuals based on the signal strength data from a wearable device.
- *Level 2 — Situation assessment*: By situation assessment relying on object clustering and relational analysis, current relationships among entities and events in the environment are described [71]. At this level, the relation of the objects and their perceived interactions with each other and with the environment suggest a leap to some form of semantics. The corresponding example in IoT would be that of "activity recognition" where the activities are fairly high—level concepts, e.g., walking, sleeping, etc. as it pertains to multi-sensor fusion for an observed individual.
- *Level 3 — Threat refinement*: By threat refinement, the current situation is projected into the future and inferences about threats, friend and foe vulnerabilities and opportunities for operations are drawn [71]. While, strictly speaking, not always a "threat" in the IoT context, this level stands above the basics of activity recognition and comes closer to describing *intent*. Here the analogy with today's civilian-grade sensor systems is less clear, but it encroaches into the analysis of trends and behavior changes across time. Rather than a threat, in IoT we would speak of "anomaly detection."
- *Level 4 — Process refinement*: It monitors the overall fusion process so that the system's performance can be assessed and improved accordingly. It addresses resource management with the target typically being that of ensuring that the right kind of resources, in the right numbers, are available at the right location, to ensure real-time performance. The corresponding IoT concepts can span from efficiency concerns to achieve real-time data fusion, to placement concerns for the sensors and/or their proper parameterization. Level 4 is a level of reflection on whether factors, such as parameters in the relevant algorithms, are tuned well to achieve the desired outcome. Equally, the choice of deployment of sensors and their parameterization can have an impact on the resulting network traffic, congestion etc.
- *Level 5 — cognitive refinement* is an optional level that targets the improvement of the interaction be-tween data fusion systems and the various

classes of users and/or analysts. Here one finds the inclusion of teamwork for decision making, and collaboration in general, as well as any cognitive assistance for the users, strategies for bias remediation, etc. In short, Level 5 represents various strategies of assisting the human users. Correspondingly, in IoT systems, one occasionally finds distinctions of various but specific user/stakeholder groups of an application. For example, a medically-related system may provide different views to clinicians as opposed to, say, patients. Nevertheless, the distinction typically does not extend beyond distinct user interface morphologies.

- *Database management system (DBMS)*: It is essential for reliable management of data and should be appropriately organized for fusion processes and application support.
- *User refinement*: It is performed to improve the interaction between a data fusion process and its users. It basically provides graphical interfaces and natural language processing tools so that the interaction between the users and fusion system is simplified [71].

In JDL model there is no clear "initialization stage", e.g., how to deploy sensors to begin with, but rather places such an operation under the, presumably dynamic, auspices of Level 4, i.e., as process refinement. A significant fraction of IoT literature either embeds some unstated a-priori decisions about placement of sensors and/or assumes a static placement, hence restricting the potential of enacting Level 4 control to just the logical domain, rather than combined logical + physical control of the sensor assets.

4.3.2 Other data fusion models

Other models have been developed, trying to abstract the notion of data fusion, albeit in drastically different terms. As shown in Table 4.2, JDL model was preferred in many research papers in literature such as data fusion [34], multi-sensor data fusion [17], biometric information fusion [22], information fusion [21,47] and wearable technologies [30]. Also, it is seen in the literature that JDL data fusion model is taken as a reference model in many of data processing applications. One of the most well-known ones of other models that is in the same category but has different features with JDL is Omnibus model proposed by Bedworth and O'Brien in 1999 [78]. It was developed to address the weaknesses of existing models by integrating the most beneficial features of the existing ones. Different from JDL, it defines an explicit ordering of processes and relies on a cyclic nature. It does not support a decomposition into modules for separate implementation, separate test and reuse in different applications. It does also not support horizontal partitioning into tasks for distributed sensing and data processing [75]. Specifically designed for distributed data fusion applications realized in ad hoc WSNs, DFuse models an application as a graph of data sources, fusion points and data sink(s) and assumes that the data sources have data available whenever required [79]. Therefore, the fusion points may realize the fusion process based on an incomplete set of data if the requested data does not arrive in time. DFuse automatically deploys fusion applications in WSNs. Being a middleware framework to support data fusion, Mr. Fusion is not exactly suitable for the communication and processing of sensor measurements but quite suitable for handling outputs from several network servers [80]. LAAS architecture handles a fusion process by mapping low-level and intermediate-level sensor fusion to a set of modules at the functional level [75]. High-level sensor fusion is represented in the decision level and there are different timing requirements at the decision level and the functional level [81]. Different from JDL, LAAS architecture guides its designers well in the implementation of reusable modules. The final model that we would like to introduce is Time-Triggered Sensor Fusion Model [82]. It works on top of the Time-Triggered Architecture [85]. The Time-Triggered Architecture offers a fully synchronous design, in which each task and communication activity is planned a priori in a static schedule [75].

4.3.3 Data fusion methods

In recent years, the development of many projects related to IoT and IoE applications has increased the importance of multi-sensor data fusion. Together with the widespread use of their applications, the importance of mathematical methods in data fusion has also increased. In this subsection, the mathematical models employed in data fusion are briefly reviewed. The models can be categorized as belonging to one of three categories: probabilistic, artificial intelligence, and evidence theory [3], as shown in Fig. 4.6.

(1) *Probabilistic based data fusion techniques*: Although their accuracy is lower than integral techniques, probabilistic techniques are commonly used in single and multi-target tracking problems and their applications [3]. Bayesian theory, Markov Chain and Monte Carlo method are examples of probabilistic techniques widely used in data fusion [71,83].

(2) *Artificial intelligence-based data fusion techniques*: Intelligent methods, especially used in the development of smart systems, are also used effectively in many different IoT and IoE applications.

Mathematical methods for data fusion

- Probabilistic based methods
  - *Bayesian analysis*
  - *Statistics*
  - *Recursive operators*
  - *Markov chain Monte Carlo*
  - *Joint probabilistic data association*
  - *Distributed joint probabilistic data association*
  - *Multiple-hypothesis tracking*
  - *Distributed multiple-hypothesis tracking*
  - *Graphical models*
  - *Optimal theory*
  - *Maximum likelihood and maximum posterior*
  - *Kalman filter*
  - *Distributed kalman filter*
  - *Particle filter*
  - *Distributed particle filter*
  - *Covariance consistency model*
- Artificial intelligence based methods
  - *Artificial neural networks*
  - *Fuzzy logic*
  - *Genetic algorithm*
  - *Support vector machine*
  - *k-Nearest neighbor*
  - *k-Means*
  - *Linear and quadratic discriminant analysis*
  - *Deep learning*
  - *Other methods in machine learning*
- Theory of evidence based methods
  - *Dempster-Shafer theory*
  - *Bayesian probability theory*
  - *Light detection and ranging*
  - *Semantic methods*
  - *Abductive reasoning*

Figure 4.6 Mathematical methods for data fusion.

(3) *Theory of evidence-based data fusion techniques*: Based on updating beliefs and a combination of evidence, they are used to represent incomplete knowledge and uncertainty explicitly [86]. Dempster-Shafer Theory (DST), also known as evidence theory, was first introduced by Arthur P. Dempster and then developed by Glenn Shafer into a framework for modeling epistemic uncertainty [72,83,87,88]. It allows combining evidence from all the available sources and arriving at a degree of belief.

As an alternative approach Luo and Kay's taxonomy [89] can be used for classifying data fusion methods. When it is followed for decision making from fusion levels of the time-varying data, features and decisions, low level fusion methods can fuse the multi-sensor data. On the other hand, medium level ones can fuse both data and feature to obtain fused feature or decision. Finally, the most sophisticated ones, namely high level fusion methods, can fuse both feature and decision to make the final decision.

4.4 Opportunities and challenges for data fusion

Single level data fusion puts a limit on the capacity and robustness of a system, because of the weaknesses in uncertainty, missing measurements, and the accuracy, precision and reliability of a single sensor system [89]. Integrating and fusing data from multiple sensors provide high robustness, reliability and flexibility. For any large-scale application, especially highly distributed ones like IoT, manipulating and understanding data is highly critical. Data science offers various techniques and methodologies for this and plays a key role in extracting insight from a given dataset, it includes many challenges [3,71,90–93]. Data has a life cycle and survives in environments that provide existential meaning as well as social context and physical support. Hence, data life cycle should be understood well so that the opportunities and challenges of data science can be explored [90].

Even in a small period in the lifetimes of IoT applications, a tremendous amount of data is created. Therefore, ensuring the quality of information is critical in the availability of such data so that highly accurate, precise and reliable results can be obtained. Data fusion is one of the most effective solutions in this regard and utilizes enormous volumes of data from many sources effectively. Essentially, multi-sensor data fusion combines data from multiple sensors with different physical characteristics and obtains inferences not feasible from a single sensor. This way, the understanding of our surroundings can be enhanced and the planning, decision-making and control of autonomous, intelligent entities can be designed.

4.4.1 Opportunities

In IoT applications, data fusion can provide several advantages as follows [3,71]:

- By data fusion, more decisive and precise information can be obtained by gathering data from multiple sources and sensors.
- Statistically, data fusion relies on the computation of N independent observations. This way the data is consolidated in an optimal manner.

- IoT environments involve a large number of very low power sensors. Since the requirement of battery replacement is a major problem in most IoT applications, sensors with very low power requirement and low accuracy can be preferred to sensors with higher power requirement and higher accuracy to improve the lifetime of the application. Data fusion helps to obtain accurate information.
- Since IoT applications involve a large number of sensors, fusing data from such large number of sensors allows obtaining more accurate and precise information.
- It can be used to hide critical information or semantics, particularly in military applications and applications realized in critical medical areas and intelligence buildings.

4.4.2 Challenges

Difficulties in the data fusion process may vary depending on the variety of applications, and sizes and type of the data. Especially, the hardware and software used in the IoT environment affect this process. The most common difficulties are explained in the light of the obtained knowledge.

- *Data imperfection*: Data obtained from some devices or sources can be inaccurate and uncertain. These data imperfections should rectify using data fusion algorithms [3,16,71].
- *Ambiguities and inconsistencies*: Inconsistency of the differently formatted data obtained from several data sources and uncertainty of various parameters in data fusion exhibit these problems known as unified calculus of uncertainty [3,71,94].
- *Conflicting or contradicting data*: Within the obtained data, conflicting data sets can be found. They are more common in evidential belief reasoning and the Dempster's rule of combination [3,71,94].
- *Data correlation and alignment*: The association of data containing parameters related to each other from different devices in the same field. The alignment problem, also known as sensor registration problem, is more common in WSNs and robotics applications and occurs when data is transformed from each sensor's local frame to a global, common frame before the fusion process [3,71].
- *Trivial data and features*: IoT environments typically involve a large number of sensors with different parameters. In some of the IoT environments such as smart cities and plants some of the sensed data may be trivial and processing of such data has an effect on the accuracy of

data fusion process. This requires selecting the most relevant features before the fusion process [3,71].

- *Iterative process*: Data fusion is a dynamic, iterative process and relies on the regular refinement of the estimates [3,71].
- *Scant algorithms*: Although there is a great deal of interest in data fusion, still there is a need for better fusion algorithms [71].
- *Overflow of data*: This situation occurs when more information is being transmitted than the hardware can handle. There are several types of overflow errors such as integer overflow, buffer overflow, heap overflow, stack overflow, and overspill [71].
- *Duplicate of data*: When data is obtained from different sensors, there may be many redundant copies. Data deduplication removes redundant copies of data, thereby can significantly reduce storage overhead. It ensures that one unique instance of data is kept on a storage media.
- *Dirty secrets and pitfalls*: There are many difficult pitfalls and dirty secrets in data fusion process and its applications [95].
- *Human-computer interface*: In order to effectively and correctly use multiple input data, explicit rules should be defined to integrate the information into the operation of the system [14].
- *Incommensurable*: Data collection by different methods, instruments, devices, software tools from the system which containing different each other multi-sensors can change the quantifiability [94].
- *Data registration*: Data registration is the task of transferring to any database of the several data sets such as images, information, numbers, parameters etc. according to identified format and the order. During the data registration in any system in IoT applications, it may be encountered different record problems due to several reasons. Therefore, it can be said that either it may be originated from incorrect information entry of the people or it may be not high quality and effective communication each other the devices in the link state [71,94,95].
- *Different resolutions*: Data collection with different resolutions is performed by using several types of data acquisition methods from a set of different data sampling devices [96]. The challenge here is that this varies according to a given task, as a result there may be diverse solutions which can be associated with various types of uncertainty in different resolution cases [94].
- *Incompatible Data Size*: *In real-world applications*, quite seldom different data sets involve the same number of samples. It may be a result of

the differences in data collection resolutions in different acquisition techniques and experimental setups [94,97,98].

- *Balancing Information from Different Origins*: In real-world applications, there may be data entries with different levels of confidence, reliability and quality. Because some sensors may provide higher-valued or provide more certain measurements [71,94].
- *Noise*: In finding and an estimate of the unknown parameters, it is expected to estimate the uncertainty in applications. However, true measurements are corrupted because of noise and other sensor errors in estimation applications. Unfortunately, some of the unavoidable reasons for noise are stated as calibration errors, quantization, finite precision and thermal noise. Sometimes the states/parameters of a system cannot be measured directly because of this situation [94].
- *Missing Values*: The challenge of missing values is commonly found in various applications including the ones relying on data fusion. It is generally seen on matrix completion in algebra [71,94].

Although data preprocessing can address most of the above-mentioned challenges, some of these still need to be studied thoroughly. For instance, database management systems used for desktop and server operating systems have been designed application flexibility in mind and are not suitable for the real-time requirements of the data processing of multi-sensor systems [99]. Database management for data fusion is still challenging and difficult to implement. Because large and varied databases with numerous records and record formats exist in traditional relational databases. Traditional database management systems do not support for rapid updates for incoming data and do not provide fusion results quickly. In addition, they do not support automated data fusion process and do not provide user-friendly and flexible interfaces for this goal. If traditional database management system will be used in IoT applications, it must be able to accept free-text message formats in addition to fixed formats under multiple protocols [99]. Finally, maintaining of real-time data integrity is a must for IoT applications if sensor data is provided in a rapid fashion and intense human interactions are encountered and in case of synchronous, out-of-order, and false sensor reports [99].

In real-world applications, most multi-sensor fusion applications need to fulfill previously set timing constraints. Although the abstract models and generic architectures presented in the preceding sections can support real-time behavior, respective means like timestamping and deterministic communication should be provided to the multi-sensor fusion system. On the other hand, the Time-Triggered Sensor Fusion Model is very rigid regarding the timing assumptions and therefore suitable for designing soft real-time fusion applications if a strict communication schedule can be provided by a tool like [100].

4.5 Future research directions

In emerging smart grid applications, various IoT devices are distributed throughout the energy chain from the generating power plants to the end-users. Because, IoT can improve the existing infrastructures of power grids by facilitating real-time control and monitoring of the grid components [101]. Nevertheless, cybersecurity is widely accepted as one of the major barriers to IoT acceptance, as because of the massive number of devices connected to the communication networks, ensuring the safety of grid-connected devices is quite challenging [101]. Therefore, research activities that focus on exploring the potential risks of IoT enabled smart grid applications are essential.

Multi-sensor fusion is still in development and constantly changing [102]. Still there is no unified fusion architecture and efficient, generalized fusion model; although, JDL is widely accepted as the standard or base model. There are still unsolved problems such as fault tolerance, robustness, association ambiguity. In view of the need for solving these problems and for improving the performance of multi-sensor fusion processes, in recent years, artificial intelligence techniques such as NN, fuzzy theory and genetic algorithms have been introduced to the multi-sensor fusion domain. While relevant prior data can be used to improve the performance of data fusion processes, sophisticated fusion methods with parallel processing capabilities should be developed by the integration of artificial intelligence techniques. However, the integration of artificial intelligence techniques can create an additional burden especially in IoT applications that involve single and multi-platform architectures and homogeneous and heterogeneous sensor arrays [102]. Therefore, testing and evaluation platforms should be developed to model computational complexity of fusion methods that are under development. Then, these platforms should be used to standardize the existing methods. In addition to this, the development of processing hardware with higher computing resources should be considered to allow running complex fusion methods in real-time data acquisition scenarios.

Different from centralized computing systems, distributed systems inherently provide redundant data and the opportunities that could be provided by such data can only be exploited with cooperative algorithms, distributed reasoning systems and novel, distributed data fusion methods [103,104]. However, these introduce challenges such as spatial and temporal alignments of the information in the fusion process, availability of out-of-sequence measurements, and data correlation problems. In Ref. [105] a model to find out the correlation between the sensor measurements by matching the parameter values for the different obtained manifolds was proposed. These challenges still need to be studied on.

Reliability and credibility of input data affects the overall success of data fusion processes [49]. In this regard, in terms of reliability and credibility the degree of confidence in input data should be calculated [107,108]. In Ref. [109] based on the use of existing knowledge measures of reliability of the source of information and measures of its credibility are evaluated then the measured reliability of the source of information is combined with its measured credibility. Although a long time passed after [109] still the current formalism about the evaluation of the reliability of an information source is still not complete. Regarding the quality of data supplied to a fusion system another important fact is the rate at which data comes [49]. This is particularly critical in applications involving decentralized systems such as IoT systems where communication is not perfect.

In IoT applications, performance evaluation has different, typically conflicting dimensions and these make it difficult to develop a unified measure for performance evaluation needs in related fusion processes [49]. Therefore, a comprehensive measurement methodology might fail to reveal all the dimensions of performance evaluation and deploying a set of measurement policies for performance evaluation may suit well [42]. In addition to this, over time or according to the given context, this set of policies should be adjusted so that the desired performance levels can be maintained. Moreover, compared to single sensor systems, multi-sensor ones have more parameters to evaluate. For such systems, metrics can be evaluated for both individual targets and a group of targets [42]. Considering the fact that most evaluations especially the ones that are associated with localization still rely on the knowledge of ground truth values, there is an ample of open research directions exist here.

In multi-sensor systems fault detection is a critical process [42]. Therefore, several innovative approaches were proposed to handle this, relying on different techniques including Kalman Filter [110], fuzzy networks [111] and virtual sensors [112]. Another sensor technology related fact is that while large sensors may offer quite good operating characteristics but their sizes limit their marketability, portability, availability to specific application scenarios and increase their manufacturing cost as expected. Therefore, in the last couple of years there has been a significant interest to reduce sensor sizes. Thanks to the advancements in semiconductor technology small and reliable processors are produced in the form of integrated circuits [89] by very large-scale integration [113]. The interest in IoT applications has created a considerable demand for smart sensors [113] and nano sensors [114].

Multi-sensor fusion methods rely on exact information about their surroundings. Nevertheless, as proven in many researches involving field tests and real world application scenarios, most of the sensors are sometimes perfectly functional and precise data about the environment is scarce [89]. In order to address various forms of uncertainties due to the use of imperfect sensors. Although some methods have been proposed for this such as [115], still there is a need for adaptive fusion methods focusing IoT applications.

Although they have mostly the same or similar spectral configuration, images obtained from two different satellites have a lot of differences in terms of solar geometry, viewing angle, atmospheric conditions, spatial resolution, spectral response function, and bandwidth [116]. Therefore, in a spatiotemporal data fusion process, precise image alignment is a must for mitigating the differences between different image types. While some methods can successfully resolve some of these problems up to a degree, considering the interest in remote sensing and satellite imaging novel spatiotemporal fusion methods with less sensitivity to the geo-registration error [116] are needed. Regarding this, another open research issue lies in the use of deep learning to fuse satellite images with different resolutions. Deep learning can be used to model complex relationships between coarse images and fine images so that the fine image can be predicted. For spatiotemporal fusion, accuracy assessment is one of the key issues, but a widely accepted accuracy assessment method has not been developed yet. In Ref. [117] a public data set for the assessment of spatiotemporal data fusion was proposed but still there is a need for abundant data sets for different applications due to the increasing interest in satellite imaging. Widely accepted assessment models typically use correlation coefficient, Root Mean Square Error, relative difference, and absolute difference. But, these typically cannot reflect pixel-wise spatial details satisfactorily [116]. A new index was proposed in Ref. [118] for thermal images. It can be extended to evaluate the accuracy of spatiotemporal data fusion process. In addition, the use of standard datasets is important to make a fair comparison between different spatiotemporal data fusion methods. A publicly available data set for this purpose is provided in Ref. [117]. A final open research issue regarding spatiotemporal data fusion process is the need for methods to deal with low computing efficiency, particularly for image sets that cover large areas [116].

There are many reasons that can cause unknown correlations in real world applications. Since unknown correlations are inevitable, dividing them into categories as complete unknown and partial unknown ones can be a useful methodology in improving the existing fusion methods and developing new ones. In addition to this, instead of the traditional way of being labeling labeled, unknown correlations should be explicitly characterized via mathematical models [119]. Since unknown correlations are inherent in most applications, an interesting research topic would be to design consensus filter or hybrid consensus filter, in case of increasing complexity of system dynamics, involving information matrices, information vectors, and optimal weights simultaneously [119]. On the other hand, although consensus is a commonly used strategy in solving distributed filtering problems, it may not be a suitable strategy in certain circumstances and instead of it the diffusion strategy proposed in Ref. [120] can be considered. Therefore, strategies that can make a trade-off between different strategies are needed, especially in case of randomly occurring nonlinearities and fading measurements.

An emerging use of multi-fusion methods is the fusing of information in learning analytics. This approach focuses on the types of collected data and integration of them in an appropriate way that allows the accurate characterization of the learning process [121]. This approach relies on the advancements in artificial intelligence techniques at the expense of additional computing load.

4.6 Conclusion

Sensing and associated processing steps are the key enablers of IoT applications in smart grids. Due to the diversity of sensors and cost-related issues, data fusion, a big trend in the fields of signal, image and data analysis, has become an inevitable process in IoT applications. Although data fusion combines and organizes numerical entities for analysis thereby providing many opportunities, due to the huge diversity of sensors, large number of measurements and big volume of data in real-time the development of sophisticated fusion techniques for extracting valuable information from IoT systems is essential. However, most IoT designers face challenges in terms of data harmonization, data interface abstraction and computation cost as well as cost-related limitations, this makes it difficult for them to implement a data fusion technique freely. Recent works conducted in the last couple of years highlighted the need for relevant information extraction via data dictionaries dedicated to IoT scenarios and data fusion and computational frameworks designed for IoT applications.

In this paper, we comprehensively reviewed relevant literature on multi-sensor data fusion and its applications, particularly focusing on its application in IoT environment. For this, we first conducted an extensive survey and created a taxonomy for the reviewed papers. Then, based on the taxonomy we created; we presented the fusion methods in the literature. Finally, we presented the opportunities and challenges encountered in data fusion applications in IoT environments and stated future research directions in this domain.

Although JDL is currently the most commonly used method in multi-sensor data fusion applications, the trend biased toward IoT will make it necessary the integration of sophisticated multi-sensing platforms and systems. It is obvious that multi-sensor data fusion is in a state of continuous development, particularly in terms of multi-level fusion process, fault detection, micro/nano smart sensors, and adaptive multi-sensor fusion. Therefore, the development of state-of-the-art data processing algorithms and novel fusion solutions that accelerate computation and provide reliability and accuracy will be unavoidable.

Acknowledgments

This work was supported by scholarships given to Dr. Resul Daş by the Scientific and Technological Research Council of Turkey (TUBITAK-BIDEB 2219) International Postdoctoral Research Scholarship Program. The author would like to thank TUBITAK-BIDEB for its financial support and Professor Ioanis Nikolaidis his guidance and support during his studies.

References

[1] J. Qi, P. Yang, G. Min, O. Amft, F. Dong, L. Xu, Advanced internet of things for personalised healthcare systems: a survey, Pervasive and Mobile Computing 41 (2017) 132−149.

[2] O.B. Sezer, E. Dogdu, A.M. Ozbayoglu, Context-aware computing, learning, and big data in internet of things: a survey, IEEE Internet of Things Journal 5 (2018) 1−27.

[3] F. Alam, R. Mehmood, I. Katib, N.N. Albogami, A. Albeshri, Data fusion and IoT for smart ubiquitous environments: a survey, IEEE Access 5 (2017) 9533−9554.

[4] A. Zanella, N. Bui, A. Castellani, L. Vangelista, M. Zorzi, Internet of things for smart cities, IEEE Internet of Things Journal 1 (2014) 22–32.

[5] P.P. Ray, A survey on internet of things architectures, Journal of King Saud University - Computer and Information Sciences 30 (3) (2016) 291–319.

[6] J. Mineraud, O. Mazhelis, X. Su, S. Tarkoma, A gap analysis of internet-of-things platforms, Computer Communications 89 (2016) 5–16.

[7] R. Ford, M. Pritoni, A. Sanguinetti, B. Karlin, Categories and functionality of smart home technology for energy management, Building and Environment 123 (2017) 543–554.

[8] L. Liu, E. Stroulia, I. Nikolaidis, A. Miguel-Cruz, A. Rios Rincon, Smart homes and home health monitoring technologies for older adults: a systematic review, International Journal of Medical Informatics 91 (2016) 44–59.

[9] A. Antonacci, F. Arduini, D. Moscone, G. Palleschi, V. Scognamiglio, Nanostructured (Bio)sensors for smart agriculture, TrAC, Trends in Analytical Chemistry 98 (2018) 95–103.

[10] S. Schreiber-Ehle, W. Koch, The JDL model of data fusion applied to cyber-defence-A review paper, in: Proceedings of 2012 Workshop on Sensor Data Fusion: Trends, Solutions, Applications, Sept. 2012, pp. 116–119.

[11] A.D. Paola, P. Ferraro, S. Gaglio, G.L. Re, S.K. Das, An adaptive bayesian system for context-aware data fusion in smart environments, IEEE Transactions on Mobile Computing 16 (2017) 1502–1515.

[12] Z. Chen, M.K. Masood, Y.C. Soh, A fusion framework for occupancy estimation in office buildings based on environmental sensor data, Energy and Buildings 133 (2016) 790–798.

[13] J. Kim, M. Kim, K. Kim, Development of a wearable HCI controller through sEMG IMU sensor fusion, in: Proceedings of 13th International Conference on Ubiquitous Robots and Ambient Intelligence, 2016, pp. 83–87.

[14] A. Corradini, M. Mehta, N.O. Bernsen, J.-C. Martin, Multimodal input fusion in human-computer interaction on the example of the on-going NICE project, in: Proceedings of the NATO-ASI Conference on Data Fusion for Situation Monitoring, Incident Detection, Alert and Response Management, Yerevan, Armenia, 2003.

[15] D. Evans, The Internet of Things: How the Next Evolution of the Internet is Changing Everything, White paper, Cisco, San Jose, CA, USA, April 2011.

[16] R. Das, I. Turkoglu, Creating meaningful data from web logs for improving the impressiveness of a website by using path analysis method, Expert Systems with Applications 36 (2009) 6635–6644.

[17] I.M. Pires, N.M. Garcia, F. Florez-Revuelta, Multi-sensor data fusion techniques for the identification of activities of daily living using mobile devices, in: Proceedings of the ECMLPKDD, 2015.

[18] S. Rawat, S. Rawat, Multi-sensor data fusion by a hybrid methodology – a comparative study, Computers in Industry 75 (2016) 27–34.

[19] B. Chandrasekaran, S. Gangadhar, J.M. Conrad, A survey of multisensor fusion techniques, architectures and methodologies, in: SoutheastCon 2017, 2017, pp. 1–8.

[20] F. Castanedo, A review of data fusion techniques, The Scientific World Journal 2013 (2013) 1–19.

[21] M. Nilsson, J.v. Laere, T. Susi, T. Ziemke, Information fusion in practice: a distributed cognition perspective on the active role of users, Information Fusion 13 (2012) 60–78.

[22] G. Slanzi, G. Pizarro, J.D. Velásquez, Bio- metric information fusion for web user navigation and preferences analysis: an overview, Information Fusion 38 (2017) 12–21.

[23] Y. Zhang, S. Huang, S. Guo, J. Zhu, Multi- sensor data fusion for cyber security situation awareness, Procedia Environmental Sciences 10 (2011) 1029–1034.
[24] D. Hall, J. Llinas, An introduction to multisensor data fusion, Proceedings of the IEEE 85 (1997) 6–23.
[25] S. Birogul, Y. Sonmez, U. Guvenc, A survey of data fusion, Politeknik Dergisi 10 (3) (2007) 235–240.
[26] Y.-B. Huang, Y.-B. Lan, W.C. Hoffmann, R.E. Lacey, Multisensor data fusion for high quality data analysis and processing in measurement and instrumentation, Journal of Bionic Engineering 4 (Mar. 2007) 53–62.
[27] N.A. Giacobe, Application of the JDL data fusion process model for cyber security, Proceedings of SPIE 7710 (2010) 77100R.
[28] I.M. Pires, N.M. Garcia, N. Pombo, F. Flórez- Revuelta, From data acquisition to data fusion: a comprehensive review and a roadmap for the identification of activities of daily living using mobile devices, Sensors 16 (2016) 184.
[29] W.-T. Sung, M.-H. Tsai, Data fusion of multi- sensor for IOT precise measurement based on improved PSO algorithms, Computers & Mathematics With Applications 64 (2012) 1450–1461.
[30] R.C. King, E. Villeneuve, R.J. White, R.S. Sherratt, W. Holderbaum, W.S. Harwin, Application of data fusion techniques and technologies for wearable health monitoring, Medical Engineering & Physics 42 (2017) 1–12.
[31] R. Tanner, N.K. Loh, A toxonomy of multi- sensor fusion, Journal of Manufacturing Systems 11 (5) (1992) 314–325.
[32] W. Elmenreich, An Introduction to Sensor Fusion, Vienna University of Technology, Austria, 2002.
[33] A.N. Steinberg, C.L. Bowman, Rethinking the JDL data fusion levels, NSSDF JHAPL 38 (2004) 39.
[34] J. Bleiholder, F. Naumann, Data fusion, ACM Computing Surveys 41 (2009) 1–41.
[35] J. Dong, D. Zhuang, Y. Huang, J. Fu, Advances in multi-sensor data fusion: algorithms and applications, Sensors 9 (2009) 7771–7784.
[36] L.J. De Vin, M. Holm, A.H.C. Ng, The information fusion JDL-U model as a reference model for virtual manufacturing, Robotics and Computer-Integrated Manufacturing 26 (2010) 629–638.
[37] L. Han, Z. Lei, Clustering methods for multi-sensor data fusion, in: Proceedings of 2012 International Conference on Industrial Control and Electronics Engineering, August 2012, pp. 1166–1169.
[38] G. Navarro-Arribas, V. Torra, Information fusion in data privacy: a survey, Information Fusion 13 (2012) 235–244.
[39] O. Sidek, S.A. Quadri, A review of data fusion models and systems, International Journal of Image and Data Fusion 3 (2012) 3–21.
[40] E. Blasch, A. Steinberg, S. Das, J. Llinas, C. Chong, O. Kessler, E. Waltz, F. White, Revisiting the JDL model for information exploitation, in: Proceedings of 16th International Conference on Information Fusion, 2013, pp. 129–136.
[41] V. Borges, W. Jeberson, Survey of context information fusion for sensor networks based ubiquitous systems, Computer Science and Information Technology 2 (3) (2014) 165–178, https://doi.org/10.13189/csit.2014.020306.
[42] B. Khaleghi, A. Khamis, F.O. Karray, S.N. Razavi, Multisensor data fusion: a review of the state-of-the-art, Information Fusion 14 (2013) 28–44.
[43] S.B. Ayed, H. Trichili, A.M. Alimi, Data fusion architectures: a survey and comparison, in: Proceedings of 15th International Conference on Intelligent Systems Design and Applications, 2015, pp. 277–282.
[44] E. Azimirad, J. Haddadnia, A. Izadipour, A comprehensive review of the multi-sensor data fusion architectures, Journal of Theoretical and Applied Information Technology 71 (1) (2015).

[45] M.M. Fouad, N.E. Oweis, T. Gaber, M. Ahmed, V. Snasel, Data mining and fusion techniques for WSNs as a source of the big data, Procedia Computer Science 65 (2015) 778–786.
[46] D. Novak, R. Riener, A survey of sensor fusion methods in wearable robotics, Robotics and Autonomous Systems 73 (2015) 155–170.
[47] L. Snidaro, J. García, J. Llinas, Context-based information fusion: a survey and discussion, Information Fusion 25 (2015) 16–31.
[48] J.A. Balazs, J.D. Velásquez, Opinion mining and information fusion: a survey, Information Fusion 27 (2016) 95–110.
[49] N.-E.E. Faouzi, L.A. Klein, Data fusion for ITS: techniques and research needs, Transportation Research Procedia 15 (2016) 495–512.
[50] K. Golestan, R. Soua, F. Karray, M.S. Kamel, Situation awareness within the context of connected cars: a comprehensive review and recent trends, Information Fusion 29 (May 2016) 68–83.
[51] J.J.A. Mendes Jr., M.E.M. Vieira, M.B. Pires, S.L. Stevan Jr., Sensor fusion and smart sensor in sports and biomedical applications, Sensors 16 (September 2016) 1569.
[52] R. Gravina, P. Alinia, H. Ghasemzadeh, G. Fortino, Multi-sensor fusion in body sensor networks: state-of-the-art and research challenges, Information Fusion 35 (2017) 68–80.
[53] B. Noack, J. Sijs, M. Reinhardt, U.D. Hanebeck, Decentralized data fusion with inverse covariance intersection, Automatica 79 (2017) 35–41.
[54] Y. Zhu, D. Liu, R. Grosu, X. Wang, H. Duan, G. Wang, A multi-sensor data fusion approach for atrial hypertrophy disease diagnosis based on characterized support vector hyperspheres, Sensors 17 (2017) 2049.
[55] G. Hernández-Peñaloza, A. Belmonte-Hernández, M. Quintana, F. Álvarez, A multisensor fusion scheme to increase life autonomy of elderly people with cognitive problems, IEEE Access 6 (2018) 12775–12789, https://doi.org/10.1109/ACCESS.2017.2735809.
[56] I.A. Ramgopal, P.V. Manivannan, Development of multi-sensor data fusion technique for the automated bus rapid transport system, in: Proceedings of 2016 International Conference on Robotics: Current Trends and Future Challenges (RCTFC), Dec. 2016, pp. 1–6.
[57] H. Medjahed, D. Istrate, J. Boudy, J.L. Baldinger, B. Dorizzi, A pervasive multisensor data fusion for smart home healthcare monitoring, in: Proceedings of 2011 IEEE International Conference on Fuzzy Systems, June 2011, pp. 1466–1473.
[58] L. Vargas-Meléndez, B.L. Boada, M.J.L. Boada, A. Gauchía, V. Díaz, A sensor fusion method based on an integrated neural network and Kalman filter for vehicle roll angle estimation, Sensors 16 (9) (2016) 1400.
[59] A.M. Sabatini, V. Genovese, A sensor fusion method for tracking vertical velocity and height based on inertial and barometric altimeter measurements, Sensors 14 (2014) 13324–13347.
[60] W.-T. Sung, Multi-sensors data fusion system for wireless sensors networks of factory monitoring via BPN technology, Expert Systems With Applications 37 (2010) 2124–2131.
[61] P.-H. Chou, Y.-L. Hsu, W.-L. Lee, Y.-C. Kuo, C.-C. Chang, Y.-S. Cheng, H.-C. Chang, S.-L. Lin, S.-C. Yang, H.-H. Lee, Development of a smart home system based on multi-sensor data fusion technology, in: Applied System Innovation (ICASI), 2017 International Conference on, IEEE, 2017, pp. 690–693.
[62] L.P.L. Billy, N. Wijerathne, B.K.K. Ng, C. Yuen, Sensor fusion for public space utilization monitoring in a smart city, IEEE Internet of Things Journal 5 (2) (2018) 473–481, https://doi.org/10.1109/JIOT.2017.2748987.

[63] E. Kanjo, E.M. Younis, N. Sherkat, Towards unravelling the relationship between on-body, environmental and emotion data using sensor information fusion approach, Information Fusion 40 (2018) 18–31.
[64] O. Banos, M.A. Toth, M. Damas, H. Pomares, I. Rojas, Dealing with the effects of sensor displacement in wearable activity recognition, Sensors 14 (2014) 9995–10023.
[65] X. Hao, P. Jin, L. Yue, Efficient storage of multi-sensor object-tracking data, IEEE Transactions on Parallel and Distributed Systems 27 (2016) 2881–2894.
[66] S. Kumar, R.M. Hegde, Multi-sensor data fusion methods for indoor localization under collinear ambiguity, Pervasive and Mobile Computing 30 (2016) 18–31.
[67] S.O.H. Madgwick, A.J.L. Harrison, R. Vaidyanathan, Estimation of IMU and MARG orientation using a gradient descent algorithm, in: 2011 IEEE International Conference on Rehabilitation Robotics, Zurich, 2011, pp. 1–7, https://doi.org/10.1109/ICORR.2011.5975346.
[68] S.A. Ludwig, K.D. Burnham, A.R. Jiménez, P.A. Touma, "Comparison of attitude and heading reference systems using foot mounted MIMU sensor data: basic, Madgwick, and Mahony", in: Proceeding SPIE 10598, Sensors and Smart Structures Technologies for Civil, Mechanical, and Aerospace Systems 2018, March 27, 2018, p. 105982L, https://doi.org/10.1117/12.2296568.
[69] R. Mahony, T. Hamel, J. Pflimlin, Nonlinear complementary filters on the special orthogonal group, IEEE Transactions on Automatic Control 53 (5) (June 2008) 1203–1218, https://doi.org/10.1109/TAC.2008.923738.
[70] R.E. Kalman, A new approach to linear filtering and prediction problems, Journal of Basic Engineering 82 (1960) 35–45, https://doi.org/10.1115/1.3662552.
[71] M.E. Liggins, D.L. Hall, J. Llinas (Eds.), Handbook of Multisensor Data Fusion: Theory and Practice, CRC Press, Boca Raton, 2009.
[72] A. Kushwah, S. Kumar, R.M. Hegde, Multi-sensor data fusion methods for indoor activity recognition using temporal evidence theory, Pervasive and Mobile Computing 21 (2015) 19–29.
[73] N.A. Tmazirte, M.E.E. Najjar, C. Smaili, Pomorski, D. Pomorski, Dynamical reconfiguration strategy of a multi sensor data fusion algorithm based on information theory, in: Proceedings of IEEE Intelligent Vehicles Symposium, June 2013, pp. 896–901.
[74] R. Das, I. Türkoglu, Extraction of interesting patterns through association rule mining for improvement of website usability, Istanbul University - Journal of Electrical & Electronics Engineering 9 (2) (2009) 1037–1046.
[75] W. Elmenreich, A review on system architectures for sensor fusion applications, in: Proceedings of the 5th IFIP WG 10.2 International Conference on Software Technologies for Embedded and Ubiquitous Systems, 2007, pp. 547–559.
[76] M. Markin, C. Harris, M. Bernhardt, J. Austin, M. Bedworth, P. Greenway, R. Johnston, A. Little, D. Lowe, Technology Foresight on Data Fusion and Data Processing, The Royal Aeronautical Society, 1997.
[77] J.R. Boyd, A Discourse on Winning and Losing, Unpublished set of briefing slides, Air University Library, Maxwell AFB, AL, USA, May 1987.
[78] M. D Bedworth, J. O'Brien, The omnibus model: a new architecture for data fusion?, in: Proceedings of the 2nd International Conference on Information Fusion Finland, Helsinki, 1999.
[79] R. Kumar, M. Wolenetz, B. Agarwalla, J. Shin, P. Hutto, A. Paul, U. Ramachandran, DFuse: a framework for distributed data fusion, in: Proceedings of the International Conference on Embedded Networked Sensor Systems, 2003, pp. 114–125.
[80] A. Franz, R. Mista, D. Bakken, C. Dyreson, M. Medidi, Mr. Fusion: a programmable data fusion middleware subsystem with a tunable statistical profiling service, in: Proceedings of the International Conference on Dependable Systems and Networks, 2002, pp. 273–278.

[81] R. Alami, R. Chatila, S. Fleury, M. Ghallab, F. Ingrand, An architecture for autonomy, The International Journal of Robotics Research 17 (4) (1998) 315–337.
[82] W. Elmenreich, S. Pitzek, The time-triggered sensor fusion model, in: Proceedings of the 5th IEEE International Conference on Intelligent Engineering Systems, Finland, Helsinki–Stockholm–Helsinki, 2001, pp. 297–300.
[83] D.L. Hall, S.A.H. McMullen, Mathematical Techniques in Multisensor Data Fusion, second ed., Artech Print on Demand, Boston, February 2004.
[84] W. Koch, Tracking and Sensor Data Fusion, Springer Berlin Heidelberg, Heidelberg, 2014.
[85] H. Kopetz, G. Bauer, The time-triggered architecture, Proceedings of the IEEE 91 (1) (2003) 112–126.
[86] G.M. Provan, The validity of Dempster-Shafer belief functions, International Journal of Approximate Reasoning 6 (3) (1992) 389–399.
[87] G. Shafer, A Mathematical Theory of Evidence, Princeton University Press, 1976.
[88] J. Kohlas, P.-A. Monney, Theory of evidence – a survey of its mathematical foundations, applications and computational aspects, Zor-Mathematical Methods of Operations Research 39 (1994).
[89] R.C. Luo, C.-C. Yih, K.L. Su, Multisensor fusion and integration: approaches, applications, and future research directions, IEEE Sensors Journal 2 (2) (2002) 107–119.
[90] F. Berman, R. Rutenbar, B. Hailpern, H. Christensen, S. Davidson, D. Estrin, M. Franklin, M. Martonosi, P. Raghavan, V. Stodden, A.S. Szalay, Realizing the potential of data science, Communications of the ACM 61 (2018) 67–72.
[91] M. Guri, Y. Elovici, Bridgeware: the air-gap malware, Communications of the ACM 61 (2018) 74–82.
[92] N. Forsgren, M. Kersten, DevOps metrics, Communications of the ACM 61 (2018) 44–48.
[93] C. Sadowski, E. Aftandilian, A. Eagle, L. Miller- Cushon, C. Jaspan, Lessons from building static analysis tools at google, Communications of the ACM 61 (2018) 58–66.
[94] D. Lahat, T. Adali, C. Jutten, Multimodal data fusion: an overview of methods, challenges, and prospects, Proceedings of the IEEE 103 (2015) 1449–1477.
[95] D. Hall, A. Steinberg, Dirty secretsin multi- sensor data fusion, in: D. Hall, J. Llinas (Eds.), Multisensor Data Fusion, 2 Volume Set, vol. 3, CRC Press, June 2001.
[96] T. Stathaki (Ed.), Image Fusion: Algorithms and Ap- Plications, Academic Press, Amsterdam, 2008. OCLC: 845386598.
[97] I. Van Mechelen, A.K. Smilde, A generic linked-mode decomposition model for data fusion, Chemometrics and Intelligent Laboratory Systems 104 (2010) 83–94.
[98] T. Wilderjans, E. Ceulemans, I. Van Mechelen, Simultaneous analysis of coupled data blocks differing in size: a comparison of two weighting schemes, Computational Statistics & Data Analysis 53 (2009) 1086–1098.
[99] D. Mcdaniel, G. Schaefer, Real-time dbms for data fusion, Proceedings of the Sixth International Conference of Information Fusion 2 (2003) 1334–1341.
[100] W. Elmenreich, C. Paukovits, S. Pitzek, Automatic generation of schedules for time-triggered embedded transducer networks, in: Proceedings of the 10th IEEE Conference on Emerging Technologies and Factory Automation, Catania, Italy, vol. 2, 2005, pp. 535–541.
[101] A. Goudarzi, F. Ghayoor, M. Waseem, S. Fahad, I. Traore, A survey on IoT-enabled smart grids: emerging, applications, challenges, and outlook, Energies 15 (19) (2022).
[102] L. Kong, X. Peng, Y. Chen, P. Wang, M. Xu, Multi-sensor measurement and data fusion technology for manufacturing process monitoring: a literature review, International Journal of Extreme Manufacturing 2 (2) (2020).

[103] F. Castanedo, A review of data fusion techniques, The Scientific World Journal 2013 (2013), https://doi.org/10.1155/2013/704504.

[104] F. Castanedo, J. García, M.A. Patricio, J.M. Molina, Data fusion to improve trajectory tracking in a cooperative surveillance multi-agent architecture, Information Fusion 11 (3) (2010) 243–255.

[105] M.A. Davenport, C. Hegde, M.F. Duarte, R.G. Baraniuk, Joint manifolds for data fusion, IEEE Transactions on Image Processing 19 (10) (2010) 2580–2594.

[107] V. Nimier, Information evaluation: a formalization of operational recommendations, in: Proceedings of the International Conference on Information Fusion, 2009, pp. 1166–1171.

[108] L. Cholvy, Modelling information evaluation in fusion, in: Proceedings of 10th International Conference on Information Fusion, Quebec, 2007, pp. 1–6, https://doi.org/10.1109/ICIF.2007.4408060.

[109] Nato, Standardization Agreements: Intelligence Reports STANAG 2022, North Atlantic Treaty Organization, 1992.

[110] M. Fernandez, H.F. Durrant-Whyte, A failure detection and isolation algorithm for a decentralised multisensor system, in: Proceedings of IEEE International Conference on Multisensor Fusion and Integration for Intelligent Systems, Las Vegas, NV, USA, 1994, pp. 27–33, https://doi.org/10.1109/MFI.1994.398476.

[111] H. Zhu, O. Basir, A novel fuzzy evidential reasoning paradigm for data fusion with applications in image processing, Soft Computing Journal 10 (12) (2006) 1169–1180.

[112] D. Rotondo, F. Nejjari, V. Puig, A virtual actuator and sensor approach for fault tolerant control of LPV systems, Journal of Process Control 24 (2014) 203–222.

[113] S. Xie, A.J.P. Theuwissen, On-chip smart temperature sensors for dark current compensation in CMOS image sensors, IEEE Sensors Journal 19 (18) (2019) 7849–7860, https://doi.org/10.1109/JSEN.2019.2919655.

[114] J. Liu, L. Pan, C. Shang, B. Lu, R. Wu, Y. Feng, W. Chen, R. Zhang, J. Bu, Z. Xiong, W. Bu, J. Du, J. Shi, A highly sensitive and selective nanosensor for near-infrared potassium imaging, Science Advances 6 (16) (2020) eaax9757, https://doi.org/10.1126/sciadv.aax9757.

[115] L. Hong, Adaptive data fusion, in: Proceedings of IEEE International Conference on Systems, Man, and Cybernetics, Charlottesville, VA, USA vol. 2, 1991, pp. 767–772, https://doi.org/10.1109/ICSMC.1991.169779.

[116] X. Zhu, F. Cai, J. Tian, T.K.-A. Williams, Spatiotemporal fusion of multisource remote sensing data: literature survey, taxonomy, principles, applications, and future directions, Remote Sensing 10 (4) (2018), https://doi.org/10.3390/rs10040527.

[117] I.V. Emelyanova, T.R. McVicar, T.G. Van Niel, L.T. Li, A.I.J.M. van Dijk, Assessing the accuracy of blending Landsat–MODIS surface reflectances in two landscapes with contrasting spatial and temporal dynamics: a framework for algorithm selection, Remote Sensing of Environment 133 (2013) 193–209.

[118] L. Gao, W. Zhan, F. Huang, X. Zhu, J. Zhou, J. Quan, P. Du, M. Li, Disaggregation of remotely sensed land surface temperature: a simple yet flexible index (SIFI) to assess method performances, Remote Sensing of Environment 200 (2017) 206–219.

[119] W. Li, Z. Wang, G. Wei, L. Ma, J. Hu, D. Ding, A survey on multisensor fusion and consensus filtering for sensor networks, Discrete Dynamics in Nature and Society 2015 (2015) 12, https://doi.org/10.1155/2015/683701.

[120] D. Franken, A. Hupper, Improved fast covariance intersection for distributed data fusion, in: Proceedings of International Conference on Information Fusion, 2005, pp. 154–160.

[121] S. Mu, M. Cui, X. Huang, Multimodal data fusion in learning analytics: a systematic review, Sensors 20 (23) (2020), https://doi.org/10.3390/s20236856.

CHAPTER 5

Impact of cybersecurity attacks on electrical system operation

Sinan Atıcı[1] and Gurkan Tuna[2]
[1]Department of Software, SECHARD Information Technologies Company, Istanbul, Türkiye;
[2]Computer Technologies, Trakya University, Edirne, Türkiye

5.1 Introduction

Undoubtedly, the rapid transition from the conventional Electrical Grid to the Smart Grid (SG) signifies a momentous paradigm shift in the domain of energy management and distribution. The assimilation of the Industrial Internet of Things (IIoT) augments the efficiency and reliability of power systems, presenting an auspicious trajectory toward a more sustainable and technologically sophisticated future [1]. The SG promises to facilitate heightened grid intelligence, optimal resource utilization, and improved grid resiliency, thereby reshaping the energy sector and aligning it with modern demands and advancements in technology and sustainability.

SG systems, evolving into cyber-physical systems (CPSs), integrate complex software-based controllers, sophisticated communication networks, and distributed cooperative control protocols. They represent an evolution of the conventional power grid, encompassing power systems, information, and communication infrastructures, with a critical emphasis on cybersecurity to ensure secure, reliable, and efficient operation. The reliance on distributed control, vital for optimal performance, makes these systems more susceptible to cybersecurity threats, particularly false-data injection (FDI) attacks, due to potential limited global situational awareness and the propagation of physical anomalies' effects [2,3].

In this chapter, we embark on an in-depth exploration of cybersecurity within the realm of electrical system operations. We begin by examining diverse threats that pose risks to the seamless functioning of these systems. We then delve into real-life case studies that shed light on the gravity of cybersecurity attacks and their implications. Moving forward, we analyze

Cyber Security Solutions for Protecting and Building the Future Smart Grid
ISBN 978-0-443-14066-2
https://doi.org/10.1016/B978-0-443-14066-2.00008-6

how these threats impact the reliability and stability of power grids. Subsequently, we thoroughly investigate the extensive economic and societal repercussions stemming from cybersecurity breaches. Furthermore, we underscore the critical implications for national security, emphasizing the roles of governmental and defense agencies. Finally, we focus on crucial cybersecurity measures and best practices essential for safeguarding the dependable functionality of electrical systems.

5.1.1 Understanding the crucial role of cybersecurity in the operation of electrical systems

Critical infrastructure, encompassing essential elements like objects, networks, services, and systems, forms the backbone of a nation's functionality. Any disruption or malfunction within these components poses a significant threat to the health, safety, and overall well-being of the populace. Notably, over half of recent recorded incidents involve severe cyber-attacks on critical infrastructure facilities, underscoring the urgent need to enhance cybersecurity measures for the stability of our society [4].

In today's digital era, electrical systems are integral to our daily lives, powering homes, industries, and economies. However, this extensive integration and reliance expose us to heightened vulnerabilities to cyber threats. Recognizing the pivotal role of cybersecurity is essential to fortify and ensure the smooth operation of these crucial systems. Cybersecurity, an ever-evolving discipline, is fundamental in preserving the integrity and stability of electrical systems. It comprises diverse practices, technologies, and protocols aimed at thwarting unauthorized access, data breaches, and malevolent attacks. Within electrical systems, cybersecurity acts as a vital defense against a range of cyber threats capable of disrupting the power grid and our routines. To address these risks, various security solutions have been proposed, taking into account physical models, operation data, side channel information, and other techniques. These measures aim to protect critical systems and ensure the secure operation of manufacturing processes. By implementing comprehensive security measures and staying vigilant against cyber threats, the integrity and functionality of modern systems can be effectively safeguarded [5,6].

At the heart of safeguarding electrical infrastructure lies a profound understanding of cybersecurity and the need to identify potential vulnerabilities. The amalgamation of digital technologies and the growing interconnectivity of devices provides fresh entry points for cybercriminals, enabling them to exploit weaknesses and gain unauthorized access to critical components, manipulate control systems, or disrupt the seamless energy flow. Proper software development is foundational to system security, demanding developers' fluency in secure coding guidelines and best practices [7]. Effectively mitigating these risks requires robust cybersecurity measures, encompassing the deployment of firewalls, intrusion detection systems (IDS), encryption protocols, and regular security updates. Additionally, educating personnel within the energy sector is paramount, cultivating a culture of awareness and vigilance against emerging cyber threats.

Furthermore, today's electrical systems play a vital role in our daily lives, ensuring the power that drives our homes, industries, and economies. However, the extensive integration and reliance on digital technologies expose us to heightened vulnerabilities to cyber threats. Understanding the significance of cybersecurity is paramount to fortify and ensure the smooth operation of these crucial systems. As we delve into this discussion, it is important to keep in mind the core principles of the Confidentiality, Integrity, and Availability (CIA) triad which underpin data security.

Confidentiality is the key principle of preventing unauthorized access to sensitive information, whether it is safeguarding personal data like vital signs collected by medical equipment in the IoT or protecting critical data on the traditional Internet, such as phone numbers and passwords. Various encryption techniques, including Rivest-Shamir-Adleman (RSA) and Data Encryption Standart (DES), are used to maintain data integrity and privacy, even in resource-constrained IoT devices [7a].

Integrity, within the context of data security, serves as the foundational principle that limits the access to and alteration of information solely to authorized individuals or processes [8].

Availability refers to the elements of the system's infrastructure that secure convenient access to information for its users [9]. These principles guide us in safeguarding against an evolving landscape of cyber threats, and it is imperative to consider them as we explore the multifaceted world of electrical infrastructure security.

In conclusion, cybersecurity stands as a linchpin in the operation of electrical systems, safeguarding against an evolving landscape of cyber threats. It is imperative to invest in research, education, and cutting-edge technologies to bolster our defenses. By doing so, we ensure the reliable, uninterrupted, and secure distribution of electricity, meeting the burgeoning needs of our modern society. Only through a deep understanding and proactive resolution of these cybersecurity challenges can we build and preserve a resilient future SG.

5.1.2 Technologies used for electrical system operations

The continuous modernization of the electrical grid is driven by a set of pivotal improvements. These encompass the reinforcement of switching capabilities within electrical substations, the incorporation of cutting-edge power electronics such as phasor measurement networks to enable precise monitoring and enhance system reliability, and the integration of advanced metering technologies. These technologies empower consumers to flexibly adapt to variable tariffs and actively participate in demand response initiatives. Furthermore, microgrids are envisaged as fundamental constituents of the future grid, with their development facilitated by advancements in microgrid management systems [10].

In their study [10a], highlighted that cyberattacks specifically target critical components, including Supervisory Control and Data Acquisition (SCADA) systems, Advanced Metering Infrastructure (AMI), Phasor Measuring Units (PMU), and Phasor Data Concentrators (PDC).

CPSs seamlessly integrate computer-based control with physical processes, and within this framework, SCADA systems assume a pivotal role in monitoring and controlling diverse industrial operations. However, the amalgamation of information technology (IT) and operational technology in SCADA systems exposes them to analogous cyber threats. This section sheds light on several noteworthy vulnerabilities that have the potential to impact SCADA systems. Foremost among these vulnerabilities are Denial of Service (DoS) attacks, which pose a substantial risk by inundating Internet-connected devices with an overwhelming influx of spurious packets, thereby causing network congestion and impeding legitimate communication. The incursion of malicious codes, such as viruses, Trojan horses, and worms, represents an additional peril as they can compromise SCADA systems that possess direct or indirect connections to the Internet. Moreover, insider threats originating from dissatisfied employees with unimpeded access to SCADA systems warrant concern, as exemplified by the notorious Maroochy Shire sewage treatment system attack. Furthermore, unpatched vulnerabilities furnish hackers with a window of opportunity to exploit weaknesses within SCADA systems, permitting them to gain unauthorized control over critical infrastructure. Additionally, non-technical attacks, particularly those leveraging social engineering techniques, can circumvent sophisticated security measures by artfully manipulating individuals into disclosing sensitive information. To effectively mitigate these potential threats, it is imperative to harness established security concepts, including management protocols, filtering mechanisms, encryption methodologies, and IDS employed in traditional IT systems, while adapting them to suit the distinctive requirements of SCADA systems [11].

In the realm of Electrical System Operations, several key technologies play pivotal roles in facilitating efficient and secure information exchange and control within industrial systems. These include IEEE 1815 DNP3 [12] an interoperable communication framework, vital for ensuring secure data exchange in systems like SCADA. Sunspec Modbus [13] extends the Modbus protocol [14] to monitor and control Distributed Energy Resources (DERs) [15] parameters, such as power and voltage, along with ancillary services. OpenADR [16], on the other hand, serves as the energy market management standard, governing demand response through signaling to DERs and controllable devices. IEEE 2030.5 [17] stands as the default protocol for DER management, providing a smart energy profile application standard. These technologies collectively form the bedrock for efficient and robust operations within the electrical systems domain [18].

5.2 Various cybersecurity threats to electrical system operation

In this section, we will explore a range of cybersecurity threats that present substantial risks to the smooth functioning of electrical systems. Grasping the nature of these threats is vital for devising proactive strategies to address potential vulnerabilities and bolster the overall security of electrical infrastructure.

5.2.1 Exploring different types of cybersecurity threats affecting electrical system operations

Traditionally centralized for cost optimization, Energy Management Systems (EMS) are now garnering increased attention due to the resilience of distributed controllers against cyber threats, while attackers focus on disrupting power generation to raise costs and decrease energy efficiency, posing a substantial economic risk for grid operators, and employing deceptive tactics to target EMS to interfere with SCADA systems [19].

Cyberattacks on SGs encompass a wide range of threats, with a particular focus on the CIA triad, which includes energy availability, system integrity, and consumer identity confidentiality. These attacks are often carried out through the use of various malicious software, such as worms, Trojans, viruses, spyware, trapdoors, and backdoors, which can be introduced via pre-installed software. Among the most extensively studied attacks are FDI attacks, DoS, distributed DoS (DDoS), and spoofing attacks. Additionally, SG communication faces disruptions caused by DoS and DDoS attacks, leading to compromised data transmission and information integrity. Moreover, prevalent spoofing attacks, including man-in-the-middle (MitM) attacks, software exploitation, message replays, and a distinct jamming attack, highlight the potential for disrupting the SG system [20,21].

5.2.1.1 State estimation (SE) model

The relationship between measurements Z and state variables x is expressed by the nonlinear equation:

$$Z = H(x) + e \tag{5.1}$$

The state vector x contains bus voltage angles and magnitudes to be estimated, while the measurements Z include power injections and bidirectional power flows for each line acquired from Remote Terminal Units (RTUs). The measurement error is represented by the vector e, typically assumed to follow an independent Gaussian distribution [22].

In the study by Ref. [23] titled "Phasor data correction and transmission system SE under MitM attack" we can represent the impact of the attack on the measurements with the following model:

$$Z_{observed} = SHv + e \tag{5.2}$$

Where S a latent variable and v is the state vector.

In their study [24,25], delved into a critical equation, known as the "classical Weighted Least Squares (WLS) method" (Eq. 5.3), central to SG SE. This equation plays a pivotal role in evaluating the accuracy and reliability of SE in SG, especially when facing challenges such as measurement inaccuracies and cybersecurity threats.

$$\begin{aligned} minH(x) &= \sum_{i=1}^{n} q_i(b_i - g_i(X) + \varepsilon_i + \alpha)^2 \\ &= [z - h(\widehat{x})]^T R^{-1}[z - h(\widehat{x})] \end{aligned} \tag{5.3}$$

Within this equation, $minH(x)$ signifies the objective function under minimization. This equation assumes a fundamental role in assessing the comprehensive expense of SE. It considers the squared discrepancies between measured values b_i and anticipated values $g_i(X)$, while also accommodating measurement errors ε_i and potential repercussions from cyber-attacks via the threat vector α. Its pivotal role lies in appraising the precision and trustworthiness of SE in SG, especially when confronted with issues like measurement inaccuracies and cybersecurity threats. By finding state vector values X that trim the expenses, it actively bolsters the robustness and security of SG operations.

The Jacobian matrix [26] is defined as follows:

$$H(\widehat{x}) = \frac{\partial h(\widehat{x})}{\partial \widehat{x}} \tag{5.4}$$

Where $H(\widehat{x})$ represents the Jacobian matrix that describes the relationship between the measurements and the state variables $\widehat{x}$. The optimization equation for state estimation is given by:

$$g(\widehat{x}) = \frac{\partial J}{\partial \widehat{x}} = -H(\widehat{x})R^{-1}[z - h(\widehat{x})] = 0 \tag{5.5}$$

This equation is used iteratively to determine the state variables vector $\widehat{x}$ where $\widehat{x}^t$ the solution at a given iteration is t. The Gain matrix $G(\widehat{x}^t)$, presented in the following equation, is used to solve the matrix system iteratively using the Gauss–Newton method. The iterative solution of the state variables is provided using the Gauss–Newton method with the following equation:

$$\widehat{x}^{t+1} = \widehat{x}^t - [G(\widehat{x}^t)]^{-1} g(\widehat{x}^t) \tag{5.6}$$

Here, $G(\widehat{x}^t)$ represents the Gain matrix at iteration t and is defined as:

$$G(\widehat{x}^t) = \frac{\partial g(\widehat{x}^t)}{d\widehat{x}} = H^T(\widehat{x}^t)R^{-1}H(\widehat{x}^t) \tag{5.7}$$

Here, R represents a diagonal matrix.

5.2.1.2 False data injection (FDI) attack model

FDI attack is a form of cyber-attack that entails the insertion of fabricated information into the power grid's SE process, which can result in erroneous actions or consequences [27]. According to a study by Ref. [28], the equation for an FDI attack is as follows:

$$a = \left[a_D; \quad a_F; \quad -a_F \right]^T \in R^{(nmx1)} \tag{5.8}$$

Here, a: FDI attack vector, which has a length of nm, where n is the number of buses in the system and m is the number of measurements. Each component of the vector represents a change in a measurement.

a_D: Vector representing changes in load demand, with a length of n.

a_F: Vector representing changes in active power flow, with a length of $2m$.

T: Transpose operator.

$R^{(nmx1)}$: Indicates that this vector is composed of nm rows and 1 column in the set of real numbers.

If we expand these parameters a_D and a_F from Eq. (5.8):

$$a_D = \left[a_{D1} \quad a_{D2} \quad \cdots \quad a_{Dn} \right] \tag{5.9}$$

$$a_F = \begin{bmatrix} a_{F1,1} & a_{F1,2} & \cdots & a_{F1,m} \\ a_{F2,1} & a_{F2,2} & \cdots & a_{F2,m} \\ \vdots & \vdots & \ddots & \vdots \\ a_{Fn,1} & a_{Fn,2} & \cdots & a_{Fn,m} \end{bmatrix} \tag{5.10}$$

Using these equations (Eqs. 5.9 and 5.10), we can expand the equation (Eq. 5.8) as follows:

$$a = \begin{bmatrix} a_{D1} & a_{D2} & \cdots & a_{Dn} \\ a_{F1,1} & a_{F1,2} & \cdots & a_{F1,m} \\ a_{F2,1} & a_{F2,2} & \cdots & a_{F2,m} \\ \vdots & \vdots & \ddots & \vdots \\ a_{Fn,1} & a_{Fn,2} & \cdots & a_{Fn,m} \\ -a_{F1,1} & -a_{F1,2} & \cdots & -a_{F1,m} \\ -a_{F2,1} & -a_{F2,2} & \cdots & -a_{F2,m} \\ \vdots & \vdots & \ddots & \vdots \\ -a_{Fn,1} & -a_{Fn,2} & \cdots & -a_{Fn,m} \end{bmatrix} \tag{5.11}$$

This expanded equation illustrates each component of the FDI attack vector. The a_{D1} component represents a change in load demand at the first bus, while the $a_{F1,1}$ component represents a change in active power flow at the first branch of the first bus, and the $-a_{F1,1}$ component represents the negative change in active power flow at the first branch of the first bus.

This expanded equation clarifies how the FDI attack works. It generates a vector that signifies alterations in load demand or active power flow, employing the measurement Jacobian matrix (Eq. 5.13), as discussed in these works by Refs. [29,30]. This vector is then incorporated into the measurements sent to the control center, distorting the system's understanding (Eq. 5.12), as discussed in these works by Refs. [31,32].

In practical power systems, the FDI attack can elude detection in the control center if the equation $a = \mathrm{H}c$ is satisfied and where c represents a vector with arbitrary nonzero elements. The measurement Jacobian matrix H is defined as:

$$\mathrm{H} = \left[A^T\mathrm{MA};\quad \mathrm{MA};\quad -\mathrm{MA}\right]^{\mathrm{T}} \tag{5.12}$$

Here, M represents diagonal branch susceptance matrix.

$$Z_{bad} = Z + a \tag{5.13}$$

Where Z represents measurement vector.

5.2.1.3 Man-in-the-middle (MitM) attack model

The MiTM attack is a form of cyber-attack in which an adversary intercepts and modifies communications between two parties in an attempt to manipulate the system's output, potentially jeopardizing the integrity of critical infrastructure systems like power grids and industrial control systems (ICS) [33]. We can utilize Eq. (5.2), where S represents a latent variable, which can be Π in the case of a MitM attack. A MitM attack leads to the reordering of PMU measurements. In their research on "Phasor data correction and transmission system SE under spoofing attacks," [34] introduced the concept of a permutation matrix, denoted as Π in Eq. (5.14). This matrix, consisting of zeros and ones, ensures that the sum of each row and column equals one, thus playing a crucial role in creating the shuffling effect caused by the attack.

$$Z_{observed} = \Pi H v + e \tag{5.14}$$

5.2.1.4 Denial-of-service (DoS) attack model

A DoS attack is a cyberattack that disrupts a system's normal operation by targeting the network's accessibility, making it inaccessible to legitimate users through the overwhelming influx of false data, ultimately resulting in network failures [35]. In CPSs, DoS attacks disrupt controller-actuator communication by consuming bandwidth, severing connections, causing feedback delays, and overloading resources, ultimately leading to system instability [36]. According to the analysis provided by Ref. [37], the state estimator's utilization of measurements at time k in the presence of DoS attacks can be represented as follows:

$$Z_k = \sum_{i=1}^{d+1} M_k(i) Z_{k-i-1} = Z_{k-\tau_k+1} \quad (5.15)$$

Here, M_k is a row vector that identifies the most recently received, and d maximum number of consecutive attacks an attacker can launch.

5.3 Case studies: Cybersecurity attacks on electrical system operation

This section explores real-life case studies, shedding light on the serious implications of cybersecurity attacks on the operational integrity of electrical systems. Analyzing these cases offers crucial insights into the true extent of cyber threats and their consequential impacts on critical infrastructure.

5.3.1 Studying to understand severity and consequences Ukraine's cybersecurity attacks on electrical operations

The international landscape witnesses an escalating trend of cyber-attacks orchestrated by highly sophisticated entities who leverage the Internet for clandestine and lucrative activities. A paramount concern is the organized cyber-attacks capable of causing debilitating disruptions in a nation's critical infrastructures, which encompass functions crucial to society, the economy, and national security. The elusive origins and motives of these cyber-attacks are further compounded by the intricate interplay of interests across national, international, public, and private domains. The rapidly evolving cyber environment adds complexity to the challenge of mitigating these multifaceted threats [38].

Since the Russian invasion, there have been several attempts to cyberattack Ukraine's electric grid. The first attack, which reportedly began on March 19, shortly after the integration of Ukraine's grid with ENTSO-E, was reminiscent of an earlier cyberattack on December 23, 2015. This 2015 attack struck three Ukrainian energy distribution companies in the Ivano-Frankivsk Oblast, leading to a temporary electricity supply disruption for about 225,000 people and involved multiple stages, compromising seven 110 kV and twenty-three 35 kV substations. Unlike the 2015 attack, the 2016 incident involved an automated assault using Industroyer malware. In the 2022 attacks, a similar malware, referred to as Industroyer2, was discovered by investigators. Furthermore, another attack, similar to the 2016 incident, reportedly occurred in early April. If successful, these attacks would have affected approximately 2 million consumers, which is around 12% of Ukraine's total of 16.8 million household consumers and 579,000 non-household consumers, potentially causing significant power outages and disruptions (Popik, 2022) [38,39].

The 2017 NotPetya malware attack encrypted systems and disrupted global businesses, commencing in Ukraine and rapidly spreading to over 60 countries. Among the hardest-hit was the global shipping company Maersk, necessitating a complete IT infrastructure rebuild. During the 9-day recovery period, Maersk struggled with operational continuity, improvising manual workarounds, ultimately incurring nearly $300 million in losses [40,40a].

Globally, the increasing frequency of cyber-attacks conducted by sophisticated actors poses a significant threat to critical infrastructure, national security, and economic stability. These attacks, often shrouded in secrecy, are further complicated by the complex web of interests in domestic, international, public, and private domains. The ever-evolving cyber landscape adds complexity to the challenge of mitigating multifaceted cyber threats. The events surrounding the cyber-attacks on Ukraine's electric grid and the NotPetya malware attack serve as stark reminders of the potential consequences, highlighting the importance of robust cybersecurity measures and international cooperation in addressing these threats.

5.3.2 Other cyber-security breaches

The North American blackout of 2003 serves as a stark reminder of the critical importance of improving power system operations. This 4-day event, triggered by a relatively small failure, inflicted a substantial economic toll, costing the United States between $4 and $10 billion. The introduction of WLS SE in power system operations marked significant progress, with SEs now being a standard in transmission control centers. Nonetheless, the lessons learned from the August 2003 blackout underscore the ongoing need to reduce susceptibility to erroneous data and guarantee a reliable solution, ensuring the resilience and security of power system operations [41,42].

In 2007, the Aurora Attack at the Idaho National Laboratory in the USA manipulated a diesel generator's circuit breaker using an FDI Attack, leading to the generator's explosion [43].

The major blackout that struck Venezuela in March 2019, lasting from March 7 to 14, had far-reaching consequences, not only in its immediate impact but also within the realm of political discourse [44]. It left countless households, businesses, and public facilities without power, causing disruptions at airports and hospitals, leading to incidents of looting and local conflicts [45], and affecting approximately 20 million people [46]. Subsequently, in 2020, Venezuela's national grid's 765 trunk line was targeted, resulting in blackouts in all 11 states, except the capital, Caracas [47].

In recent events within Europe, the power infrastructure responsible for providing electricity to the United Kingdom and Ireland experienced a cyberattack in July 2017. This specific attack had the primary objective of infiltrating power management systems, with the aim of potentially disrupting a portion of the energy grid. The attackers deployed a tactic that involved sending deceptive emails to senior executives within the power industry [48].

Moreover, in April 2022, another significant cybersecurity incident occurred, impacting the monitoring and control communication systems for approximately 2000 wind turbines in Germany. These incidents shed light on the growing challenges and vulnerabilities faced by Europe's critical infrastructure. They emphasize the pressing need to bolster cybersecurity measures to better defend against such threats [49].

5.4 Impact on grid reliability and stability

In this section, we will delve into the effects of cybersecurity breaches on the reliability and stability of power grids. Understanding these impacts is vital for developing strategies to safeguard power supply reliability, voltage and frequency stability, and overall grid resilience in the face of cyber threats.

5.4.1 Investigating how cybersecurity breaches affect power supply reliability, voltage, frequency stability, and overall grid stability

The potential harm stemming from a threat event involves unauthorized access, alteration, destruction, or loss of information and system functionality, impacting various stakeholders including agency heads, business proprietors, information stewards, and individuals both within and outside the organization [50].

Additionally, as per [51], impacts typically manifest as costs resulting from ICS component failures and replacements, reputation management due to customer dissatisfaction, or compliance with new post-attack regulatory requisites.

The power system undergoes evaluations to ensure reliability, adequacy, and security for delivering high-quality power, encompassing normal operations and resilience to disruptions like short circuits, with a careful analysis of frequency and voltage stability during contingencies [52]. This robust evaluation is crucial for maintaining a steady and secure power supply, ensuring the system can withstand unexpected events. Additionally, these assessments aid in formulating proactive strategies to enhance the overall resilience of the power infrastructure.

Power system functions vulnerable to cyberattacks encompass electricity supply (generation) stability and reliability, electricity transmission and distribution stability and reliability, communication between systems or equipment, information regarding the operating status of generation, transmission, or distribution equipment, black start capability, as well as equipment performance and its capacity to recover, particularly involving backup systems [53] (Table 5.1).

Asset management is not only essential for optimizing electrical power system operations and making decisions regarding equipment maintenance, repair, and replacement but also for ensuring long-term economic success and addressing the challenges of aligning strategy with stakeholder values and managing performance-based rates. This comprehensive approach involves categorizing asset management into three dimensions: technical, which centers around asset condition and maintenance; economical, focusing on financial aspects like maintenance costs and inventory management; and societal, which considers the broader societal and environmental impact of resource utilization, potentially affecting the status of distribution companies due to disturbances in various locations such as schools, government offices, and convention centers [54].

Table 5.1 Electrical system cybersecurity risks and precautions [53].

References	Functions	Affected risks	Risk level	Precautions
[60]	Electricity supply (generation) stability and reliability	Disruption or damage through cyberattacks	High	Implement strong encryption, access controls, regularly update and patch systems.
[54]	Electricity transmission and distribution stability and reliability	Power outages and reduced reliability	Medium	Asset management.
[55]	Communication between systems or equipment	Data manipulation or interruptions	High	Implementing strong cybersecurity protocols such as SSL and 128-bit encryption.
[58]	Information on the operating conditions of generation, transmission, or distribution equipment	Misuse of deceptive data and malfunctions	Medium	Utilize equipment monitoring and diagnostic systems.
[57,59]	Black start capability	Loss of the critical black start capability for emergencies	High	Implement redundant black start systems.

Efficient advanced distribution automation relies on the establishment of robust communication links connecting controlled devices to a central control unit. However, the prominent use of Power Line Carrier (PLC) technology as a communication channel introduces significant challenges, notably open circuit vulnerabilities. In the domain of cybersecurity, the Electric Power Research Institute underscores the vital importance of implementing precautionary measures, including the utilization of robust security protocols like secure socket layers (SSL), 128-bit encryption, and frame relays to mitigate potential risks [55].

Black start involves re-establishing the power supply using self-starting black start units [56]. As the number of blackstart-capable power plants in the United States has dwindled due to the retirement of conventional generation assets and the growing reliance on renewable energy resources, there is a pressing demand for enhanced grid resilience during blackouts. These blackouts typically result from various factors, including transformer damage, criminal attacks, tree growth, electrical faults, severe weather, and cyberattacks, to name a few [57].

The increasing power demand and grid expansion elevate the demand for reliable power supply. Therefore, analyzing the distribution network's status is crucial for ensuring this reliability. Utilizing monitoring and diagnostics to assess equipment conditions becomes essential for enabling efficient energy transmission within the power system. However, assessing this equipment poses challenges due to its widespread distribution and the absence of uniform standards [58].

At a legislative level, the U.S. Congress was proactive. The Secure Energy Infrastructure Act (SEIA) introduced a pilot program to reduce grid digitization and enhance manual backups for grid operators. This aims to minimize blackout durations and draws inspiration from the 2015 Ukrainian grid attack, which used backup manual controls successfully [59].

5.4.2 The cybersecurity domino effect

The domino effect is initiated by a primary event in one unit and propagates to nearby units through escalation vectors, causing consequential events and potential damage based on event type, distance, and unit characteristics [61]. In a parallel context, cybersecurity breaches in power grids can act as a catalyst, initiating a chain of events that compromise the reliability and stability of the network, much like the high-impact phenomena seen in various chemical and process plants worldwide, where secondary incidents caused by primary ones have resulted in catastrophic damage.

Ref. [62] suggests that the article emphasizes the shift from isolated to connected critical infrastructure in electrical grids, highlighting the role of smart substations in the evolving SG. It underscores the vulnerability of connected smart substations to cyber threats, which can trigger a chain reaction of blackouts and faults, necessitating a comprehensive review of architectural, communication, and cybersecurity standards along with defense mechanisms to address this critical issue. Additionally, frequency fluctuations resulting from load changes or cyber-attacks in one area can have cascading effects, imperiling the stability of the entire interconnected network [63].

5.5 Economic and social consequences of cybersecurity breaches

This section thoroughly explores the significant influence of cybersecurity breaches on both the economy and broader society. Gaining insights into the financial losses, societal disruptions, and enduring effects of cyber-attacks on electrical system operations is imperative for formulating strategies to alleviate these outcomes and bolster overall resilience.

5.5.1 Important of social responsibility

In today's dynamic business environment, the emphasis on social responsibility is undeniably evident, and overlooking it can have far-reaching consequences for both businesses and society. Focusing solely on short-term economic gains not only risks causing long-term environmental and public health issues but also erodes trust, undermines reputation, and hinders sustainable growth. In other words, when businesses neglect their social and environmental responsibilities, it is not just their profitability that is at stake; it is also the trust and the carefully built reputation they have established over the years. Furthermore, the adoption of SG technology in the power industry reveals a compelling connection between SG and social responsibility, as it naturally enhances social responsibility through renewable energy integration, proactive consumer engagement in demand management, and the establishment of self-healing energy networks. As a result, this not only propels the energy sector toward a more sustainable and environmentally responsible future but also empowers communities and organizations to address the pressing concerns of climate change and resource conservation with innovative solutions, thereby strengthening the sustainability of the energy sector and promoting a more environmentally-conscious and socially responsible energy ecosystem. Consequently, it becomes clear that the integration of SG technology goes beyond mere technological advancement, contributing significantly to our broader goals of social responsibility and environmental stewardship [64].

5.5.2 Assessing the economic losses and social disruptions caused by cybersecurity attacks on electrical system operations

Cyber-attacks present not only technical and physical threats but also substantial financial risks, particularly in smart microgrids with extensive renewable energy integration in grid-connected mode. These attacks can influence economic operations, especially in deregulated electricity markets encompassing day-ahead and real-time markets, thereby affecting load forecasting and power flow calculations [65].

Expanding on this [66], carried out experiments to assess the impact of DoS and MitM attacks on SG networks, monitored smart meters for load fluctuations, extended the attacks to an IEEE-30 bus system using MATPOWER simulations, and explored protective measures including cryptographic data security and the importance of firewalls and antivirus databases for enhancing cybersecurity. The study also yielded significant economic findings, highlighting the potential monetary losses faced by service providers due to these cyberattacks. The results of the conducted experiments shed light on the critical vulnerabilities and economic implications associated with cyberattacks on SG systems. In the case of DoS attacks, we observed significant fluctuations in load output over a 3-day period, resulting in an average load loss of approximately 10.423%. These fluctuations were attributed to changes in the attacker's bandwidth and attack environment, with a considerable monetary loss estimated at around Rs. 9,00,00,000 for a service provider serving nearly 400,000 customers. Similarly, MitM attacks led to a 3.5% reduction in load output, causing a substantial economic impact, with an estimated loss of approximately Rs. 3,10,00,000 for the same service provider. Furthermore, the extension of these cyberattacks to an IEEE-30 bus system emphasized the need for robust protective measures, as changes in load and attack power were observed. These findings underscore the urgency of enhancing the cybersecurity infrastructure for SGs to safeguard both the grid's operational stability and the financial interests of service providers in the face of evolving cyber threats.

According to Ref. [67], The impact of DoS attacks on local electricity markets (LEM) can be summarized as follows: Increasing DoS attack severity leads to greater price dispersion in both neighborhoods, resulting in the convergence of median prices, causing significant discrepancies in true transmission that may exceed circuit capacity. Furthermore, the distribution of net quantities demanded widens and shifts higher as apparent prices align with the rising DoS severity, increasing the likelihood of transmission levels surpassing the circuit's capacity and overloading. Even when the circuit is not overloaded, DoS attacks introduce supply-demand imbalances, necessitating the market maker's intervention to compensate for these disparities and raising financial risk as costs increase with the heightened DoS severity associated with supplying extra power.

5.5.3 Understanding the long-term economic impacts on affected regions

According to Ref. [68], the escalating cyber-attacks on critical infrastructures, leading to disruptions and data breaches with a projected 1100 major attacks causing over USD 1 million in damages in the next 5 years, are exacerbated by the global cybercriminal economy offering diverse cyber-attack services, including training, software development, network scanning, DoS, phishing, and money laundering. Furthermore, the increasing issue of electricity theft in the electrical infrastructure has emerged as a significant contributor to nontechnical losses in the distribution systems of the SG, highlighting the importance of addressing this challenge. This necessitates a comprehensive understanding of the long-term economic impacts on affected regions as they grapple with these multifaceted cybersecurity threats.

5.6 National security implications

In this segment, we explore the pivotal aspect of national security concerning cyber threats targeting vital electrical infrastructure. Evaluating the implications on national security and understanding the functions of government and defense agencies are essential steps in devising a holistic strategy to safeguard the critical electrical systems of a nation.

5.6.1 Analyzing the national security implications of cyber threats on critical electrical infrastructure

Analyzing the national security implications of cyber threats on critical electrical infrastructure is of paramount concern, given the multifaceted risks they pose to finance, reputation, and politics, extending beyond mere system disruptions to potential harm to human lives, organizational integrity, and even national security. Neglecting this issue could result in catastrophic consequences, including economic paralysis at various scales, making the adoption of business continuity and disaster recovery plans (BCDRPs) essential for fortifying defenses and raising preparedness levels [69].

Additionally, as society's dependence on information and communication technologies grows, safeguarding these technologies becomes vital for national interests. National cybersecurity strategies, adopted by various countries, are key tools in addressing these challenges. They outline high-level visions, goals, and priorities, clarify stakeholders' roles, and describe measures to protect national cyber infrastructure. These strategies acknowledge the evolving cyber threat landscape and provide a framework to secure critical infrastructure and national security in our digital age [70].

5.6.2 Discussing the roles of government and defense agencies in safeguarding the electrical system

The influence of cyberattacks on public confidence remains uncertain. Some studies suggest that exposure to cyberattacks targeting critical infrastructure has a similar impact on public confidence in the government's ability to counter future attacks, akin to traditional terrorism [71]. On the contrary, other research indicates that personal exposure to cyberattacks may increase individuals' apprehension about future targeting and lead to doubts about the effectiveness of government intervention.

To address these challenges, governments are increasingly emphasizing the importance of enhancing cybersecurity awareness through education, training, and best practices among their citizens. They are also taking concrete steps to mitigate cybersecurity risks, including the implementation of legislation and regulations that not only reduce cybersecurity risks but also enforce penalties for cybercrimes. These regulations are designed to remain adaptable to the ever-evolving technology landscape, ensuring they effectively tackle emerging cybersecurity issues [72].

Complementing these governmental initiatives, private sector-government collaboration has become a pivotal aspect of safeguarding the nation's critical infrastructure. This collaboration was initially set in motion through the National Infrastructure Protection Plan (NIPP) and has been further refined by Presidential Policy Directive-21 (PPD-21). Under this framework, the nation's critical infrastructure is categorized into 16 sectors, with Sector-Specific Agencies (SSAs) assigned to each. The result is mandated partnerships among the Federal Government, critical infrastructure stakeholders, and state, local, tribal, and territorial (SLTT) government entities. This sector-based partnership structure involves the formation of Sector Coordinating Councils (SCCs), Government Coordinating Councils (GCCs), SSAs, and cross-sector councils, all of which foster collaboration on critical infrastructure security and resilience programs, ultimately working together to achieve national objectives [73]. As depicted in Fig. 5.1, the National Partnership Structure and key elements of critical infrastructure coordination illustrate the comprehensive framework established by the [73].

Furthermore, the emphasis on cybersecurity in safeguarding critical infrastructure extends beyond the private sector and government collaboration. As highlighted by Ref. [74], the imperative to enhance cybersecurity aligns with global efforts aimed at reducing greenhouse gas emissions, compelling the U.S. federal government to prioritize energy efficiency. This commitment extends to various sectors, including U.S. Department of Defense (DoD) installations worldwide. The significance of sustainability is underscored by the ongoing security concerns related to the U.S. electricity grid, accentuating the vital role that cybersecurity plays in ensuring the resilience and security of this crucial infrastructure. While DoD's combat assets are exempt from these energy efficiency measures, the primary focus lies in enhancing the energy efficiency of fixed installations and non-tactical vehicle fleets, which constitute integral components of federal facilities and fleets.

Moreover, the overarching commitment to cybersecurity and its role in infrastructure resilience encompasses not only energy efficiency but also privacy protection. As outlined in the ePrivacy regulation, stringent privacy rules have been proposed for all electronic communications, addressing the needs of our increasingly interconnected digital world. These regulations extend to various facets of online activities, including communications content and metadata, cookies, and protection against spam. They also involve enhanced enforcement mechanisms to ensure compliance. This comprehensive approach to privacy, as detailed by Ref. [75], is designed to safeguard individuals' digital interactions while upholding the integrity of critical infrastructure. Thus, in the ever-evolving landscape of technology and security, the holistic strategy extends to both cybersecurity and digital privacy.

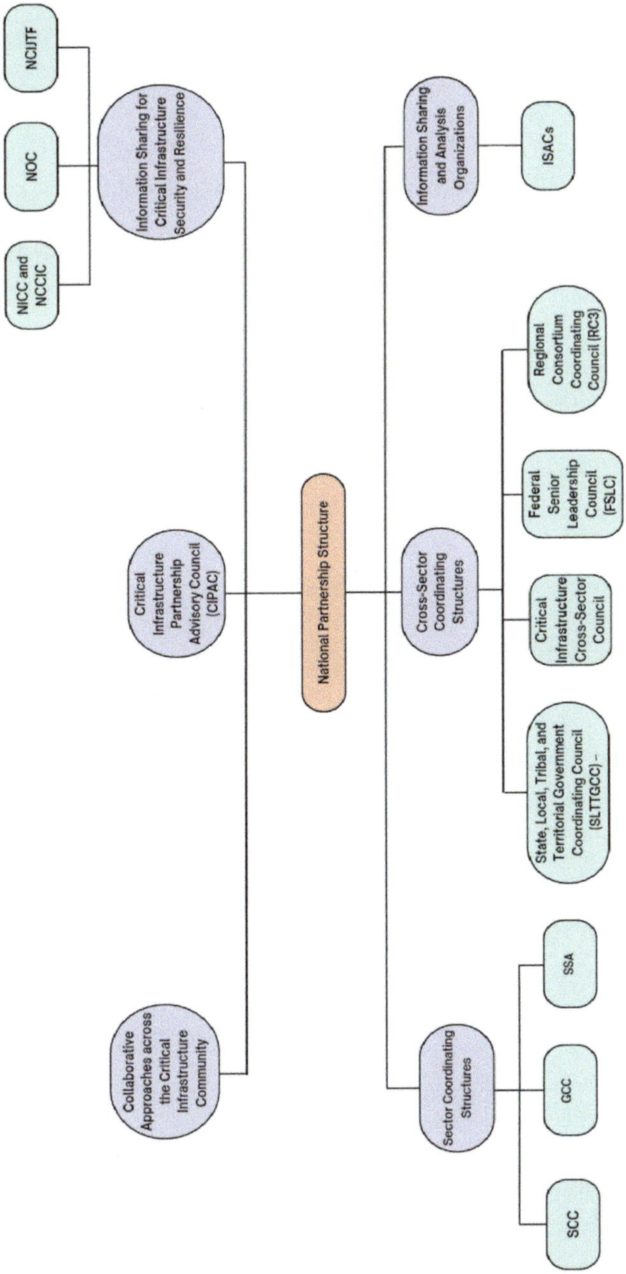

Figure 5.1 National partnership structure and key elements of critical infrastructure coordination [73].

5.7 Cybersecurity measures and best practices for electrical system operation

This section thoroughly explores vital cybersecurity measures and optimal practices meticulously crafted to guarantee the secure functionality of electrical systems. Grasping the nuances of cybersecurity frameworks, employee training, secure design, and security hardening is paramount in reinforcing the steadfastness and dependability of electrical infrastructure.

5.7.1 Exploring cybersecurity frameworks, standards, and best practices tailored for securing electrical system operations

Multiple U.S. laws impose cybersecurity requirements on utility companies, applying to specific organizations; some, like federal power marketing administrations, must comply with multiple regulations, which can encompass cybersecurity standards, guidance, and regulatory mandates, occasionally leading to ambiguity among different entity categories [76]. In response to this complex landscape, North American Electric Reliability Corporation (NERC) addresses the situation by developing Reliability Standards through an American National Standards Institute (ANSI) accredited process, which emphasizes openness, transparency, consensus, and timely development, as outlined by the NERC [77]. These standards define reliability requirements for the North American bulk power system, with a focus on performance and risk management. As shown in Table 5.2, various cybersecurity frameworks, standards, and best practices play a crucial role in defining essential cybersecurity objectives, controls, and guidelines for securing electrical system operations.

5.7.2 Holistic approach to cybersecurity: Employee training, secure design, security hardening and assessment

Employee Training and Awareness: Regular training programs are crucial in educating employees about cybersecurity threats, best practices, and effective risk identification. It is vital for employees to promptly recognize and report any suspicious activities they encounter. Understanding that cybersecurity incidents resemble a chain of behavioral reactions, each influenced by a diverse range of psychological factors, emphasizes the importance of comprehensive training. Cyber adversaries often strategically exploit and manipulate their targets' psychological processes, highlighting the vulnerability of humans as the weakest link in cybersecurity. Despite this, cybersecurity education and training programs

Table 5.2 Cybersecurity frameworks, standards, and best practices for electrical system operations.

References	Cybersecurity frameworks, standards, and best practices	Definitions
[78]	National Institute of Standards and Technology (NIST) cybersecurity framework (CSF)	Defines essential cybersecurity objectives that can aid an organization in reducing its cybersecurity risks.
[79]	Center for Internet security (CIS) controls	Encompasses fundamental controls to combat cyber-attacks and guides organizations in adopting the most effective defense strategies. It presents crucial controls for the cybersecurity of electrical & electronical systems.
[80]	ISA/IEC 62443 Series	International society of automation (ISA), under the ISA99 initiative, formed collaborative working groups, leading to the development of the IEC 62443 standards. This standard encompasses organizational policies, procedures, risk assessment, and security measures for hardware and software components.
[81]	NERC CIP standards	NERC has developed the critical infrastructure protection (CIP) standards. These standards are designed to ensure the safeguarding of assets essential for operating North America's bulk electric systems (BES).
[82,83]	[82,83]	ISO/IEC 27001 outlines requirements for establishing and maintaining an information security management system tailored to any organization, offering a proven methodology to enhance resilience and efficiently address security risks. ISO/IEC 27002 provides generic information security controls and guidance for implementing internationally recognized best practices within the framework of an information security management system (ISMS) based on ISO/IEC 27001.

Continued

Table 5.2 Cybersecurity frameworks, standards, and best practices for electrical system operations.—cont'd

References	Cybersecurity frameworks, standards, and best practices	Definitions
[84]	[84]	ISO/IEC 27005 serves as a global benchmark for information security risk management, assisting organizations in creating and upgrading their risk management procedures and aligning with ISO/IEC 27001 provisions for evaluating and managing information security risks.

may sometimes fall short in adequately addressing the intricate psychological aspects, leaving a critical facet insufficiently covered [85].

Security by Design Principles: Incorporating security at the initial design stages of electrical systems ensures that security features and considerations are an integral part of the system architecture, rather than added as an afterthought.

Concepts such as Industry 4.0, Cyber-Security, CPS, and the IIoT are significant areas that represent the transformation of industrial processes and systems in today's evolving technological landscape. SG, power systems, and electrical systems signify the formation of intelligent networks through digitalization, automation, and data integration in energy production and distribution processes. Additionally, cyber-security is a crucial focus in this digital environment, ensuring the protection of systems and data within the power sector. Furthermore, mechatronic system design aims to develop intelligent and interactive systems by integrating mechanical, electronic, and software engineering principles. CPS refers to systems where physical processes merge with cyber components, emphasizing real-time interaction and control, particularly relevant in the context of SGs and power systems. Moreover, IIoT enables the connection and data sharing of industrial processes, creating a more efficient and flexible energy production and distribution environment. These concepts represent key components of the transformation of the power sector and will shape the future direction of system design and operation [86].

According to the findings of [87], it has been observed that current proposals for designing IoT applications often neglect security considerations, employ specialized approaches that deviate from general distributed application design, and do not seamlessly integrate with secure development methodologies. As a result, the adaptation of established secure systems development methodologies for distributed systems, which leverage patterns as key artifacts, is deemed more effective than creating entirely new methodologies.

According to the study conducted by the authors [88], IIoT projects often prioritize functional aspects over security, resulting in the integration of security measures at later stages of the development process. As a solution, the article introduces IgniteSec, a novel method that combines the Ignite System Development Method with the industrial security standard NIST 800-82. The case study demonstrates that by integrating these two fronts from the initial stages, the project successfully incorporates both functional and security elements into the software design, ensuring compliance with the security standard.

Threat Modeling: Threat modeling is a methodical process aimed at enumerating, analyzing, and prioritizing threats and vulnerabilities within an information system. The outcomes of this process serve as a basis for decision-making regarding high-risk threats and vulnerabilities, as well as the cybersecurity measures essential for their mitigation. The application of threat models is imperative for conducting systematic vulnerability assessments on IoT devices connected to the grid [89,90].

Multiple factors and User Authentication: User authentication enhances security through multiple factors like what you know (passwords), what you have (tokens or devices), and what you are (biometrics). These factors can replace or complement passwords, bolstering protection. Two-factor and multi-factor authentication require verification via different methods, increasing security, albeit with potential usability considerations. MFA, which combines user-device and device-web verification, significantly strengthens security by requiring multiple forms of verification before granting access to critical systems and sensitive data [91].

Patch Management (PM): Patch Management (PM) presents a highly intricate process in the realm of information technology, necessitating a profound grasp of system components and their associated patches. Its intricacy is compounded by the uncertainties surrounding system responses to patches and potential issues arising from vendor-released patches, which can inadvertently introduce new vulnerabilities. Ensuring up-to-date software with the latest security patches and following secure configurations for hardware, software, and network devices based on industry best practices stand as paramount tasks. Regular updates to address evolving threat landscapes form the bedrock of a resilient security stance [92].

Zero Trust Security: Zero trust constitutes a security framework based on the core tenet of "never trust, always verify." It operates under the assumption that no entity, whether it is a user, device, or application, can be inherently considered trustworthy. Instead, their identity, authorization, and legitimacy must undergo continuous validation before granting access to a system or sensitive information. This model plays a pivotal role in contemporary security strategies, especially in light of evolving threats and the vulnerabilities associated with conventional username and password-based access. The zero-trust framework encompasses six essential pillars:

- Identity security
- Device and endpoint protection
- Data security
- Application and API governance
- Infrastructure security
- Network protection

These pillars collectively establish a comprehensive approach to enhancing security in today's digital landscape [93].

Security Hardening: Security hardening is the proactive practice of reducing system and network vulnerabilities to minimize their attack surface and mitigate potential risks. This comprehensive process involves actions such as patching vulnerabilities, resetting default passwords, eliminating vulnerable services, deactivating unused system applications, configuring firewalls, implementing access controls, and encrypting network traffic [94]. In the context of CPS, security hardening is of utmost importance for enhancing resilience against a wide range of threats. It requires integrated security approaches that bridge both physical and cyberspace components, recognizing their interdependencies to effectively counter traditional and emerging security challenges. By doing so, security hardening plays a crucial role in ensuring the safety and reliability of CPS in our increasingly interconnected digital world [95].

Privileged Access Management (PAM): Privileged Access Management (PAM) serves as a vital subset of IT security, focusing on supervising how privileged users, such as system administrators, interact with essential systems and resources, particularly high-value assets like domain controllers and production servers. While PAM offers valuable functions for password and access management, along with aiding in security, compliance, and audit requirements, it does not replace a comprehensive Zero Trust platform. However, integrating PAM with a Zero Trust system can enhance the overall security posture. Organizations possessing PAM should prioritize safeguarding the PAM server itself. Robust access controls and effective privilege management mechanisms are crucial elements in restricting access to authorized individuals, minimizing the potential attack surface, and enhancing overall security [96].

Security Assessment: According to Ref. [10], the key forms of cybersecurity assessment, as detailed in Table 5.3, encompass compliance checking, vulnerability identification, vulnerability analysis, penetration testing, simulation or emulation-based testing, formal analysis, and reviews.

Risk Assessment: According to Ref. [10], the key forms of cybersecurity assessment encompass compliance checking, vulnerability identification, vulnerability analysis, penetration testing, simulation or emulation-based testing, formal analysis, and reviews.

5.7.3 Intrusion detection systems (IDS)

Industrial robots, Programmable Logic Controllers (PLCs), and SCADA systems are essential components of the SG. However, the integration and interdependence of these systems can lead to increased vulnerabilities and security challenges. Cyberattacks targeting these interconnected systems can result in production disruptions, manipulation of control commands, and unauthorized access to critical infrastructure. Therefore, robust security measures are necessary to maintain the integrity and functionality of these systems.

Effective IDS play a significant role in enhancing the security of interconnected systems like industrial robots, PLCs, and SCADA systems. Snort3 and Suricata are two widely accepted open-source IDS solutions used in network security. Snort3, considered one of the most popular IDS, utilizes a rule-based approach to detect malicious network activities by comparing network packets against predefined criteria. It detects packets that match these criteria and rapidly generates alerts for users. Snort3 also offers flexibility by allowing users to create custom rules, providing a more personalized and robust detection system [104].

Similarly, Suricata is an advanced IDS with a multi-threaded architecture, efficiently managing high-volume network traffic. Through comprehensive network-based functions such as deep packet inspection,

Table 5.3 Cybersecurity assessment methods for electrical system operations.

References	Cybersecurity assessment methods	Description	Techniques/tools used	Advantages
[97]	Compliance checking	The process of verifying and ensuring adherence to established laws, regulations, policies, and standards.	Reference to specific standards or regulations.	- Validates compliance with established standards and regulations. - It effectively prevents insider violations and errors by utilizing nested business processes for compliance checking.
[98]	Vulnerability identification	Vulnerability identification is the process of systematically recognizing and assessing vulnerabilities within an organization's critical infrastructure, involving the utilization of systematized knowledge, relevant standards, and acquired experience.	It encompass industry standards, practical experience, active inspection and testing, asset mapping, the common vulnerability scoring system (CVSS), and structured threat information eXpression (STIX) for cyber threat intelligence, all aimed at systematically identifying and assessing vulnerabilities in energy-related critical infrastructures.	- Identifies and highlights vulnerabilities that may be exploited.

[47,99]	Vulnerability analysis	Vulnerability analysis is the process of systematically assessing and identifying weaknesses or susceptibilities within a system, be it related to security risks in CPSs or potential hazards in industrial processes, to evaluate potential threats and their impact, ultimately enhancing defense strategies or safety measures.	Security vulnerability analysis utilizes a range of techniques and tools, including layered analysis, specific metrics (e.g., CP-SAM), data-driven modeling, deep learning, network analysis, and SG standards, to assess and mitigate security risks in diverse critical infrastructure systems.	- Vulnerability analysis in complex critical CPSs is crucial to identifying weaknesses, enhancing defense strategies, and leveraging advanced techniques like data-driven models and deep learning for proactive threat management, with the potential for efficient integration and industry collaboration.
[100]	Penetration testing	Penetration testing comprises the utilization of detected system vulnerabilities to evaluate the security of its IT infrastructure.	Methods and tactics used by cyber intruders	- Offers a realistic perspective on possible cyberattacks.

Continued

Table 5.3 Cybersecurity assessment methods for electrical system operations.—cont'd

References	Cybersecurity assessment methods	Description	Techniques/tools used	Advantages
[10,101]	Simulation or emulation-based testing	Utilizes simulation and emulation techniques to model or simulate the context or environment of the assessed object.	Simulation and emulation software like DIgSILENT PowerFactory, GridLAB-D, openSCADA, GridSpice, DEVS-CAS.	- Provides a cost-effective alternative to physical deployment. - Enables evaluation without system disruption. - Allows for testing of various architectural scenarios.
[10,102]	Formal analysis	Formal analysis in cybersecurity employs mathematical methods and formal notations to create precise models of the system and its context, enabling rigorous examinations similar to mathematical theorem proving, often applied to assess less complex system elements due to resource-intensive requirements.	Formal analysis relies on mathematical methods and formal notations to build accurate system models, and these models are then verified through manual or automated examinations, often employing theorem provers to ensure their correctness.	- Formal cybersecurity analysis provides a robust and detailed method to assess system security, offering unambiguous representations of systems and contexts to identify vulnerabilities, especially suitable for evaluating simpler elements, instilling confidence in their robustness.

[103]	Reviews	Encompasses passive, typically manual, review of documentation related to the evaluated object.	Technical specifications, logs, rules, configuration files.	- Assesses system components and configurations from a documentation perspective. - Identifies potential weaknesses through documentation review.

protocol analysis, and traffic pattern matching, Suricata can detect and respond to various types of network attacks, contributing to the overall security of interconnected systems.

With the use of solutions like Snort3 and Suricata, organizations can effectively detect and prevent cyberattacks on industrial robots, PLCs, and SCADA systems. These IDS continuously monitor network traffic, analyze packet behaviors, and generate alerts or take necessary actions against suspicious activities. They serve as critical components in preserving the integrity and functionality of modern systems while protecting against production disruptions, control command manipulations, and unauthorized access to critical infrastructure. Additionally, by combining the strength of open-source IDS solutions with machine learning techniques, organizations can efficiently detect and prevent cyberattacks, safeguarding the integrity and functionality of interconnected modern systems.

Concepts such as Industry 4.0, Cyber-Security, Mechatronic System design, CPSs, and the IIoT are significant areas that represent the transformation of industrial processes and systems in today's evolving technological landscape. Industry 4.0 signifies the formation of smart factories through digitalization, automation, and data integration in production processes. Additionally, Cyber-Security is a crucial focus in this digital environment, ensuring the protection of systems and data. Furthermore, Mechatronic System design aims to develop intelligent and interactive systems by integrating mechanical, electronic, and software engineering. CPS refers to systems where physical processes merge with cyber components, emphasizing real-time interaction and control. Moreover, IIoT enables the connection and data sharing of industrial processes, creating a more efficient and flexible manufacturing environment. These concepts represent key components of the industrial sector's transformation and will shape the future direction of production and system design [86].

5.8 Future challenges and emerging threats in electrical system operation

In this segment, we explore the envisaged future challenges and emerging cyber threats that are expected to affect the functioning of electrical systems. Grasping these evolving risks is pivotal for taking proactive cybersecurity measures and fortifying the resilience of electrical infrastructure.

5.8.1 Anticipating future challenges and emerging cyber threats specific to electrical system operations

Addressing the future challenges and emerging cyber threats specific to electrical system operations is of utmost importance, as the reliance on advanced technologies, such as Artificial Intelligence (AI), Federated Learning (FL), and Computer Vision (CV), becomes integral to ensuring uninterrupted power supply, infrastructure security, and energy efficiency in the context of smart cities and evolving energy systems.

5.8.1.1 Artificial intelligence (AI)

Smart cities' reliance on technology underscores the challenge of meeting their energy needs, with solar energy emerging as a sustainable solution; advanced AI and ML techniques are vital for power generation and infrastructure security to ensure uninterrupted supply and prevent disruptions [105].

While AI and ML based IoT security introduces both increased efficiency and enhanced protection against cyber threats, they simultaneously bring forth potential risks, with privacy concerns, algorithmic bias, and data breaches at the forefront. Ensuring that IoT devices employing artificial intelligence do not unintentionally collect personal information without user consent is essential for safeguarding individual privacy rights, including compliance with General Data Protection Regulation (GDPR) regulations. Addressing algorithmic bias necessitates extensive testing, independent audits, and ongoing evaluations. Organizations must make substantial efforts to establish secure configurations for connected devices and maintain regular patching and updating processes to mitigate potential vulnerabilities. These considerations are of paramount importance to avert adverse consequences that could emerge if adequate precautions are not taken, despite the numerous advantages provided by AI and ML-driven IoT security systems [106].

5.8.1.2 Federated learning (FL)

FL is a decentralized machine learning approach that prioritizes privacy by training models on local client devices, thus reducing the need for centralized data transmission. This approach enhances the security and efficiency of various systems, including applications like Smart Grid, while mitigating concerns related to connectivity, bandwidth, latency, and data privacy [107,108]. Although FL provides protection for the centralized server against various attack vectors, it raises user-related concerns, particularly regarding the risk of information leakage, as attackers can potentially compromise FL systems through methods like poisoning and presentation. Moreover, algorithm vulnerabilities, including those in gradient descent, may lead to information leakage, potentially exposing sensitive user data to malicious actors, emphasizing the importance of strengthening the defense mechanisms of FL systems to safeguard both centralized and distributed systems against a spectrum of security threats [109].

5.8.1.3 Computer vision (CV)

CV, a mature AI technology, is revolutionizing the energy sector by supporting cost-effective, large-scale AI vision systems in applications ranging from energy inspection and infrastructure monitoring to rooftop photovoltaic panel identification, enhancing operational efficiency. Industry leaders like ABB, Siemens, and GE increasingly adopting AI-based CV to improve energy services. However, as CV integrates rapidly into the energy industry, it introduces new cybersecurity vulnerabilities, including data privacy concerns, device security, and the risk of cyberattacks targeting energy infrastructure. To ensure the reliability and resilience of electrical system operations, the energy sector must proactively address these evolving cybersecurity threats and vulnerabilities [110].

5.8.1.4 5G, 6G and beyond: Advancing communication networks

The multifaceted challenges of 5G technology demand comprehensive solutions that address issues such as security, cost, reliability, and efficiency. These solutions encompass innovations in virtualization, spectrum utilization, energy efficiency, and AI-driven resource management. While progress has been made, the evolution of 5G and subsequent networks must contend with research gaps and security concerns, necessitating continuous advancement to ensure robust technologies and prevent performance degradation. Looking ahead, future mobile communication networks, including 6G, aspire to achieve comprehensive connectivity, high-speed data transfer, and advanced strategies. Yet, persistent challenges revolve around ensuring full connectivity, maintaining high-quality service, enhancing security, and enabling efficient network management. These key concerns encompass security, scalability, energy optimization, quality of service, and the integration of machine learning, all of which demand ongoing research efforts to overcome and advance network technologies [111]. Moreover, it is vital to seamlessly integrate these advanced mobile communication networks with the evolving electrical grid infrastructure, fostering the efficient and secure management of future energy systems.

5.8.2 Discussing strategies to integrate advanced technologies securely into the electrical grid

Current protocols lack sufficient security, putting confidentiality, privacy, integrity, and responsibility at risk, necessitating the development or enhancement of new security protocols for SG networks based on specific application requirements [112].

On the other hand [113], highlight the essentiality of robust data protection for the security of advanced electrical grid technologies. Legal frameworks, e.g., the Consumer Data Protection Act, Data Care Act, and GDPR, have contributed to data protection. Nonetheless, the rapidly increasing data volume requires the integration of ML. Traditional centralized ML models have raised concerns regarding digital security. In response to these challenges, Federated Learning (FL) has emerged as an alternative approach, enabling local data storage and training. Despite the increasing interest in FL, multiple challenges persist, particularly within large-scale networks characterized by diverse IoT environments. Achieving secure integration of technology into the electrical grid requires the implementation of privacy-preserving techniques and communication optimization. Ongoing pursuit is needed to effectively address the challenges and enhance security in this context.

To tackle the growing challenges associated with managing and integrating advanced technologies within the electrical grid, the approach put forth by Ref. [114] involves the implementation of the Hypervision interface framework (depicted in Fig. 5.2). This framework serves as a centralized hub for real-time business events, furnishing operators with a structured decision-making platform in the form of digital cards. These cards facilitate the prioritization and organization of tasks, offering vital information, actionable steps, and potential recommendations. Automation based on forecasts or contextual data can generate these cards, or operators have the option to manually create them to address unique situations. This adaptable framework fosters efficient coordination among operators and finds relevance across a spectrum of industrial domains, playing a pivotal role as a fundamental tool for the secure integration of advanced technologies into the electrical grid.

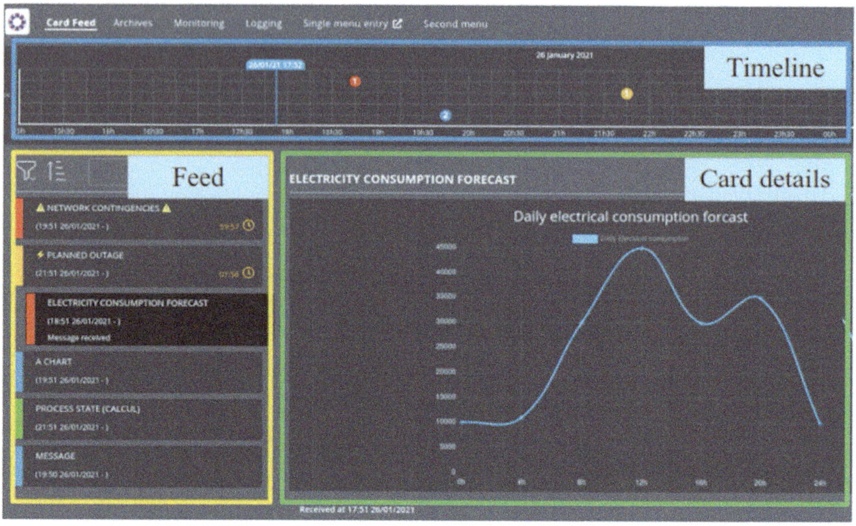

Figure 5.2 Hypervision interface of OperatorFabric [114].

5.9 Conclusion

The intricate and interdependent landscape of electrical system operations places a high premium on robust cybersecurity measures. In an era marked by relentless technological advancement and digital transformation, the challenges and implications associated with securing our electrical infrastructure have never been more pronounced. The extensive reliance on electrical systems in our daily lives, industries, and critical infrastructures makes it essential to comprehend the pivotal role of cybersecurity in safeguarding the power grids and operations we depend on.

As highlighted in this comprehensive exploration, the significance of securing electrical systems is multi-faceted. Beyond the obvious need for reliability and continuity in power distribution, cybersecurity is inexorably linked to public safety, infrastructure integrity, and economic stability. Disruptions resulting from cyberattacks not only compromise the provision of essential services but can also result in substantial economic losses. Maintaining data integrity and privacy is essential to building trust with consumers and complying with regulatory mandates. Furthermore, the security of electrical infrastructure holds paramount importance in the context of national security.

The threats facing electrical systems are diverse, ranging from malware and phishing to social engineering, DoS, and DDoS attacks. Real-life case studies have illustrated the serious implications of cybersecurity attacks, revealing the dire need to address vulnerabilities and enhance the security of critical infrastructure.

The impact of cybersecurity breaches extends to power supply reliability, voltage and frequency stability, and overall grid resilience. The cybersecurity domino effect, akin to the phenomenon seen in chemical plants, underscores the potential ripple effects of cyber incidents. Economic and social consequences resulting from cybersecurity breaches have a profound influence on our economy and society, emphasizing the imperative role of social responsibility in securing critical electrical systems.

To mitigate these risks, cybersecurity measures and best practices must be implemented comprehensively. A holistic approach encompasses employee training, secure design, security hardening, and continuous security assessment. Leveraging established cybersecurity frameworks and standards tailored for electrical system operations is fundamental. These standards provide a roadmap for organizations to assess risks, implement protective measures, detect threats, and minimize their impact.

Anticipating future challenges and emerging threats is a critical aspect of securing electrical systems. Rapid technological advancements, IoT proliferation, AI threats, supply chain vulnerabilities, and insider threats necessitate a proactive approach. Integrating advanced technologies securely into the electrical grid requires a focus on cyber-physical security, continuous monitoring and adaptation, collaborative information sharing, and resilience and redundancy planning.

In conclusion, safeguarding electrical system operations through comprehensive cybersecurity measures is essential to ensure the reliability, stability, and security of critical infrastructure. It is an ongoing endeavor that requires a deep understanding of the evolving threat landscape and a commitment to proactive security practices. By addressing these challenges and embracing emerging technologies securely, we can fortify our electrical systems and maintain their dependability in an increasingly interconnected and digital world.

References

[1] I. Siniosoglou, P. Radoglou-Grammatikis, G. Efstathopoulos, P. Fouliras, P. Sarigiannidis, A unified deep learning anomaly detection and classification approach for smart grid environments, IEEE Transactions on Network and Service Management 18 (2) (2021) 1137–1151, https://doi.org/10.1109/TNSM.2021.3078381.

[2] A. Mohammed, G. George, Vulnerabilities and strategies of cybersecurity in smart grid—evaluation and review, in: 2022 3rd International Conference on Smart Grid and Renewable Energy (SGRE), 2022, pp. 1–6, https://doi.org/10.1109/SGRE53517.2022.9774038.

[3] S. Gyawali, O. Beg, Cyber attacks detection using machine learning in smart grid systems, in: IEEE INFOCOM 2022 - IEEE Conference on Computer Communications Workshops (INFOCOM WKSHPS), 2022, pp. 1–2, https://doi.org/10.1109/INFOCOMWKSHPS54753.2022.9797941.

[4] I. Fursov, K. Yamkovyi, O. Shmatko, Smart Grid and wind generators: an overview of cyber threats and vulnerabilities of power supply networks, Radioelectronic and Computer Systems 4 (2022) 50–63, https://doi.org/10.32620/reks.2022.4.04.

[5] P. Hehenberger, M. Habib, D. Bradley (Eds.), EcoMechatronics: Challenges for Evolution, Development and Sustainability, Springer International Publishing, 2022, https://doi.org/10.1007/978-3-031-07555-1.

[6] H. Pu, L. He, P. Cheng, M. Sun, J. Chen, Security of industrial robots: vulnerabilities, attacks, and mitigations, IEEE Network 37 (1) (2023) 111–117, https://doi.org/10.1109/MNET.116.2200034.

[7] T.E. Gasiba, U. Lechner, M. Pinto-Albuquerque, A. Porwal, Cybersecurity Awareness Platform with Virtual Coach and Automated Challenge Assessment (arXiv:2102.10430). arXiv, 2021. http://arxiv.org/abs/2102.10430.

[7a] A.N. Lone, S. Mustajab, M. Alam, A comprehensive study on cybersecurity challenges and opportunities in the IoT world, Security and Privacy 6 (6) (2023) e318, https://doi.org/10.1002/spy2.318.

[8] A.A. Baybulatov, V.G. Promyslov, Industrial control system availability assessment with a metric based on delay and dependency, IFAC-PapersOnLine 54 (13) (2021) 472–476, https://doi.org/10.1016/j.ifacol.2021.10.493.

[9] E.M. Maseno, Z. Wang, H. Xing, A systematic review on hybrid intrusion detection system, Security and Communication Networks (2022) 1–23, https://doi.org/10.1155/2022/9663052.

[10] R. Leszczyna, Cybersecurity in the Electricity Sector: Managing Critical Infrastructure, Springer International Publishing, 2019, https://doi.org/10.1007/978-3-030-19538-0.

[10a] L.S. Gajanan, M. Kirar, M. Raju, Cyber-attacks on smart grid system: a review, 2022 IEEE 10th Power India International Conference (PIICON), 2022, pp. 1–6, https://doi.org/10.1109/PIICON56320.2022.10045208.

[11] A. Almalawi, Z. Tari, A. Fahad, X. Yi, SCADA Security: Machine Learning Concepts for Intrusion Detection and Prevention, Wiley, 2021.

[12] IEEE Standard for Electric Power Systems Communications—Distributed Network Protocol (DNP3), Institute of Electrical and Electronics Engineers, 2012.

[13] A. SunSpec, SunSpec Modbus Conformance Test Procedures Results Reporting, 2021.

[14] L. Arathi, B. Jacob, Design and implementation of Modbus serial line server RTU VIP, in: 2023 International Conference on Control, Communication and Computing (ICCC), IEEE, 2023, pp. 1–6.

[15] L. Strezoski, Distributed energy resource management systems—DERMS: state of the art and how to move forward, WIREs Energy and Environment 12 (1) (2023) e460, https://doi.org/10.1002/wene.460.

[16] A. Augello, P. Gallo, E.R. Sanseverino, G. Sciume, M. Tornatore, A coexistence analysis of blockchain, SCADA systems, and OpenADR for energy services provision, IEEE Access 10 (2022) 99088–99101, https://doi.org/10.1109/ACCESS.2022.3205121.

[17] IEEE Standard for Smart Energy Profile Application Protocol, IEEE, 2018, https://doi.org/10.1109/IEEESTD.2018.8608044.

[18] I. Zografopoulos, N.D. Hatziargyriou, C. Konstantinou, Distributed energy resources cybersecurity outlook: vulnerabilities, attacks, impacts, and mitigations, IEEE Systems Journal (2023) 1–15, https://doi.org/10.1109/JSYST.2023.3305757.

[19] N.D. Tuyen, N.S. Quan, V.B. Linh, V. Van Tuyen, G. Fujita, A comprehensive review of cybersecurity in inverter-based smart power system amid the boom of renewable energy, IEEE Access 10 (2022) 35846–35875, https://doi.org/10.1109/ACCESS.2022.3163551.

[20] K. Arora, S.L. Tripathi, P. Sanjeevikumar (Eds.), Smart Electrical Grid System: Design Principle, Modernization, and Techniques, CRC Press, 2023.

[21] U. Inayat, M.F. Zia, S. Mahmood, T. Berghout, M. Benbouzid, Cybersecurity enhancement of smart grid: attacks, methods, and prospects, Electronics 11 (23) (2022) 3854, https://doi.org/10.3390/electronics11233854.

[22] M. Du, G. Pierrou, X. Wang, M. Kassouf, Targeted false data injection attacks against AC state estimation without network parameters, IEEE Transactions on Smart Grid 12 (6) (2021) 5349–5361, https://doi.org/10.1109/TSG.2021.3106246.

[23] A. Tharzeen, B. Natarajan, B. Srinivasan, Phasor data correction and transmission system state estimation under Man-in-the-Middle attack, in: 2023 IEEE Power & Energy Society Innovative Smart Grid Technologies Conference (ISGT), 2023, pp. 1–5, https://doi.org/10.1109/ISGT51731.2023.10066426.

[24] M. Ghiasi, T. Niknam, Z. Wang, M. Mehrandezh, M. Dehghani, N. Ghadimi, A comprehensive review of cyber-attacks and defense mechanisms for improving security in smart grid energy systems: past, present and future, Electric Power Systems Research 215 (2023) 108975, https://doi.org/10.1016/j.epsr.2022.108975.

[25] D. Agnew, N. Aljohani, R. Mathieu, S. Boamah, K. Nagaraj, J. McNair, A. Bretas, Implementation aspects of smart grids cyber-security cross-layered framework for critical infrastructure operation, Applied Sciences 12 (14) (2022) 6868, https://doi.org/10.3390/app12146868.

[26] J.S. Mingorança, I.D. Melo, A.B.D. Santos, Critical buses identification for voltage stability assessment considering the application of modal analysis and a robust state estimation with bad data suppression, Electrical Engineering 105 (1) (2023) 335–348, https://doi.org/10.1007/s00202-022-01677-3.

[27] Z. Zhang, R. Deng, D.K.Y. Yau, P. Cheng, M.-Y. Chow, Security enhancement of power system state estimation with an effective and low-cost moving target defense, IEEE Transactions on Systems, Man, and Cybernetics: Systems 53 (5) (2023) 3066–3081, https://doi.org/10.1109/TSMC.2022.3222793.

[28] M. Tian, Z. Dong, X. Wang, Analysis of false data injection attacks in power systems: a dynamic Bayesian game-theoretic approach, ISA Transactions 115 (2021) 108–123, https://doi.org/10.1016/j.isatra.2021.01.011.

[29] B. Liu, H. Wu, Q. Yang, H. Zhang, Y. Liu, Y. Zhang, Matrix-completion-based false data injection attacks against machine learning detectors, in: IEEE Transactions on Smart Grid, 1–1, 2023, https://doi.org/10.1109/TSG.2023.3308339.

[30] R. Bobba, K. Davis, Q. Wang, H. Khurana, K. Nahrstedt, T. Overbye, Detecting False Data Injection Attacks on DC State Estimation, 2010.

[31] J. Sakhnini, H. Karimipour, A. Dehghantanha, Smart grid cyber attacks detection using supervised learning and heuristic feature selection, in: 2019 IEEE 7th International Conference on Smart Energy Grid Engineering (SEGE), 2019, pp. 108–112, https://doi.org/10.1109/SEGE.2019.8859946.

[32] M. Higgins, W. Xu, F. Teng, T. Parisini, Cyber–physical risk assessment for false data injection attacks considering moving target defences: best practice application of respective cyber and physical reinforcement assets to defend against FDI attacks,

International Journal of Information Security 22 (3) (2023) 579–589, https://doi.org/10.1007/s10207-022-00621-7.
[33] F.H. Alotaibi, H. Ibrahim, J. Kim, P. Enjeti, A method to detect man in the middle attack (MiTM) on a grid following PV inverter, in: 2023 IEEE 24th Workshop on Control and Modeling for Power Electronics (COMPEL), 2023, pp. 1–6, https://doi.org/10.1109/COMPEL52896.2023.10221187.
[34] A. Tharzeen, B. Natarajan, B. Srinivasan, Phasor data correction and transmission system state estimation under spoofing attacks, Electric Power Systems Research 221 (2023b) 109435, https://doi.org/10.1016/j.epsr.2023.109435.
[35] T.T. Khoei, N. Kaabouch, Densely connected neural networks for detecting denial of service attacks on smart grid network, in: 2022 IEEE 13th Annual Ubiquitous Computing, Electronics & Mobile Communication Conference (UEMCON), 2022, pp. 0207–0211, https://doi.org/10.1109/UEMCON54665.2022.9965631.
[36] L. Cao, X. Jiang, Y. Zhao, S. Wang, D. You, X. Xu, A survey of network attacks on cyber-physical systems, IEEE Access 8 (2020) 44219–44227, https://doi.org/10.1109/ACCESS.2020.2977423.
[37] X. Li, C. Jiang, D. Du, W. Li, M. Fei, L. Wu, A novel state estimation method for smart grid under consecutive denial of service attacks, IEEE Systems Journal 17 (1) (2023) 513–524, https://doi.org/10.1109/JSYST.2022.3171751.
[38] M. Lehto, P. Neittaanmäki (Eds.), Cyber Security: Critical Infrastructure Protection, vol 56, Springer International Publishing, 2022, https://doi.org/10.1007/978-3-030-91293-2.
[39] P.H. O'Neill, Russian Hackers Tried to Bring Down Ukraine's Power Grid to Help the Invasion, MIT Technology Review, May 2022. https://www.technologyreview.com/2022/04/12/1049586/russian-hackerstried-to-bring-down-ukraines-power-grid-to-help-the-invasion/.
[40] M. Coden, M. Reeves, K. Pearlson, S. Madnick, C. Berriman, An action plan for cyber resilience, MIT Sloan Management Review 64 (2) (2023) 1–6.
[40a] M. Willett, Lessons of the SolarWinds hack, Survival 63 (2) (2021) 7–26, https://doi.org/10.1080/00396338.2021.1906001.
[41] A. Farooq, K. Shahid, Y. Gui, R.L. Olsen, Impact of cyber-attack on coordinated voltage control in low voltage grids, IET Renewable Power Generation 17 (2023) 2887–2894, https://doi.org/10.1049/rpg2.12571.
[42] B. Gou, D. Shue, Advances in algorithms for power system static state estimators: an improved solution for bad data management and state estimator convergence, IEEE Power and Energy Magazine 21 (1) (2023) 16–25, https://doi.org/10.1109/MPE.2022.3219168.
[43] R.V. Yohanandhan, R.M. Elavarasan, P. Manoharan, L. Mihet-Popa, Cyber-physical power system (CPPS): a review on modeling, simulation, and analysis with cyber security applications, IEEE Access 8 (2020) 151019–151064, https://doi.org/10.1109/ACCESS.2020.3016826.
[44] L.F. Ribas Monteiro, Y.R. Rodrigues, A.C. Zambroni De Souza, Cybersecurity in cyber—physical power systems, Energies 16 (12) (2023) 4556, https://doi.org/10.3390/en16124556.
[45] J. Devanny, L.R.F. Goldoni, B.P. Medeiros, The 2019 Venezuelan Blackout and the consequences of cyber uncertainty, Revista Brasileira de Estudos de Defesa 7 (2) (2021), https://doi.org/10.26792/rbed.v7n2.2020.75204.
[46] A. Solat, G.B. Gharehpetian, M.S. Naderi, A. Anvari-Moghaddam, On the control of microgrids against cyber-attacks: a review of methods and applications, Applied Energy 353 (2024) 122037, https://doi.org/10.1016/j.apenergy.2023.122037.
[47] D. Du, M. Zhu, X. Li, M. Fei, S. Bu, L. Wu, K. Li, A review on cybersecurity analysis, attack detection, and attack defense methods in cyber-physical power

systems, Journal of Modern Power Systems and Clean Energy 11 (3) (2023) 727–743, https://doi.org/10.35833/MPCE.2021.000604.
[48] A. Goudarzi, F. Ghayoor, M. Waseem, S. Fahad, I. Traore, A survey on IoT-enabled smart grids: emerging, applications, challenges, and outlook, Energies 15 (19) (2022) 6984, https://doi.org/10.3390/en15196984.
[49] V.S. Rajkumar, A. Stefanov, A. Presekal, P. Palensky, J.L.R. Torres, Cyber attacks on power grids: causes and propagation of cascading failures, IEEE Access 11 (2023) 103154–103176, https://doi.org/10.1109/ACCESS.2023.3317695.
[50] Joint Task Force Transformation Initiative, Guide for Conducting Risk Assessments (NIST SP 800-30r1; 0 ed., P. NIST SP 800-30r1), National Institute of Standards and Technology, 2012, https://doi.org/10.6028/NIST.SP.800-30r1.
[51] E. Covert, Case study: conducting a risk assessment for an electrical utility, International Conference on Cyber Warfare and Security 18 (1) (2023) 49–56, https://doi.org/10.34190/iccws.18.1.948.
[52] P.-S. Woo, B.H. Kim, Contingency analysis to evaluate the robustness in large-scale smart grids: based on information security objectives and frequency stability, Energies 13 (23) (2020) 6267, https://doi.org/10.3390/en13236267.
[53] A. Dagoumas, Assessing the impact of cybersecurity attacks on power systems, Energies 12 (4) (2019) 725, https://doi.org/10.3390/en12040725.
[54] A.E. Olajuyin, E. Olubakinde, Reliability and maintenance of assets in electric power distribution network, Current Journal of Applied Science and Technology (2022) 8–17, https://doi.org/10.9734/cjast/2022/v41i931686.
[55] M.W. Mufana, A. Ibrahim, Monitoring with communication technologies of the smart grid, Idosr Journal of Applied Sciences 7 (1) (2022) 102–112.
[56] S. Verma, R. Kumari, T.R. Chelliah, Black start capability of synchronous and asynchronous hydro generating units, in: 2022 IEEE International Conference on Power Electronics, Drives and Energy Systems (PEDES), 2022, pp. 1–6, https://doi.org/10.1109/PEDES56012.2022.10080038.
[57] H.N. Villegas Pico, V. Gevorgian, Blackstart capability and survivability of wind turbines with fully rated converters, IEEE Transactions on Energy Conversion 37 (4) (2022) 2482–2497, https://doi.org/10.1109/TEC.2022.3173903.
[58] N. Wang, F. Zhao, An assessment of the condition of distribution network equipment based on large data fuzzy decision-making, Energies 13 (1) (2020) 197, https://doi.org/10.3390/en13010197.
[59] B. Monarch, Black start: the risk of grid failure from a cyber attack and the policies needed to prepare for it, Journal of Energy and Natural Resources Law 38 (2) (2020) 131–160, https://doi.org/10.1080/02646811.2020.1744368.
[60] K.J. Rahul, Cybersecurity and confidentiality in smart grid for enhancing sustainability and reliability, Recent Research Reviews Journal 2 (2) (2023) 215–241, https://doi.org/10.36548/rrrj.2023.2.001.
[61] R. Arief, N. Khakzad, W. Pieters, Mitigating cyberattack related domino effects in process plants via ICS segmentation, Journal of Information Security and Applications 51 (2020) 102450, https://doi.org/10.1016/j.jisa.2020.102450.
[62] J. Gaspar, T. Cruz, C.-T. Lam, P. Simões, Smart substation communications and cybersecurity: a comprehensive survey, in: IEEE Communications Surveys & Tutorials, 1–1, 2023, https://doi.org/10.1109/COMST.2023.3305468.
[63] A.M. Mohan, N. Meskin, H. Mehrjerdi, A comprehensive review of the cyberattacks and cyber-security on load frequency control of power systems, Energies 13 (15) (2020) 3860, https://doi.org/10.3390/en13153860.
[64] M. Farhadkhani, Social responsibility aspects of smart power grids, in: 2019 Smart Grid Conference (SGC), 2019, pp. 1–6, https://doi.org/10.1109/SGC49328.2019.9056630.

[65] F. Nejabatkhah, Y.W. Li, H. Liang, R. Reza Ahrabi, Cyber-security of smart microgrids: a survey, Energies 14 (1) (2020) 27, https://doi.org/10.3390/en14010027.
[66] S. Mudgal, S. Pranjale, T. Balaji, S.A.A. Ahmed, N.K. Singh, P.K. Gupta, V. Mahajan, Impact of cyber-attacks on economy of smart grid and their prevention, U.Porto Journal of Engineering 8 (2) (2022) 51–64, https://doi.org/10.24840/2183-6493_008.002_0005.
[67] J. Dedrick, K.A. Perrin, E. Sabaghian, P.J. Wilcoxen, Assessing cyber attacks on local electricity markets using simulation analysis: impacts and possible mitigations, Sustainable Energy, Grids and Networks 34 (2023) 100993, https://doi.org/10.1016/j.segan.2023.100993.
[68] H. Riggs, S. Tufail, I. Parvez, M. Tariq, M.A. Khan, A. Amir, K.V. Vuda, A.I. Sarwat, Impact, vulnerabilities, and mitigation strategies for cyber-secure critical infrastructure, Sensors 23 (8) (2023) 4060, https://doi.org/10.3390/s23084060.
[69] N.R. Vajjhala, K.D. Strang, Cybersecurity for Decision Makers, first ed., CRC Press, 2023, https://doi.org/10.1201/9781003319887.
[70] Z. Oruj, Cyber security: contemporary cyber threats and national strategies, Distance Education in Ukraine: Innovative, Normative-Legal, Pedagogical Aspects 2 (2023) 100–116, https://doi.org/10.18372/2786-5495.1.17309.
[71] R. Shandler, M.A. Gomez, The hidden threat of cyber-attacks – undermining public confidence in government, Journal of Information Technology & Politics 20 (4) (2023) 359–374, https://doi.org/10.1080/19331681.2022.2112796.
[72] A.M. Alelayani, F.M.A. Zahrani, A.M. Munshi, R.M. Monshi, S.A. Al, Cybersecurity Regulation and Governance, 2020.
[73] Department of Homeland Security, NIPP 2013: Partnering for Critical Infrastructure Security and Resilience, 2013. Available online at: https://www.cisa.gov/sites/default/files/2022-11/national-infrastructure-protection-plan-2013-508.pdf.
[74] C. Samaras, W.J. Nuttall, M. Bazilian, Energy and the military: convergence of security, economic, and environmental decision-making, Energy Strategy Reviews 26 (2019) 100409, https://doi.org/10.1016/j.esr.2019.100409.
[75] J. Srinivas, A.K. Das, N. Kumar, Government regulations in cyber security: framework, standards and recommendations, Future Generation Computer Systems 92 (2019) 178–188, https://doi.org/10.1016/j.future.2018.09.063.
[76] D. Dolezilek, L. Hussey, Requirements or recommendations? Sorting out NERC CIP, NIST, and DOE cybersecurity, in: 2011 64th Annual Conference for Protective Relay Engineers, 2011, pp. 328–333, https://doi.org/10.1109/CPRE.2011.6035634.
[77] North American Electric Reliability Corporation NERC, NERC Standards and Criteria, October 28, 2023. https://www.nerc.com/pa/Stand/Pages/default.aspx.
[78] G.M. Nist, The NIST Cybersecurity Framework 2.0 (NIST CSWP 29 Ipd; P. NIST CSWP 29 Ipd), National Institute of Standards and Technology, 2023, https://doi.org/10.6028/NIST.CSWP.29.ipd.
[79] CIS Controls, CIS Controls TM V8, 2021 [Online], https://www.cisecurity.org/controls/v8.
[80] C. Tremlet, The IEC 62443 Series of Standards: How to Defend against Infrastructure Cyberattacks, 2023.
[81] G.A. Francia III, E. El-Sheikh, NERC CIP standards: review, compliance, and training, in: G. Francia III, J. Zanzig (Eds.), Global Perspectives on Information Security Regulations: Compliance, Controls, and Assurance, IGI Global, 2022, pp. 48–71, https://doi.org/10.4018/978-1-7998-8390-6.ch003.
[82] ISO/IEC, ISO - ISO/IEC 27001:2022 - Information Security, Cybersecurity and Privacy Protection — Information Security Management Systems — Requirements, 2022. https://www.iso.org/standard/82875.html.

[83] ISO/IEC, ISO/IEC 27002:2022, ISO, 2022, 2022b, https://www.iso.org/cms/render/live/en/sites/isoorg/conte nts/data/standard/07/56/75652.html. (Accessed 28 October 2023).

[84] ISO/IEC, ISO/IEC 27005:2022, ISO, 2022. https://www.iso.org/standard/80585.html.

[85] J. Taylor-Jackson, J. McAlaney, J.L. Foster, A. Bello, A. Maurushat, J. Dale, Incorporating psychology into cyber security education: a pedagogical approach, in: M. Bernhard, A. Bracciali, L.J. Camp, S. Matsuo, A. Maurushat, P.B. Rønne, M. Sala (Eds.), Financial Cryptography and Data Security, vol 12063, Springer International Publishing, 2020, pp. 207–217, https://doi.org/10.1007/978-3-030-54455-3_15.

[86] P. Kotuszewski, K. Kukielka, P. Kluk, A. Ordys, K. Bieńkowski, J.M. Kościelny, M. Syfert, P. Wnuk, J. Możaryn, B. Fajdek, Cyber-security assessment of industry 4.0 enabled mechatronic system, Complexity 2021 (2021) 1–10, https://doi.org/10.1155/2021/6670625.

[87] E.B. Fernandez, H. Washizaki, N. Yoshioka, T. Okubo, The design of secure IoT applications using patterns: state of the art and directions for research, Internet of Things 15 (2021) 100408, https://doi.org/10.1016/j.iot.2021.100408.

[88] S.B. Brandao Filho, C.D.A.C. Cesar, A secure method for industrial IoT development, SN Computer Science 3 (2) (2022) 173, https://doi.org/10.1007/s42979-022-01061-9.

[89] B. Bokan, J. Santos, Threat modeling for enterprise cybersecurity architecture, in: 2022 Systems and Information Engineering Design Symposium (SIEDS), 2022, pp. 25–30, https://doi.org/10.1109/SIEDS55548.2022.9799322.

[90] J.A. Font, J. Jarauta, R. Gesteira, R. Palacios, G. López, Threat models for vulnerability analysis of IoT devices for manipulation of demand attacks, in: 2023 JNIC Cybersecurity Conference (JNIC), 2023, pp. 1–8, https://doi.org/10.23919/JNIC58574.2023.10205781.

[91] P.C. Van Oorschot, Computer Security and the Internet: Tools and Jewels from Malware to Bitcoin, Springer International Publishing, 2021, https://doi.org/10.1007/978-3-030-83411-1.

[92] V.A. Mehri, P. Arlos, E. Casalicchio, Automated patch management: an empirical evaluation study, in: 2023 IEEE International Conference on Cyber Security and Resilience (CSR), 2023, pp. 321–328, https://doi.org/10.1109/CSR57506.2023.10224970.

[93] M. Dunkerley, M. Tumbarello, Mastering Windows Security and Hardening: Secure and Protect Your Windows Environment from Cyber Threats Using Zero-Trust Security Principles, second ed., Packt Publishing, 2022.

[94] S.Y. Enoch, J. Mendonça, J.B. Hong, M. Ge, D.S. Kim, An integrated security hardening optimization for dynamic networks using security and availability modeling with multi-objective algorithm, Computer Networks 208 (2022) 108864, https://doi.org/10.1016/j.comnet.2022.108864.

[95] I. Traore, I. Woungang, S. Saad (Eds.), Artificial Intelligence for Cyber-Physical Systems Hardening, vol. 2, Springer International Publishing, 2023, https://doi.org/10.1007/978-3-031-16237-4.

[96] J. Garbis, J.W. Chapman, Zero Trust Security: An Enterprise Guide, Apress, 2021, https://doi.org/10.1007/978-1-4842-6702-8.

[97] Q. Chen, M. Zhou, Z. Cai, S. Su, Compliance checking based detection of insider threat in industrial control system of power utilities, in: 2022 7th Asia Conference on Power and Electrical Engineering (ACPEE), 2022, pp. 1142–1147, https://doi.org/10.1109/ACPEE53904.2022.9784085.

[98] N. Nikolaou, A. Papadakis, K. Psychogyios, T. Zahariadis, Vulnerability identification and assessment for critical infrastructures in the energy sector, Electronics 12 (14) (2023) 3185, https://doi.org/10.3390/electronics12143185.

[99] D.P. Nolan, Adaptation to security vulnerability analysis, in: Safety and Security Review for the Process Industries, Elsevier, 2015, pp. 14–16, https://doi.org/10.1016/B978-0-323-32295-9.00004-5.
[100] Q.S. Qassim, N. Jamil, M. Daud, A. Patel, N. Ja'affar, A review of security assessment methodologies in industrial control systems, Information & Computer Security 27 (1) (2019) 47–61, https://doi.org/10.1108/ICS-04-2018-0048.
[101] S. Kara, S. Hizal, A. Zengin, Design and implementation of a DEVS-based cyber-attack simulator for cyber security, International Journal of Simulation Modelling 21 (1) (2022) 53–64, https://doi.org/10.2507/IJSIMM21-1-587.
[102] A. AlQadheeb, S. Bhattacharyya, S. Perl, Enhancing cybersecurity by generating user-specific security policy through the formal modeling of user behavior, Array 14 (2022) 100146, https://doi.org/10.1016/j.array.2022.100146.
[103] S.F. Ahmed, N. A. Hikal, A review of cyber-security measuring and assessment methods for modern enterprises, JOIV: International Journal on Informatics Visualization 3 (3) (2019), https://doi.org/10.30630/joiv.3.3.241.
[104] L. Manases, D. Zinca, Automation of network traffic monitoring using Docker images of Snort3, Grafana and a custom API, in: 2022 21st RoEduNet Conference: Networking in Education and Research (RoEduNet), 2022, pp. 1–4, https://doi.org/10.1109/RoEduNet57163.2022.9921063.
[105] A.R. Javed, F. Shahzad, S.u. Rehman, Y.B. Zikria, I. Razzak, Z. Jalil, G. Xu, Future smart cities: requirements, emerging technologies, applications, challenges, and future aspects, Cities 129 (2022) 103794, https://doi.org/10.1016/j.cities.2022.103794.
[106] U. Tariq, I. Ahmed, A.K. Bashir, K. Shaukat, A critical cybersecurity analysis and future research directions for the Internet of Things: a comprehensive review, Sensors 23 (8) (2023) 4117, https://doi.org/10.3390/s23084117.
[107] H. Bousbiat, R. Bousselidj, Y. Himeur, A. Amira, F. Bensaali, F. Fadli, et al., Crossing Roads of Federated Learning and Smart Grids: Overview, Challenges, and Perspectives. arXiv Preprint, 2023, https://doi.org/10.48550/arXiv.2304.08602.
[108] J. Jithish, B. Alangot, N. Mahalingam, K.S. Yeo, Distributed anomaly detection in smart grids: a federated learning-based approach, IEEE Access 11 (2023) 7157–7179, https://doi.org/10.1109/ACCESS.2023.3237554.
[109] S. Pandya, G. Srivastava, R. Jhaveri, M.R. Babu, S. Bhattacharya, P.K.R. Maddikunta, S. Mastorakis, M.J. Piran, T.R. Gadekallu, Federated learning for smart cities: a comprehensive survey, Sustainable Energy Technologies and Assessments 55 (2023) 102987, https://doi.org/10.1016/j.seta.2022.102987.
[110] N. Zhao, H. Zhang, X. Yang, J. Yan, F. You, Emerging information and communication technologies for smart energy systems and renewable transition, Advances in Applied Energy 9 (2023) 100125, https://doi.org/10.1016/j.adapen.2023.100125.
[111] F. Salahdine, T. Han, N. Zhang, 5G, 6G, and beyond: recent advances and future challenges, Annals of Telecommunications 78 (9–10) (2023) 525–549, https://doi.org/10.1007/s12243-022-00938-3.
[112] S.J. Pinto, P. Siano, M. Parente, Review of cybersecurity analysis in smart distribution systems and future directions for using unsupervised learning methods for cyber detection, Energies 16 (4) (2023) 1651, https://doi.org/10.3390/en16041651.
[113] B. Ghimire, D.B. Rawat, Recent advances on federated learning for cybersecurity and cybersecurity for federated learning for Internet of Things, IEEE Internet of Things Journal 9 (11) (2022) 8229–8249, https://doi.org/10.1109/JIOT.2022.3150363.
[114] A. Marot, A. Kelly, M. Naglic, V. Barbesant, J. Cremer, A. Stefanov, J. Viebahn, Perspectives on future power system control centers for energy transition, Journal of Modern Power Systems and Clean Energy 10 (2) (2022) 328–344, https://doi.org/10.35833/MPCE.2021.000673.

CHAPTER 6

Cyber security: Emerging trends best practices

Divya Asija and R.K. Viral
Department of Electrical and Electronics Engineering, Amity School of Engineering and Technology, Amity University Uttar Pradesh, Noida, India

6.1 Introduction

In an increasingly huge data-driven world, where information security and data privacy are constantly at danger, consider social networks, online transactions, information saved or managed over the internet, and automated procedures carried out by means of IT systems. Cybercrime is continuously growing in terms of the quantity of attacks and severity of harm done to its users due to the invention of new tools and methods. Attackers seek to undermine the confidentiality, integrity, and availability of information by creating new methods of gaining unauthorized access to networks, programs, and data. They can target anyone, from individuals to small and medium-sized businesses to the largest businesses in the world of trade. On an annual basis, there appears to be an increase in both the total number of attacks and the number of attacks that circumvent the security measures on an annual basis, there appears to be an increase in both the total number of cyberattacks and the number of attacks that breach the security of very large businesses, thereby impacting information security, business continuity, and confidence among consumers. The year 2014, widely referred to as "the year of cyber-attacks," witnessed a breakthrough in this escalating trend, but the researchers predict that this is not the turning point provided significant solutions are put into effect [1]. This chapter aims to present countermeasures that ought to be carried out to support an enhancement of security and the reduction of global cybercrime. It also illustrates the results, trends, and structures that the contributors witnessed during the study of the attacks that have been disclosed over the preceding three or 4 years. The article is structured in three main parts: it begins by

Cyber Security Solutions for Protecting and Building the Future Smart Grid
ISBN 978-0-443-14066-2
https://doi.org/10.1016/B978-0-443-14066-2.00011-6

presenting the general view of cyber-crime from the perspective of specialized literature, international law, as well as historical facts and continues through revealing the main results and interpretation of the study performed over the last 3 years' reported attacks. It ends by drawing some of the main countermeasures that companies may undertake to ensure improvement of controls covering the information confidentiality, integrity and availability, while decreasing the security breaches.

6.1.1 Application domains for cyber attacks

Cyberattacks target a wide range of sectors and industries, including financial services, healthcare, critical infrastructure, retail and e-commerce, manufacturing and industrial control systems, education, media and entertainment, and nonprofit organizations. It also targets government and military organizations. These attacks demonstrate the varied effects of cyber threats across various domains as they attempt to steal confidential data, interfere with operations, seek ransom payments, or spread widespread chaos. The key target locations for cyberattacks are depicted in the figure, along with the subdomains inside these domains where potential attackers could undermine security and integrity. Fig. 6.1 indicates the major application domain of cyberattacks [2].

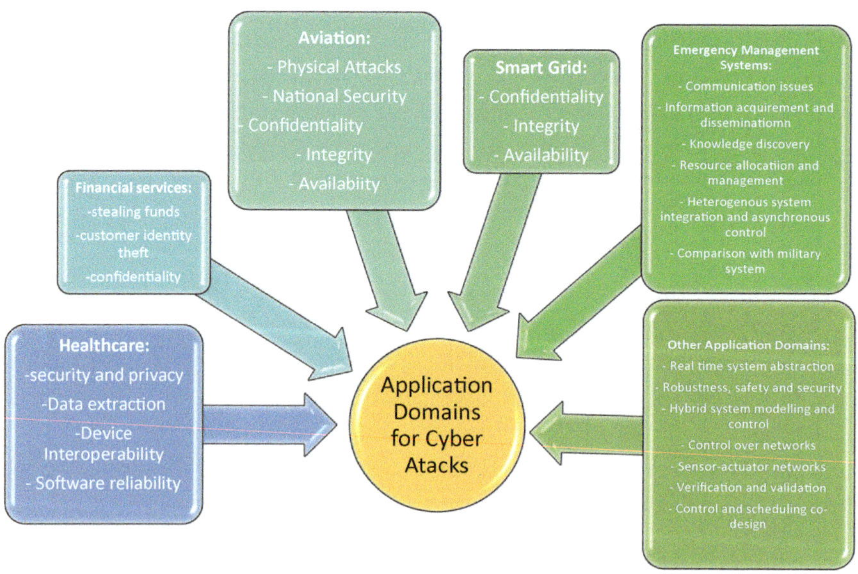

Figure 6.1 Application domain of cyberattacks.

6.2 Types of cyberattacks/threats

Cyber threats are classified based on different traits or factors. Fig. 6.2 shows the classification of such attacks effecting the system security and authenticity [3].

6.2.1 Based on purpose

i. *Access attack*

In this specific type of attack, the assailant manages to breach the security of a host's computer system without proper authorization, with

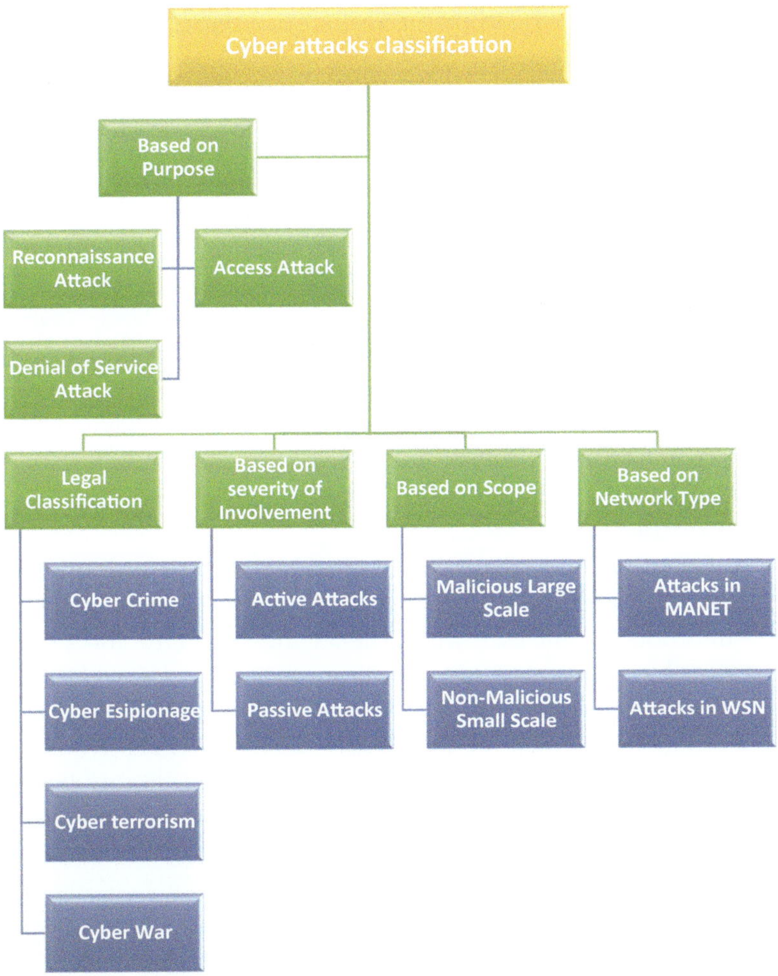

Figure 6.2 Attacks classification in cyber system.

the intention of altering data. As a result of vulnerabilities in web application services and file transfer mechanisms, hackers can exploit these weaknesses, thereby infiltrating electronic accounts, databases, and other sensitive information.

Access attacks consist of the following [3]

- *Attacks on Secret Code*

  This method is commonly referred to as a "dictionary attack," wherein an individual without proper authorization endeavors to gain access to an account by analytically trying out every possible password combination within a defined scope. It is further reconstituted into two types of attacks i.e., password prediction, and password reset.
- *Utilization of Trust Port*

  A reputable host is breached by a hacker who exploits it to execute attacks against other dependable hosts.
- *Port Redirection*

  An attacker utilizes a trusted host to enter other hosts within the network that are safeguarded by firewalls.
- *Man-in-the-middle Attacks.*

  This method is alternatively referred to as a Janus attack or a bucket brigade attack. It falls under the category of active eavesdropping, wherein the attacker individually establishes connections with the victims and then relays messages between them. This tactic is employed to create the illusion of private communication between the victims.
- *Social Engineering*

  This method utilizes the technique of SQL injection, malicious code is inserted into social engineering websites, leading to the infection of any user who accesses the site. This also introduces the potential for content modification.
- *Phishing*

  Under this method, individual is sending a deceptive email under the guise of a legitimate company's identity aims to deceive the recipient into sharing personal information, which is then exploited for the purpose of identity theft.

ii. *Reconnaissance attack*

An attack when the culprit connects to the system being attacked to look for any kind of machine vulnerabilities and acquire information. In other terms, we might characterize it as reconnaissance attacks, which include unauthorized identification, machine mapping process, and operations. This is a situation like committing theft, for instance, in a home where weak locks, gates, and linked windows are easily broken.

Reconnaissance attacks can comprise of the following:

- *Packet Sniffers*

  A special device is used to eavesdrop upon traffic between networked computers and it will capture data addressed to other machines saving it for later analysis.

- *Scanning the Port*

  A series of messages sent by an attacker attempting to break into a computer to learn which computer services each associated with a well-known port number.

- *Sweeping the Ping*

  As scanning method used by the attacker to determine the range of IP addresses mapped to live hosts.

- *Queries Regarding Internet Information*

  An attacker can use DNS Queries to learn who owns a domain and what addresses have been assigned to that domain.

iii. *Denial of Service Attack*

Denial-of-service attacks, which are outlined in Refs. [4,5], attempt to either crash or render the system ineffective by purposefully halting the system's operations. Data loss or manipulation may potentially be part of these assaults. Further, in accordance with reference [6], attackers can totally destroy the network or mess with its operations with the goal of preventing legitimate users from accessing it.

6.2.2 Based on legal classification

i. *Cyber crime*

According to Canadian law enforcement entities, cybercrime is defined as a criminal offense in which a computer is either the main target or an essential instrument in the execution of the offense. Cybercrime's objective is to utilize a computing device to perpetrate a crime, with the device itself frequently acting as an assisting element. As reference [7] points out, operating system shortcomings, massive computer storage capacities, the degree of privacy offered by digital surroundings, and a lack of user knowledge are the primary causes of such crimes.

ii. *Cyber Espionage*

Cyber espionage is the illegal deployment of cracking techniques and malicious software, such as Trojan horses and spyware, to get private information from individuals, entities, or authorities in order to serve one's own interest. Cyber espionage is another name for this activity, which can be done completely online and often carried out by highly skilled people working from faraway locations. On the other hand, it might involve intrusion through conventional spies or computer-trained moles. At times, amateur hackers or software developers with malicious intent can also be the cause of cyber spying.

iii. *Cyber terrorism*

Terrorist acts frequently employing Internet-based attacks, such as deliberate large-scale disruption of computer networks via the employment of instruments like computer viruses, as described in Ref. [8].

iv. *Cyber war*

Attacking a network or computer system of a different nation with the intent to cause harm or disruption is known as a cyber war.

6.2.3 Based on severity of involvement

i. *Active attack*

An intermediary hack that modifies content and interferes with processes during data transfer causes significant damage.

ii. *Passive attack*

The database is not accessed by the hacker, nor is it modified. Rather, their sole objective is to keep an eye on the target and retrieve data while it's being transmitted. Basically, the hacker uses a variety of techniques to listen in on host conversations in order to collect data.

6.2.4 Based on scope

i. *Malicious Large Scale*

Malicious behavior is defined by a conscious desire to do harm. A malicious large-scale attack is one that is planned and carried out by a person or group with the intention of causing widespread chaos or disturbance. These attacks impact thousands of systems worldwide, causing system breakdowns that cause massive data loss and harm to the company's reputation.

ii. *Non-Malicious Small Scale*

These events are usually the consequence of unintentional assaults or mishandling, which frequently happens as a result of operational errors committed by people with insufficient training. System crashes or slight data loss could result from them. In most cases, data recovery is possible in these situations because there are just a few infected systems on the network. Usually, the accompanying expenses are negligible.

6.2.5 Based on network type

i. *Attacks in Mobile Adhoc Networks (MANET)*

The attacks in MANETs are [9]:

- Byzantine attack
- The Black Hole Attack
- Flood Rushing Attack
- Byzantine Wormhole Attacks
- Byzantine Overlay Network Wormhole Attacks

ii. *Attacks in Wireless Sensor Networks (WSN)*

The attacks on WSN are [10]:

- Cryptography Attack:

  Type of attack in which the adversary breaks the cryptography, pragmatically, to discover the shortcoming in an exceeding protocol, code, or ciphers to retrieve the plaintext without the key.
- Non-Cryptography related Attacks
- Attacks based on the Network Layers

  Based on the layers, methods, and attack domains, the attacks detected in the Wireless Sensor Network will be categorized.

6.2.6 Malware attack

An attack happens when a hacker deliberately installs malicious software on a target computer with an objective of infecting and destroying the system in addition to spreading viruses and acquiring private information.

6.2.7 Attack on quantum key distribution

Attacks can be carried out by counterfeiting one or more photons, altering the timing of pulses, or by launching an attack during the transfer of data across a quantum channel. Table 6.1 Outline few examples of cyberattacks across different categories [11].

Table 6.1 Examples of cyberattacks across different categories [11].

Name of the attacks	Description	Examples
Reconnaissance attacks	Type of attack which involves unauthorized detection mapping and services to steal data	Packet snillers, port scanning, ping sweeps and DNS (distributed network services) queries
Access attacks	An attack where intruder gains access to a device to which he has no right for access	Port trust utilization, port redirection, dictionary attacks, man-in-the-middle attacks, social engineering attacks and phishing
Denial of service	Intrusion into a system by disabling the network with the intent to deny service to authorized users.	Smurf, SYN flood, DNS attacks, DDoS (distributed denial of services)
Cyber crime	The use of computers and the internet to exploit users for materialistic gain.	Identity theft, credit card fraud
Cyber espionage	The act of using the internet to spy on others for gaining benefit.	Tracking cookies, RAT controllable
Cyber terrorism	The use of cyber space for creating large scale disruption and destruction of life and property.	Crashing the power grids by al-queda via a network, poisoning of the water supply
Cyberwar	The act of a nation with the intention of disruption and destruction of life and property	Russia's war on Estoria (2007), Russia's war on Georgia (2008)
Active attacks	An attack with data transmission to all parties thereby acting as a liaison enabling severe compromise.	Masquerade, reply, modification of message
Passive attacks	An attack which is primarily eaves dropping without meddling with the database	Traffic analysis, release of message contents
Malicious attacks	An attack with a deliberate intent to cause harm resulting in large scale disruption.	Spear phishing attack

Table 6.1 Examples of cyberattacks across different categories [11].—cont'd

Name of the attacks	Description	Examples
Non-malicious attacks	Accidental attack due to mishandling or operational mistakes with minor loss of data	Registry corruption, accidental crashing of hard disk
Attacks in MANET	Attacks which aim to slow or stop the flow of information between the nodes.	Byzantine attacks, black hole attack, flood rushing attack, byzantine wormhole attack
Attacks on WSN	An attack which prevents the sensors from detecting and transmitting information through the network.	Application layer attacks, transport layer attacks, network layer attacks, multi-layer attacks

6.3 Learning from the past cyber attacks

Enhancements in incident response and general security enhancements were separated into two parts of the intervention or action phase to go deeper into the research scope of cybersecurity. The purpose of this division was to distinguish research that only looked at strengthening an organization's incident response capability from studies that focused on how insights led to more comprehensive security architecture modifications. As a result, five unique categories emerged from this classification: involvement in the learning process, causal study, improvements to incident response, security in general, and assessment of the learning process.

The organization and involvement of different sectors within the organization are included in the learning process. event reporting, event classification, and cause analysis are all elements of causal analysis. The identification and prioritization of activities, the distribution of accountability, the distribution of resources, and the tracking of actions through to completion are all included in these two improvement categories. The scope of causes to be addressed is influenced by the learning approach used and the participation of different teams, and the type of modifications implemented is influenced by the reasons that have been found. The organization's security posture may be greatly impacted by the adoption of these upgrades.

The last stage of learning evaluation involves making sure that activities are carried out and confirming that they are effective in achieving expected improvements. It also evaluates the organization's ability to draw lessons from events. The learning process is refined organizationally based on the insights gained from this step. Recurring occurrences with identical causes or outcomes may be a sign of a poor learning process, showing that the organization has not adjusted its detection, response, and protection capabilities [12].

6.4 Standard security framework

i. *NIST Special Publication 800-41: Comprehensive* instructions for configuring firewalls are provided in this article, guaranteeing excellent traffic filtering and access control as well as efficient network security. These recommendations, for instance, can be used by organizations to set up firewall rules that prevent unauthorized access to confidential information kept on internal servers (NIST SP 800-41, 2009) [2].

ii. *ISO/IEC 27032:* Essential cybersecurity recommendations are provided by ISO/IEC 27032, which focuses on delivering cybersecurity support and successfully handling cyber threats across organizations. To lessen the effects of cyberattacks, for instance, organizations ought to utilize these guidelines for establishing incident response strategies (ISO/IEC 27032:2012) [1].

iii. *CIS Controls:* The Center for Internet Security established the CIS Controls, which provide a set of best practices for protecting IT networks and systems that include both fundamental and sophisticated security procedures. By using these controls, organizations may set an effective security baseline across their infrastructure and prioritize their cybersecurity efforts (CIS Controls, 2015) [4,7].

iv. *NIST Special Publication 800-53:* To ensure adherence to strict security norms and regulations, NIST Special Publication 800—53 describes security and privacy controls specifically designed for government information systems and organizations. These controls are a tool that government agencies can utilize to preserve public trust and safeguard sensitive government information (NIST SP 800-53, 2020) [8].

v. *Payment Card Industry Data Security Standard (PCI DSS):* PCI DSS lays out specifications for securing payment card data, including security protocols to defend cardholder data and stop data theft. To secure online transactions and safeguard customer payment information, retailers and e-commerce businesses can adopt PCI DSS compliance (PCI DSS, 2018) [10].

vi. *FIPS Publication 140-2:* By using cryptographic operations to ensure the integrity and confidentiality of sensitive data, FIPS Publication 140-2 lays forth security standards for cryptographic units. FIPS-compliant cryptographic units can be used by financial institutions and government organizations to encrypt communications and safeguard secret data. (FIPS 140-2, 2002) [11].

vii. *ISO/IEC 27001:* Information security management systems can be established, implemented, maintained, and continuously improved by organizations with the help of the framework for information security management systems provided by ISO/IEC 27001. Organizations across diverse sectors, including healthcare and finance, can exhibit their dedication to safeguarding confidential information and guaranteeing adherence to regulations by obtaining ISO/IEC 27001 certification (ISO/IEC 27001, 2022) [5].

viii. *NIST Special Publication 800-63:* NIST SP 800-63 provides digital identity management rules that provide safe authentication procedures and efficient digital identity management. These recommendations can be used by online service providers and governmental organizations to set up multi-factor authentication systems and safeguard user accounts against illegal access (NIST SP 800-63, 2017) [12].

ix. *ISO/IEC 27002:* A code of practice for information security controls is provided by ISO/IEC 27002, which offers recommendations for placing security controls into effect as well as best practices for successfully managing information security risks. Businesses can create thorough information security policies and procedures using ISO/IEC 27002 to safeguard their vital assets from online attacks (ISO/IEC 27002, 2022) [1].

x. *NIST Cybersecurity Framework:* Organizations can evaluate and strengthen their cybersecurity posture with the help of the NIST Cybersecurity Framework, which offers a customizable framework and a comprehensive approach to risk management. The NIST Cybersecurity Framework can be used by critical infrastructure sectors, such energy and transportation, to prioritize cybersecurity threats and strengthen their defenses against cyberattacks (NIST Cybersecurity Framework, 2024) [9].

6.5 Emerging trends in cyber security

In order to create defending technologies next to cyber risks, enhance cyber threat detection, automate cyber security procedures, and stop cyberattacks, advanced artificial intelligence (AI) and machine learning (ML) approaches are needed. In order to improve cyber security, AI and ML technologies must be used to automate the cyber defense system as cyberattacks become more sophisticated and complicated [13].

Cyber threat recognition systems are strengthened by real-time AI, which also helps cyber specialists analyze large amounts of data, spot trends in cyberattacks, identify dangers and abnormalities, and automate security actions. This makes it possible for cyber experts to react to cyberattacks more skillfully [14]. Furthermore, biometric authentication comes out as a very successful security strategy for specific businesses when combined with AI and ML technology. By using distinctive biological characteristics like fingerprints or face recognition to confirm a user's identity, biometric authentication greatly strengthens security protocols. Future cryptography researchers should concentrate on developing quantum-resistant algorithms that can fend off quantum computer assaults and guarantee the foreseeable security of confidential encrypted data. In the cyber business, automating security operations to respond automatically to cyber-attacks is critical. Also, one of the major issues to improve an organization's cyber security is the advancement of block chain technology. Aside from this, current developments involve creating unified structures and platforms for exchanging cyberthreats and presenting cooperative cyber defense strategies for industries, governments, and organizations. Next subsection will elaborate more on emerging trends in cyber security [15].

6.5.1 Rise of AI and machine learning in threat detection and response

6.5.1.1 Applications of AI/ML in cyber security

The integration of artificial intelligence and cybersecurity is becoming more and more important in handling the dynamic terrain of cyberthreats. The need for researchers, policymakers, and business professionals to arm people with human intelligence for security is become more and more apparent as con artists continue to refine their techniques. Many surveys propose improving the most advanced simulated intelligence approaches to counter artificial intelligence-motivated attacks. Such systems will be the basis of a preventive security approach against the evolving cyber threat environment, in addition to having the ability to identify and mitigate emerging threats [16]. Network security has long been based on conventional techniques for threat identification and response. Using rule sets, defined examples, and signature-based techniques to identify and mitigate potential risks. Signature-based detection methods struggle to keep up with the increasing volume and sophistication of attacks, especially when it comes to 0-day threats that take advantage of vulnerabilities prior to they are discovered. The limits of these conventional tactics are made worse by the growing threat surface caused by the diversification of cyber-attacks in terms of people, goals, and means.

Conventional approaches are successful in adhering to predetermined rules, but AI-driven solutions impact computer-based intelligence and a profound understanding of how to analyze large datasets effectively, discern patterns, and adapt to ever-changing dangers.

Numerous reviews of the literature examine the challenges posed by traditional approaches, emphasizing the necessity [14,17–19]. Based on the literature available, potential applications of AI in cybersecurity can be seen in Fig. 6.3 [20]. Several important uses of AI/ML in cybersecurity include [14,18,21,22]:

i. Analysis and identification of threats
ii. Identification of anomalies
iii. Predictive Analytics
iv. User and Entity Behavior Analytics (UEBA)
v. Automated Incident Reaction
vi. Fraud Identification
vii. Phishing Detection

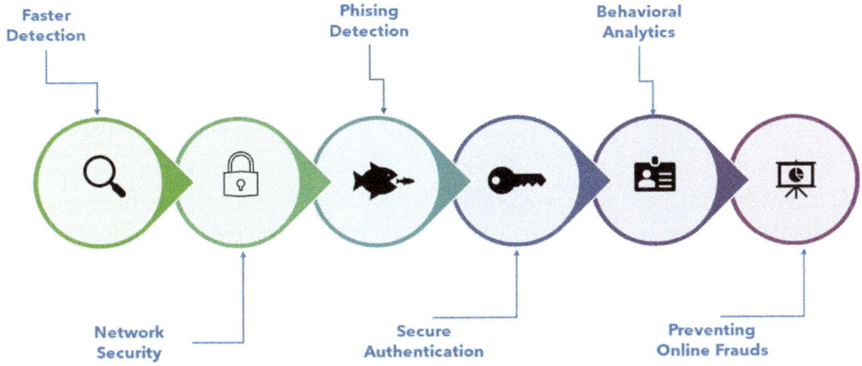

Figure 6.3 Potential applications of AI in cybersecurity.

viii. Malware Identification and Avoidance
ix. Security Analytics
x. Identity and Access Management (IAM)

6.5.1.2 Benefits and challenges

Though AI and ML have a great deal of scope to improve cybersecurity, they also bring special difficulties that businesses must overcome in order to fully utilize them and guarantee the privacy of their data and systems. Key advantages of AI/ML for cybersecurity include [13,19,22]:

i. *Improved Security Identifying:* Compared to previous approaches, algorithms based on machine learning and AI can evaluate large volumes of data fast and correctly, which helps firms to identify and react to cyber threats quicker.
ii. *Quick Reaction*
iii. *Enhanced Precision:* More precise recognition of threats and a lower rate of false positives result from ML models' ability to learn from previous occurrences and constantly enhance their detection skills.
iv. *Preemptive Defense:* Given past data as well as trends, AI/ML can forecast possible security threats and weaknesses, enabling businesses to proactively bolster their safety measures prior to an attack.
v. *Automation:* Identification of threats and other repetitious security operations may be automated with AI/ML technology.

The never-ending war of cat and mouse involving defenders and attackers is a major obstacle to using AI/ML for security. AI and ML's challenges for cybersecurity [13,15–18,22]:

i. *Data Accessibility and Quality:* Large volumes of high-quality training data are required for AI/ML models to perform well. Nevertheless, obtaining tagged data for cybersecurity-related tasks may be challenging since security-related data is sensitive.
ii. *Adversarial Maneuvers:* Cybercriminals may be able to take advantage of weaknesses in ML and AI models by using adversarial techniques such as introducing fraudulent information or changing inputs to the model to avoid identification.
iii. *Explainability and Interpretability:* Deep neural network algorithms especially are sometimes seen as "black boxes" in AI/ML, which makes it difficult to grasp the way they make judgments and undermines diffusion and confidence.

iv. *Privacy Issues:* Primarily in uses like user behavior analytics and information mining, AI and ML models may unintentionally violate people's rights to privacy, giving rise to moral and legal questions.

v. *Complexity and Scalability:* Large-scale analysis of data, training models, and implementation are complicated processes that call for a high level of experience and funding when implementing AI/ML solutions in cybersecurity.

6.5.2 Internet of things (IoT) security considerations

It is inappropriate to apply threat prevention based on typical IT networking techniques in an Internet of Things system because of the wide variety of equipment, several protocols for communication, multiple interfaces, and services that are available. In actuality, the level of safety protections in place in traditional networks might not be enough. The components that comprise the communications connection, and interfaces/service levels of an Internet of Things system are all three layers that are targeted by attack vectors, according to the Open Web Application Security Project (OWASP). Therefore, a security architect at every IoT level should be included in the implementation of internet of things security mitigation. An Internet of Things network includes Wireless Sensor Networks (WSN) and Radio Frequency Identification (RFID). Thus, it is usual for these two systems to be targets of assaults [23].

IoT system security comes with a number of special difficulties, including erratic communication, adversarial settings, and insufficient information and privileged protection [24]. There are additional security issues at the perception layer, as Fig. 6.4 illustrates. This might be due to a

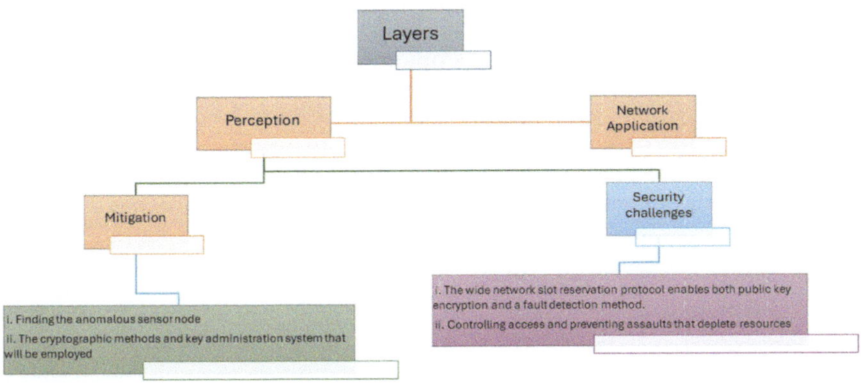

Figure 6.4 A summary of the state of IoT security issues and suggested solutions.

number of factors, including insecure network connectivity, easily accessible final nodes physically, and weak web interfaces on devices. Therefore, it may be said that the primary attack surface for adversary in Internet of Things systems is devices that are physical, or end nodes [25].

6.5.2.1 Vulnerabilities in IoT devices

There are several unique challenges associated with IoT security of systems, such as inconsistent communication, hostile environments, inadequate data, and elevated security [24]. As Fig. 6.2 indicates, there are more safety issues at the perception layer. Numerous variables, such as unsecured connection to the network, physically reachable end nodes, and objects with inadequate web interactions, could be to blame for this. Consequently, actual devices, or endpoint nodes, might be considered as the primary attack surface for adversaries in IoT [25,26].

i. *Passwords that are programmed, weak, or guessed:*

The simplest approach for intruders to infiltrate IoT gadgets and start massive networks of bots and other infections is by using fragile, standard, or programmed passwords.

ii. *Network services that are not secure:*

Adversaries seek to disable and violate critical or private data shared among the gadget & the server by taking advantage of flaws in the connection protocol and applications operating on IoT gadgets.

iii. *Unsecured ecosystems connections:*

Here, a robust permission and authentication method is required. A number of approaches have been developed to protect IoT device identification. Every time a server interacts there with an Internet of Things gadgets, it is going to be capable to distinguish both a legitimate terminal and a malicious one by requiring that endpoint to identify oneself thanks to the implementation of an efficient device identification method.

iv. *Unsecure storage of information and exchange:*

Absence of control over access or encrypting for private information at all points in the ecosystem—while it's being processed, in motion, or at repose.

v. *Insufficient handling of devices:*

Absence security-related assistance, such as management of assets, upgrade management, safe demolition, networks surveillance, and reaction abilities, on devices put into operation.

vi. Unsecure default configuration:

Computers or devices that are pre-configured with unsafe standard configurations or that are incapable of increasing system safety by preventing users from changing setups.

vii. *Absence of rigidity on a physical level:*

Absence of actual hardening initiatives, which might provide way to future remote assaults or permit hackers to acquire local control of the gadget and get crucial data.

viii. *Inadequate safeguards for security:*

Personal data of an individual that is misused, incorrectly utilized, or kept on the gadget or in the ecosystem as a whole.

6.5.2.2 Strategies for securing IoT ecosystems

Implementing solid authorization as well as access control procedures, encrypting information, updating the firmware and software on constantly, and keeping an eye out for any unusual activity on devices and network connections are all necessary to secure network integrity in the Internet of Things. A few strategies that can secure the IoT ecosystems are discussed below [25,26]:

i. *Secure Design:* Implement security principles from the start.
ii. *Control of Authentication and Access:* Employ robust methods for authentication.
iii. *Privacy of Data & Encryption:* When information is in route or at ease, encrypt it.
iv. *Safe Interaction:* When communicating, utilize safe protocols.
v. *Devices Administration:* Put remote upgrades and tracking into practice.
vi. *Occupational safety:* Safe physical entry to the gadgets.
vii. *Segmenting a network:* Employ firewall and divide up your gadgets.
viii. *User Instructions:* Inform users of potential cybersecurity hazards.
ix. *Chain of Supply Security:* Make sure that the supply chain is secure.
x. *Adherence to regulations:* Continue to abide with the rules.

6.5.3 Cloud security and its evolving landscape

Deployment of security measures for the cloud and growing reliance on cloud-based services exposes enterprises to new risks and weaknesses. Information theft, illicit use of uncertain APIs, divulged facilities that present dangers to security, loss of information, and service interruption that results in an interruption of important services, among additional safety risks and obstacles are among the difficulties associated with cyber security experienced by enterprises that rely on cloud-based services [27].

6.5.3.1 Cloud-based threats and attacks

Cloud security is an ever-evolving field. It is always changing to counter emerging and advanced threats. Following are a couple of significant advancements and trends influencing the field of research [25,27]:

- **i.** *Breach of Data:* Important information can be accessed without authorization.
- **ii.** Hijacking of Accounts: Illicit access to individual credentials in the public cloud.
- **iii.** DoS Assaults: disrupting performance by overcrowding the cloud's resources.
- **iv.** Unsecured APIs: Usability flaws with cloud applications.
- **v.** Loss of Data: Data corruption or loss, whether intentional or unintentional.
- **vi.** Inadequate Safety Measures: inadequate use of safety precautions.
- **vii.** Injection of Malware: Spyware that is executed in the cloud or uploaded to cloud stores.
- **viii.** Attacks of MitM: Influencing and listening in on conversations.
- **ix.** Dangers from Within: Illegal conduct by anyone with permission.
- **x.** Vulnerabilities in Sharing Technologies: Taking advantage of holes in common infrastructure.

Corporations ought to set a high priority on strong safety precautions such as encrypting it, entry limits, surveillance, and recurring reviews in order to mitigate these threats.

6.5.3.2 Ensuring data privacy and integrity in the cloud

The use of cloud computing has completely changed the way organizations function in the current digital world. Because of its exceptional cost-effectiveness, expansion, and adaptability, it is the go-to option for businesses of all kinds. But, enormous power is paired with enormous accountability, and the field of security for the cloud is always changing to cope with the new threats that the globe is facing. To ensure robust cloud security, organizations must adopt standards of excellence in this ever-changing environment. Following are some of the significant advancements and trends influencing the field of study [28]:

- **i.** *Put Zero Trust in Place:* Presume that there are dangers within and beyond the internet. Employ ongoing oversight and strict access limits.

ii. *Ongoing Audits and Surveillance:* Make sure to regularly check as well as evaluate the cloud infrastructure for irregularities and risks. Make use of automatic technologies to keep an eye out for breaches of security instantly.
iii. *Cryptography:* Protect information from illicit use by encrypting it while it's in transit and at rest.
iv. *Useful training for security update:* To lower the possibility of human mistakes contributing to safety concerns, provide your staff with security education.

6.6 Best practices in cyber security

To lower the possibility of mistakes by humans contributing to safety concerns, provide your staff with security education. Putting in place solid safety precautions is crucial to protecting online assets, protecting confidential information, and ensuring that business continues in the face of the increasing prevalence of cyber threats like ransomware, malicious software, and information breaches.

Organizations may reduce the likelihood of attacks via the internet, limit any harm, and guarantee the safety, reliability, and accessibility of data and systems by following recognized cybersecurity guidelines. In the midst of constantly changing online threats, an integrated approach to cybersecurity is essential, ranging from training staff members about potential hazards to putting in place strong technological safeguards and emergency reaction protocols.

i. *Frequent Security Education:* Employee education should include dangers associated with cybersecurity standards of conduct, and ways to spot and handle scams like phishing texts and fraudulent websites.
ii. Encryption and Robust Passwords: To add even more protection, promote the usage of difficult passwords and, when practical, utilize multiple factor authentication (MFA).
iii. Maintain Software Updates: To guard toward identified flaws, make sure every piece of software—including computer systems, apps, and privacy solutions—is routinely upgraded and repaired.
iv. Put Minimum Privilege into Practice: To lessen the possibility of illicit access and escalation of authority, restrict user access privileges to what is essential for their positions and duties.
v. Encryption of Data: Whether it's being kept on systems or being sent across associations, sensitive data should always be encrypted to prevent unwanted access.

vi. Continual Backups: To make sure that important data is able to be restored in the case of a virus infection, information theft, or system crash, implement routine data changes and test reinstatement methods.

vii. Segmenting a network: Division the connections enable restricted access across various network segments, hence lowering the risk of security failures and restricting the ability of hackers to relocate laterally.

viii. Track and Record Activity: Employ monitoring and recording tools to keep an eye on system activities, internet traffic, and user conduct for indications of illicit activity or compromises to security.

ix. Incident Response Plan: Develop and regularly test an incident response plan to ensure a coordinated and effective response to security incidents, minimizing their impact on operations.

x. Risk Management for Vendors: Evaluate and control the risks to security that come from outside suppliers and suppliers of services, such as cloud computing services and counterparties in the chain of supply.

xi. Carry out routine security: Perform routine security evaluations, including unauthorized access and vulnerability examinations to find and fix holes in fortifications and networks.

xii. Culture of Security: Encourage workers to demonstrate ownership of their security duties by fostering a workplace atmosphere of cybercrime knowledge and awareness.

Corporations may improve their safety on the internet and fortify oneself toward a variety of cyber threats and assaults by adhering to these recommended procedures.

6.6.1 Multifactor authentication (MFA) and strong authentication methods

Passwords safeguard information stored online, yet they're insufficient. Skilled hackers frequently search for passwords. If someone finds out your password, they could be able to access different accounts that you may have employed it for. In the case that the login information is compromised, multi-factor authentication provides an extra degree of protection to keep unwanted people from obtaining these profiles [28].

In addition to enabling digital start and reducing the danger of human mistake, forgotten login credentials, and lost devices, multi-factor authentication also enhances protection response. Business regulations and customer information are used by adaptive multi-factor authentication to decide which authentication considerations to apply.

AI and ML are used by adaptive authorization systems to spot questionable actions during access to systems by analyzing trends. These software programs have the ability to track user activities continuously in order to see trends, create basic profiles for users, and spot strange behaviors like the following [29]:

i. Trying to log in during unexpected hours
ii. Trying to log in via strange places
iii. Trying to log in through unfamiliar devices

In accordance with company policies, machine learning algorithms modify several authentication indicators in real-time and provide risk ratings to questionable situations.

6.6.2 Regular software patching and updates

Frequent updates to software and patch management are essential for mitigating cyberattack since they fix software's identified security flaws, which constitutes one of its main benefits.

Cybercriminals frequently take advantage of these weaknesses to breach the organization's security and pilfer confidential data. Organizations may seal these holes while making it more harder for attackers to penetrate their IT infrastructure by routinely upgrading and updating software.

Ensuring that the systems of a company are operating on the most recent edition of the software is an additional advantage of frequent software upgrades and management of patches. Software developers frequently provide security patches along with fresh functions and improvements when they launch the latest versions of their products. Organizations may benefit from these fresh capabilities and improvements as well as remain safe from recognized security weaknesses by utilizing the latest available software version.

Organizations have to first determine every piece of computer software that use and maintain account of its variants, covering operating systems (OSes), programs, and various additional programs, before putting in place a routine software update and patch administration procedure.

After that, they ought to set out a timetable for patching and updating. While it is best practice to scan for modifications and install fixes immediately when they appear accessible, businesses might require testing updates prior to deploying them to live systems [30].

Let's get started first examine a few programs which are useful for both software and managing patches.

i. *System Center Configuration Manager (SCCM):* Firmware patches and updates on Windows-based machines may be managed with the help of a program produced by Microsoft called SCCM. It is an effective solution for handling software upgrades and patches on a wide scale since it enables enterprises to push upgrades and fixes to numerous systems at once.

ii. *Windows Server Update Services (WSUS):* Yet another Windows tool for managing fixes and updates to software on machines running Windows is called Windows System Updates (WSUS). It may be utilized to provide fixes and updates to a restricted quantity of systems and is intended for smaller enterprises.

iii. *Ansible:* Software updates and patches may be managed on a variety of platforms, notably Windows, Linux, and macOS, with the help of Ansible, which is a freely available automated tool. It is an excellent solution for handling patches and updates to software on an extensive basis since it enables enterprises to streamline the tasks of testing for updates and installing fixes.

To put it briefly, managing patches and routine software upgrades are essential for thwarting intrusions.

Businesses may significantly reduce the ease with which attackers can breach their computer networks by fixing programs known security flaws.

Organizations may benefit from improved capabilities and improvements in addition to maintain protection against recognized security weaknesses by using the most recent version of the program. The organizations may reduce the risk of a data breach and maintain the security of their IT infrastructure by utilizing technologies like SCCM, WSUS, and Ansible, as well as by having a routine updated software and patch management procedure.

6.7 Case studies: Real-world examples

Throughout the years, there have been a number of noteworthy incidents involving cybersecurity that have had a substantial influence on the impacted firms, their clients, and occasionally even the worldwide cybersecurity environment.

6.7.1 Notable cyber security breaches and their impact

Globally, there have been a number of noteworthy incidents related to cybersecurity that have had a big influence on people, governments, and businesses.

One of the earliest examples of a cyberattack physically destroying vital infrastructure was Stuxnet, an advanced cyberweapon that was found in 2010 and was directed against Iranian nuclear capacities [31]. It caused centrifuges to literally break. Hundreds of thousands of machines were affected by the 2017 ransomware called WannaCry outbreak, which encrypted information and demanded payments for ransom [29]. NotPetya, which surfaced in 2017, was a ransomware-disguised cyberattack that caused extensive damage to companies throughout the globe. 2020s SolarWinds supplier chain assault disrupted software upgrades, giving hackers unauthorized access to a large number of businesses and government organizations [31]. The JBS assault, which affected one of the biggest meat processing corporations in the world, and the Colonial Pipeline ransomware attack from 2021, which interrupted petroleum supply, are examples of more recent incidents. These incidents serve as a reminder of the growing risk that cyberattacks represent and the necessity of strong cybersecurity defenses to lessen their effects.

6.7.2 Successful cyber security implementations and lessons learned

Strong defenses, ongoing knowledge from previous events, and proactive actions are necessary to ensure effective cybersecurity deployments. Prioritizing tasks, risk evaluation, and capable management are essential components. Training for staff promotes knowledge about safety, and detection of threats and ongoing monitoring assist in quickly identifying and responding to events. A defense-in-depth strategy is ensured by the implementation of numerous levels of safeguarding and controls, together with frequent testing and inspections to find weaknesses. Conformance to compliance and legal guidelines, emergency response strategies, and cooperation among peers all help to minimize consequences. Constant resilience and preparedness against cybersecurity issues are ensured by constant development and response to new threats [13–16].

6.8 Future outlook and challenges

The prospects for cybersecurity are bright, but they are also fraught with difficulties as attacks get more complex and technology advances. These are some major developments and difficulties that are expected in cybersecurity:

- **i.** *Quickest Technological Progress:* The integration of cutting-edge technologies like computing via the cloud, ML, AI, and the IoT not only opens up new avenues for breakthroughs but also poses new security risks and attack methods.
- **ii.** *Advanced Online Threats*: The hackers are using more sophisticated tactics to target businesses of all kinds, including supply chain assaults, ransomware, and nation/state sponsored cyberattacks. The ubiquity of cybercrime-as-a-service models facilitates the initiation of complex assaults by adversaries.
- **iii.** *Zero-Day Attacks and Weaknesses*: Organizations are exposed to considerable risks when 0-day attacks addressing flaws that have not yet been patched or mitigated are released. It is difficult to find and fix vulnerabilities in software due to its growing intricacy and interconnectedness.
- **iv.** *Expertize Gap in Cybersecurity*: The number of cybersecurity experts without the knowledge and experience needed to handle new threats is steadily declining. Business entities face challenges in attracting and retaining skilled cybersecurity personnel, which results in deficiencies in their defensive posture and response to incident capabilities.
- **v.** *Data protection laws and privacy concerns*: Organizations have to set up effective information security strategies and assure that they comply with legal requirements due to increased worries about the security and confidentiality of data and the enforcement of laws like the California Consumer Privacy Act (CCPA) and the General Data Protection Regulation (GDPR) [1,29].
- **vi.** *Chain of Supply Security*: Attacks on supply chains that target collaborators and third-party suppliers present serious threats to businesses, as prominent examples like the SolarWinds hack have shown. In order to successfully manage threats, supply chain security demands more coordination, careful investigation, and monitoring.
- **vii.** *Cyberwarfare and Tensions in Geopolitics*: Cyberwarfare and reconnaissance by nation/state actors are still ongoing, and its targets include vital infrastructure, governmental institutions, and businesses. Cyberwarfare and reconnaissance by nation/state actors are still ongoing, and its targets include vital infrastructure, governmental institutions, and businesses.

viii. *IoT Security Difficulties*: The increased number of IoT devices poses novel threats to security because of their intrinsic weaknesses, insufficient security measures, and possible influence on vital infrastructure. Strong authentication, the use of encryption, and handling of device capabilities are necessary for IoT ecosystem security.

ix. *AI and ML in Cybersecurity*: Adversarial assaults and the possibility for driven by AI cyber threats are only two of the new difficulties. Machines that use AI and ML face, even while they have the ability to improve security measures through automated warning and mitigation.

x. The Human Factors and User Understanding: In cybersecurity events, human error continues to play a crucial role despite breakthroughs in technology. minimizing risks associated to humans requires educating individuals about cybersecurity hazards, raising knowledge about cybersecurity issues, and putting in place efficient security training activities.

A cooperative and multidisciplinary strategy including governments, industry clients, academic circles, and cybersecurity experts is needed to address these issues. Organizations may successfully manage evolving risks and better prepare for a changing cybersecurity environment by being proactive, putting money into skill development and cutting-edge technologies, and cultivating a mindset of cybersecurity knowledge and adaptability.

6.9 Conclusion

Cybersecurity elucidates the fundamental elements of cybersecurity, such as safeguarding computer networks, systems, and data against diverse cyberattacks. Information systems' security, reliability, and accessibility may be jeopardized by these kinds of risks, underscoring the necessity of strong cybersecurity defenses against possible dangers and weaknesses. Cybersecurity faces a wide range of difficulties from a variety of sources, including as the insiders, hackers, government agencies, and cyber criminals. Cybersecurity faces challenges from advanced cyberattacks, malware with AI capabilities, and defensive AI and machine learning. The complexity of encryption, cloud security concerns, and IoT technology vulnerabilities are further sources of the difficulty.

In the future, developing technologies like quantum computing for enhanced AI and ML, biometric authentication, and quantum-secure cryptography will be beneficial to information security. In the ever-changing cyber scene, these developments provide potential answers to pressing problems by strengthening defenses and enhancing detection and response powers.

References

[1] J. Kaur, R. Kumar K.R, The recent trends in cybersecurity: a review, Journal of King Saud University - Computer and Information Sciences 34 (8) (2021), https://doi.org/10.1016/j.jksuci.2021.01.018.

[2] P. Dixit, S. Silakari, Deep learning algorithms for cybersecurity applications: a technological and status review, Computer Science Review 39 (100317) (2021) 100317.

[3] T.D. Wagner, E. Palomar, K. Mahbub, A.E. Abdallah, A novel trust taxonomy for shared cyber threat intelligence, Security and Communication Networks 2018 (2018) 1–11.

[4] K. Pipyros, C. Thraskias, L. Mitrou, D. Gritzalis, T. Apostolopoulos, A new strategy for improving cyber-attacks evaluation in the context of Tallinn Manual, Computers and Security 74 (2018) 371–383.

[5] J. Srinivas, A.K. Das, N. Kumar, Government regulations in cyber security: framework, standards and recommendations, Future Generation Computer Systems 92 (2019) 178–188.

[6] R. Leszczyna, A review of standards with cybersecurity requirements for smart grid, Computers and Security 77 (2018) 262–276.

[7] M.N.A. Mhiqani, R. Ahmad, Z.Z. Abidin, N.S. Ali, K.H. Abdulkareem, Review of cyber-attacks classifications and threats analysis in cyber-physical systems, International Journal of Internet Technology and Secured Transactions 9 (3) (2019) 282.

[8] D.G. Padmavathi, M.D. Shanmugapriya, A Survey of Attacks, Security Mechanisms and Challenges in Wireless Sensor Networks, 2009 arXiv [cs.CR].

[9] M. Uma, G. Padmavathi, A survey on various cyber attacks and their classification, International Journal on Network Security 15 (2013) 390–396.

[10] A. Bendovschi, Cyber-attacks – trends, patterns and security countermeasures, Procedia Economics and Finance 28 (2015) 24–31.

[11] V. Gazis, A survey of standards for machine-to-machine and the internet of things, IEEE Communications Surveys and Tutorials 19 (1) (2017) 482–511. Firstquarter.

[12] C.M. Patterson, J.R.C. Nurse, V.N.L. Franqueira, Learning from Cyber Security Incidents: A Systematic Review and Future, 2023.

[13] W. Sema Admass, Y. Yayeh Munaye, A. Abeshu Diro, Cyber security: state of the art, challenges and future directions, Cyber Security and Applications 2 (2024) 100031.

[14] H. Kavak, J.J. Padilla, D. Vernon-Bido, S.Y. Diallo, R. Gore, S. Shetty, Simulation for cybersecurity: state of the art and future directions, Journal of Cybersecurity 1 (2021) 1–13, https://doi.org/10.1093/cybsec/tyab005.

[15] J. Kaur, K.R. Ramkumar, The recent trends in cyber security: a review, Journal of King Saud University of Computer, Information and Science 34 (8) (2022) 5766–5781, https://doi.org/10.1016/j.jksuci.2021.01.018.

[16] H.B. Mahajan, A.S. Rashid, A.A. Junnarkar, N. Uke, S.D. Deshpande, P.R. Futane, A. Alkhayyat, B. Alhayani, Integration of healthcare 4.0 and blockchain into secure cloud-based electronic health records systems, Applied Nanoscience 13 (3) (2023) 2329–2342, https://doi.org/10.1007/s13204-021-02164-0.

[17] M. Humayun, M. Niazi, N. Jhanjhi, M. Alshayeb, S. Mahmood, Cyber security threats and vulnerabilities: a systematic mapping study, Arabian Journal for Science and Engineering 45 (4) (2020) 3171–3189, https://doi.org/10.1007/s13369-019-04319-2.
[18] V.V. Vegesna, Enhancing cyber resilience by integrating AI-driven threat detection and mitigation strategies, Transactions on Latest Trends in Artificial Intelligence 4 (4) (2023).
[19] C. Sjöblom, Artificial Intelligence in Cybersecurity and Network Security, 2021.
[20] https://www.orangemantra.com/blog/wp-content/uploads/2022/08/AI-in-cybersecurity-scaled.jpg. (Accessed on April 2024).
[21] A. Yaseen, AI-driven threat detection and response: a paradigm shift in cybersecurity, International Journal of Integrated Care 7 (12) (December 2023) 25–43.
[22] Mardiana binti Mohamad Noor, W.H. Hassan, Current research on internet of things (IoT) security: a survey, Computer Networks 148 (2019) 283–294, https://doi.org/10.1016/j.comnet.2018.11.025. ISSN 1389-1286.
[23] H.Z. Yuchen Yang, L. Wu, G. Yin, L. Li, A survey on security and privacy issues in internet-of-things, in: 2015 10th International Conference of Internet Technology, and Secured Transactions, vol. 4, 2015, pp. 202–207.
[24] Y. Yang, L. Wu, G. Yin, L. Li, H. Zhao, A survey on security and privacy issues in internet-of-things, in: IEEE Internet of Things Journal, vol. 4, Oct. 2017, pp. 1250–1258, https://doi.org/10.1109/JIOT.2017.2694844.
[25] K. Chen, S. Zhang, Z. Li, Y. Zhang, Q. Deng, S. Ray, Y. Jin, Internet-of-Things security and vulnerabilities: taxonomy, challenges, and practice, Journal of Hardware and Systems Security 2 (2018) 97–110.
[26] https://venafi.com/blog/top-10-vulnerabilities-make-iot-devices-insecure/, (Accessed on May 2024).
[27] K. Thakur, M. Qiu, K. Gai, M.L. Ali, An investigation on cyber security threats and security models, in: Proceedings – 2nd IEEE International Conference on Cyber Security and Cloud Computing (CSCloud 2015), IEEE International Symposium of Smart Cloud, IEEE SSC 2015, 2016, pp. 307–311, https://doi.org/10.1109/CSCloud.2015.71.
[28] https://www.linkedin.com/pulse/evolving-landscape-cloud-security-aopartnersng-6wj9f/, (Accessed on May 2024).
[29] https://aws.amazon.com/what-is/mfa/#:~:text=Multi%2Dfactor%20authentication%20(MFA),question%2C%20or%20scan%20a%20fingerprint, (Accessed on May 2024).
[30] https://www.linkedin.com/pulse/importance-regular-software-updates-patch-management-celso-matule/, (Accessed on May 2024).
[31] https://em360tech.com/top-10/top-10-most-notorious-cyber-attacks-history, (Accessed on May 2024).

PART II

Case studies and practical techniques

CHAPTER 7

Frameworks for smart grid cyber security analysis

Muhammed Zekeriya Gündüz[1], Doygun Demirol[1], Resul Daş[2] and Kazım Hanbay[3]
[1]Department of Computer Science and Technology, Bingol University, Bingöl, Türkiye; [2]Software Engineering, Technology Faculty, Firat University, Elazığ, Türkiye; [3]Software Engineering, Inonu University, Malatya, Türkiye

7.1 Introduction

The integration of communication technologies into power systems has led to the emergence of smart grid applications [1]. However, this integration presents numerous cyber security challenges, highlighting the necessity for robust frameworks to safeguard critical infrastructure from evolving threats. Devices, communication networks, and network protocols are increasingly interconnected. So, ensuring data integrity, confidentiality, and availability is essential for smart grid security [2]. Therefore, developing holistic cyber security approaches is significant for widespread using smart grid applications.

Authentication and access control mechanisms are essential for regulating communication flows throughout the smart grid. By enforcing stringent authentication protocols and access restrictions, unauthorized access and malicious activities can be mitigated, strengthening the resilience against cyber threats. By deploying advanced detection technologies and proactive countermeasures, anomalies can be promptly determined and counteracted, thus preserving the integrity of grid operations [3].

By integrating lightweight cryptographic algorithms, the computational overhead on resource-constrained devices is minimized without compromising security robustness and safeguarding data integrity and confidentiality at the device level. Examining protocol vulnerabilities and implementing secure communication protocols are pivotal in mitigating risks associated with data manipulation, interception, and unauthorized

Cyber Security Solutions for Protecting and Building the Future Smart Grid
ISBN 978-0-443-14066-2
https://doi.org/10.1016/B978-0-443-14066-2.00002-5

access across network layers, spanning from the application layer to the MAC layer. Security testbed platforms play a significant role in facilitating real-world simulations and vulnerability assessments of smart grid applications. Testbed environments enable stakeholders to proactively identify and address vulnerabilities, thereby enhancing the overall resilience of smart grid systems [4]. Practical strategies combined with relevant case studies demonstrating real-world cybersecurity challenges and solutions in the smart grid are essential for understanding and application.

This chapter presents as a guide for stakeholders engaged in deploying smart grid systems, offering main insights into the developing robust cyber security frameworks essential for safeguarding critical infrastructure within an increasingly interconnected global landscape. Additionally, the chapter focuses on directing the attention toward designing and implementing frameworks tailored to address the unique security demands inherent to smart grid environments.

7.2 Authentication and access control for all communication flow throughout the system

Smart grids represent the next generation of electricity power grids, integrating digital technology to manage electricity flow more efficiently and securely. This integration signifies the modernization of traditional electricity systems. The modernization of these systems requires ensuring the confidentiality, integrity, and availability of data, often referred to as the CIA triad [5]. Authentication and access control emerge as critical security requirements for all communication flows across the system, playing a pivotal role in ensuring the CIA triad against potential cyberattacks [6].

Authentication involves confirming the identity of a system, device, or user before allowing access to resources. Authentication ensures that only authorized entities, such as users, devices, or applications, can access critical components and perform specific actions in smart grid applications. General authentication methods include passwords, cryptographic keys, biometrics, and multi-factor authentication [7]. The choice of the method depends on the level of security required for different components within the smart grid. Table 7.1 summarizes various types of authentication mechanisms with their pros and cons.

Access control involves defining and managing permissions to restrict or grant access to specific resources based on the authenticated identity and the authorization level. Access control is crucial for limiting access to sensitive

Table 7.1 Types of authentication mechanisms along with pros and cons.

Authentication mechanisms	Pros	Cons
Password-based authentication	Easy to implement and use Familiar to most users	Vulnerable to brute force attacks Users may choose weak passwords
Digital certificates and PKI	High level of security through encryption Scalable across large networks	Complex to implement Requires ongoing management
Biometric authentication	Highly secure, using unique physical characteristics Difficult to forge or replicate	Can be expensive to implement Raises privacy concerns
Two-factor authentication	Adds an extra security layer More secure than password-only systems	Can be inconvenient for users Dependent on a second device or factor
Token-based authentication	Tokens can be easily revoked, enhancing security Useful for temporary access	Tokens can be lost or stolen Requires additional infrastructure

data, control systems, and infrastructure. It prevents unauthorized manipulation of operations and protects against cyberattacks. Access control policies can be granular, specifying who can access which resources under which conditions [8]. This level of granularity ensures that different entities have the appropriate level of access for their respective roles. Table 7.2 shows the types of access control mechanisms.

Smart grids often involve various devices and technologies from different vendors. Ensuring interoperability of authentication and access control mechanisms is essential for seamless integration. Smart grid systems need to accommodate a growing number of connected devices and users. Authentication and access control solutions must be scalable to handle the increasing complexity of the grid infrastructure [9]. Robust authentication and access control are critical for the resilience of smart grids, especially in the face of potential cyber threats. Continuous monitoring and adaptive security measures are essential for a resilient system.

Table 7.2 Types of access control mechanisms.

Access control mechanisms	Pros	Cons
Role-based access control (RBAC)	Simplifies managing user permissions Easy to understand and implement	Lack of flexibility in dynamic environments Role creep can occur over time
Attribute-based access control (ABAC)	Highly flexible and granular control Adaptable to changing conditions	Can be complex to configure Higher risk of misconfiguration
Mandatory access control	Extremely secure, used in high-security environments Centralized control over permissions	Inflexible, can hinder operational efficiency Complex to implement and manage

Various standards and frameworks, such as those developed by the NIST, the International Electrotechnical Commission (IEC), and other industry bodies, provide guidelines for implementing authentication and access control in smart grid systems [10]. Compliance with these standards helps ensure that security measures are consistent and effective across different smart grid implementations. The smart grid's dynamic and evolving nature requires authentication and access control solutions that can scale to accommodate a growing number of connected devices and users [11]. Fig. 7.1 shows a typical authentication and access control system.

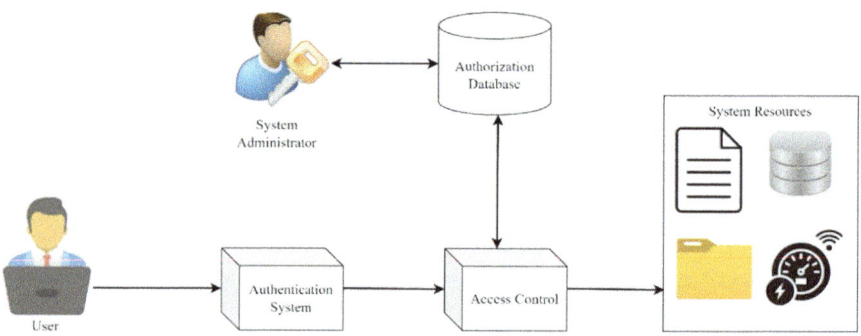

Figure 7.1 Typical authentication and access control system.

7.3 Attack detection and countermeasures everywhere in smart grid

Cyber threats stemming from various forms of malware, such as viruses and trojans present significant risks to the fundamental principles of the CIA triad within smart grid applications. Deliberately injected malware like logic bombs and backdoors represent potential vulnerabilities for future attacks. Denial of Service (DoS) attacks disrupt the flow of information while spoofing attacks exploit weaknesses in communication lines. Zero-day attacks target previously unknown vulnerabilities, adding complexity to security challenges. As a passive attack, eavesdropping compromises the confidentiality of sensitive information. On the other hand, threats such as data tampering, flooding, and other forms of manipulation undermine the integrity and availability of smart grid applications [12].

Cyber attacks on smart grids can be categorized under several headings [13]. Physical attacks involve physical harm to the infrastructure, such as sabotage of power lines or substations. Network-based attacks disrupt the communication networks, affecting data transmission and grid operations. Data integrity attacks involve tampering with data, leading to incorrect decisions and potential grid instability. The last two types of attacks can be categorized under the heading of cyber attacks. Combined attacks are sophisticated and involve a blend of physical and cyber tactics, posing a significant threat to grid security. Addressing these cyber security challenges requires the implementation of robust and adaptive security measures. There are many attack detection mechanisms developed against attacks in smart grids [14].

Anomaly detection in smart grids involves monitoring grid operations to identify deviations from normal patterns [15]. These deviations could signal potential attacks or system failures. Statistical methods and machine learning algorithms are employed for more accurate and real-time anomaly detection [16]. The statistical analysis utilizes historical data to establish a baseline of normal behavior and flag significant deviations. Machine learning models are used to automatically detect anomalies in real-time by training models on data patterns [17]. Signature-based detection relies on identifying known attack patterns or signatures within the system. It is effective against previously encountered threats. Maintaining and regularly updating a database of known attack signatures is a fundamental practice in cyber security. Real-time monitoring continuously scans network traffic and system activities with this database. In addition to these mechanisms, hybrid detection methods are also used in smart grids. Hybrid systems integrate anomaly and signature-based approaches, leveraging the advantages of each to establish a more comprehensive and efficient detection framework.

On the other hand, smart grid security can be categorized based on the CIA triad principles. The principles, traditionally critical in information technologies (IT), are equally essential in the context of smart grid application. While basic security requirements in IT are expressed as CIA, this order is AIC in the context of smart grid. Table 7.3 summarizes cyberattacks affecting the CIA triad, protocols, applications, and components while offering possible countermeasures in smart grid applications [18].

A smart grid application employs various cyber security strategies to defend against cyberattacks. Encryption and authentication are pivotal in safeguarding data integrity and confidentiality during transmission. Technical solutions such as firewalls, antiviruses, encryption systems, Demilitarized Zone (DMZ), Virtual Private Networks (VPN), Intrusion Detection Systems (IDS), early warning systems, and dynamic reconfiguration systems help prevent suspicious activities. Regular software updates and patch management ensure that all components remain current with the latest security patches. From a security management perspective, comprehensive solutions should encompass risk assessments for assets during and after cyberattacks, facilitate the exchange of security policies, manage cryptographic keys, report on security incidents, and conduct vulnerability assessments. An effective defense strategy integrates diverse security methodologies, leveraging advanced technologies such as controlled wireless propagation, machine learning, robust authentication mechanisms, certification processes, network segmentation, and the proactive deployment of real-time IDS/IPS [19]. Moreover, these solutions should encompass flexible, robust, and scalable security measures while ensuring seamless smart grid operations [20].

A secure framework necessitates the strict implementation of crucial elements:

- Robust access control and authentication should be implemented for all communication flow across the system.
- Attack detection and mitigation techniques should be integrated throughout the infrastructure to enhance overall security.
- It should be ensured that lightweight cryptographic functions should be incorporated into each node.
- Detailed attention should be paid to the design of network protocol security, covering aspects from the MAC to the application layer.

Table 7.3 Effect of cyberattacks and countermeasures in smart grid [18].

Cyber attack	Compromised protocol, application, component	Compromised security requirements	Possible countermeasures
Social engineering, traffic analysis	DNP3, Modbus	Confidentiality	SSL/TLS, authentication, secure DNP3, PKI, encryption, machine learning based IDS
Scanning service/ Port/IP vulnerabilities	DNP3, Modbus	Confidentiality, availability	Automated security compliance checks, SIEM, IDS
Trojan, worm, virus	Control device, PMU, SCADA	CIA triad	IDS, IPS, SIEM, antivirus, DLP, signature-based prevention
DoS	SCADA, PMU, RTU, AMI, state estimator, communication channel	Availability	IDS, IPS, sensing time measurement, reconfiguration methods, pushback, signal strength, flow entropy, transmission failure count, SIEM
Man-in-the-middle	DNP3, SCADA, AMI, PLC, HMI	Integrity, confidentiality	SSL/TLS, authentication, secure DNP3, encryption, PKI
Jamming channel	Communication channel, PMU, RTU, AMI, PLC, smart devices, data concentrator	Availability	Anti-jamming

Continued

Table 7.3 Effect of cyberattacks and countermeasures in smart grid [18].—cont'd

Cyber attack	Compromised protocol, application, component	Compromised security requirements	Possible countermeasures
Replay attack	PLC, authentication schemes within AMI, SCADA, IED	Integrity, confidentiality	SSL/TLS, authentication, secure DNP3, encryption, PKI, neural network based IDS
Popping the HMI	Substations, SCADA, EMS	CIA triad	Automated security compliance checks, SIEM, DLP, antivirus, IDS
Violation of integrity	RTU, EMS, SCADA, AMI, smart meter, state estimator, data concentrator, PLC	Availability, integrity	SSL/TLS, secure DNP3, authentication, encryption, PKI, IDS
Masquerade	PLC	CIA triad	PKI, secure DNP3, IDS, DLP, SIEM, SSL/TLS, authentication, encryption
Violation of privacy	Smart meter, DMS	Confidentiality	SSL/TLS, secure DNP3, authentication, encryption, PKI
Backdoor	SCADA	CIA triad	SIEM, antivirus, machine learning based IDS, IPS

AMI, Advanced Metering Infrastructure; *DLP*, Data Loss Prevention; *DMS*, Distribution Management System; *DNP*, Distributed Network Protocol; *EMS*, Energy Management System; *HMI*, Human-Machine Interface; *IDS*, Intrusion Detection System; *IED*, Intelligent Electronic Device; *IP*, Internet Protocol; *IPS*, Intrusion Prevention System; *PKI*, Public Key Infrastructure; *PLC*, Programmable Logic Controller; *PMU*, Phasor Measurement Unit; *RTU*, Remote Terminal Unit; *SCADA*, Supervisory Control and Data Acquisition; *SIEM*, Security Information and Event Management; *SSL*, Secure Sockets Layer; *TLS*, Transport Layer Security.

- Cyber security oriented testbed platforms should be applied systematically to identify and understand vulnerabilities within energy infrastructures with clarity and precision.

Ensuring comprehensive security for all aspects of the smart grid against cyberattacks poses a significant challenge. The communication system should actively identify and detect abnormal scenarios triggered by cyberattacks, continuously monitoring network flow. Monitoring methods like comparison, testing, and profiling can facilitate this process. Furthermore, the network architecture should incorporate self-healing capabilities to maintain network operations even in a cyberattack. Given the critical role of energy systems, the stability of communication networks is vital for ensuring system availability. Implementing effective security measures requires analytical frameworks capable of accurately modeling cyber and physical environments. These frameworks must account for the complex interdependencies among decision-making processes and network components. Analytical tools play a crucial role in modeling and assessing the cyber security aspects of smart grids, addressing areas such as information security, energy system security, and reliability assessment [21].

Vulnerability assessment methodologies rely on probabilistic risk assessment (PRA) and play a pivotal role in determining security levels within energy control systems. These levels are quantified by assessing the probability of cyber security events and the resulting energy loss, drawing on historical events and statistical patterns. PRA facilitates a quantitative evaluation of the impact of cyberattacks on energy systems. However, assessing the potential for large-scale DoS attacks against smart grid applications poses challenges due to insufficient historical data for profiling, and distinct DoS attacks may have varying priorities within the system. Techniques based on security metrics and graph theory encounter scalability issues, making them better suited for smaller-scale energy systems such as substation networks. In security metric-based assessments, each Intelligent Electronic Device (IED) is assigned a score based on recognized cyber threat identifications and corresponding countermeasures. Holistic security evaluation of a substation network is derived through the aggregation of scores attributed to individual IEDs [22].

A graph-based analysis is employed to depict how attacks impact energy systems. This approach determines common relationships among attack targets, outputs, and defense tactics graphically, using decision-making mechanisms to evaluate the probability and impact of attacks on energy systems. Graph-based techniques are advantageous for designing network connectivity among smart grid devices. In this context, nodes represent devices such as generators, routers, smart meters, and transformers, while edges depict the connections between these components. Modeling the connectivity can represent physical or logical links between devices. Thus, graph-based methodologies offer valuable insights into the characteristics of smart grid applications, exploring connectivity patterns and assessing the propagation of cyberattacks within smart grid environments [23].

7.4 Node wise basic and light-weight cryptographic functions

Cryptography is vital to ensure secure data transmission in smart grid environments. However, the resource limitations inherent in smart grid devices, encompassing constraints in power, memory, and battery capacity, necessitate focusing on "lightweight cryptography" within the Advanced Metering Infrastructure (AMI). Lightweight cryptographic algorithms are designed to secure data efficiently while imposing minimal demands on resources. Traditional cryptographic algorithms, designed for more resource-constrained environments, are not conducive to implementation in the constrained devices prevalent in smart grid applications [24].

The primary attribute of lightweight cryptography is its low resource requirement, including less processing power, memory, and energy consumption. Cryptographic functions can be easily integrated into various devices within the smart grid, regardless of their processing capabilities. Despite their minimalistic design, lightweight cryptographic functions do not compromise security. They provide robust protection against common cyber threats [25].

The advent of lightweight cryptographic functions is rooted in the imperative for algorithms that deliver security with an optimal utilization of resources. This resource utilization encompasses block size, number of rounds, memory usage, execution time, structure, and key size. Lightweight algorithms aim to balance performance, resource efficiency, and cryptographic robustness [26]. This balance is crucial for effective operation in resource-constrained environments while maintaining security. Diverse lightweight algorithms, including both stream and block ciphers as well as hash functions such as SIMON, TWINE, QARMA, SPECK, and PRESENT have been proposed as viable choices to their conventional counterparts like AES, RSA, and SHA3.

Light-weight cryptography involves algorithms designed for environments with limited computational resources, typical of many smart grid nodes. Essential criteria include low computational complexity, minimal memory requirements, and energy efficiency. Symmetric key algorithms like AES, LIGHT, SIMON, or PRESENT are popular in smart grid applications and require less computational power for encryption and decryption processes. Lightweight hash functions, like QUARK or SIPHASH, are employed for data integrity checks. They balance speed and security, ensuring data integrity during transmission. While public key cryptography is conventionally resource-intensive, lightweight alternatives like Elliptic Curve Cryptography (ECC) provide a balance between sufficient security and resource constraint [27].

Cryptographic functions are chosen based on the specific capabilities and roles of each node within the smart grid. A node-specific approach ensures optimal performance and security. Lightweight cryptographic functions are integrated with existing smart grid protocols to enhance security without significant alterations to the system architecture. Node-wise cryptographic functions encompass the implementation of security algorithms at the granularity of individual nodes within a smart grid, playing a fundamental role in ensuring secure communication and protecting data against potential cyber threats [28].

The primary challenge lies in balancing the trade-off between security and computational efficiency. Ongoing research is focused on developing more advanced lightweight cryptographic methods that do not compromise on security. As cyber threats evolve, cryptographic functions must also evolve. Continuous development and updates are necessary to maintain the integrity of smart grid networks.

7.5 Security of network protocols from the application layer to the MAC layer

Ensuring the security of network protocols within smart grids is essential, encompassing aspects from the application layer down to the Medium Access Control (MAC) layer. Protocols for smart grid applications can vary depending on specific requirements and technologies. Authentication systems are essential to ensure that only authorized entities can access and manage critical applications. Encryption techniques are often employed to secure the integrity and confidentiality of data exchanged between smart grid components. Moreover, adhering to secure application programming practices helps reduce vulnerabilities and potential exploits [3].

At the application layer, various security measures are implemented to safeguard the smart grid's software and communication processes. Smart grid application layer protocols define how devices and systems communicate and exchange information within the smart grid. These practices operate above lower-level communication protocols, concentrating on the interpretation and organization of the data.

In transitioning to the transport layer, the implementation of secure communication protocols becomes paramount. This layer ensures the integrity and confidentiality of data during transmission. Moreover, anomaly detection mechanisms may be integrated at this level to identify and mitigate potential threats, enhancing the overall resilience of the smart grid against malicious activities.

The network layer in smart grids focuses on routing and addressing, and security measures are applied to safeguard the routing infrastructure and prevent unauthorized access to critical network resources. Firewalls and IDS/IPS systems are implemented to monitor and control traffic, ensuring that only legitimate communication occurs within the smart grid network. Network segmentation and isolation strategies further enhance security by limiting the impact of a potential breach.

At the MAC layer, which manages access to the shared communication medium, security protocols are designed to prevent unauthorized access. Using secure key exchange protocols and authentication mechanisms helps ensure that only authorized devices can participate in the communication network. Additionally, secure MAC layer protocols contribute to the prevention of various attacks, including replay attacks and man-in-the-middle attacks.

In smart grids, some important protocols stand out for ensuring robust and secure communication. Open Smart Grid Protocol (OSGP) offers a structured approach for various applications. IEC 61850 focuses on substation automation and protection communication. DLMS/COSEM enables meter data management. IEEE 2030.5 enables interoperability between distributed energy resources and utility systems. DNP3 is used for Supervisory Control and Data Acquisition (SCADA) communication in distribution systems. Transport Layer Security (TLS) is utilized to establish encrypted and authenticated connections between devices within the smart grid. OpenADR enables utilities to manage energy demand by sending price signals to smart devices. Zigbee is used for low-power communication between smart meters and home energy management systems. Modbus is a simple protocol for communication between industrial devices often used in microgrids. Secure Hash Algorithm (SHA) is used for digital signatures and data integrity verification. MQTT, a lightweight messaging protocol, is increasingly being adopted for machine-to-machine communication within smart grid applications. Many other protocols can be tailored to specific requirements and functionalities. Furthermore, there is a growing trend in leveraging blockchain technology to achieve secure and transparent data management [21–23].

Providing the security of network protocols involves a multi-layered approach that spans from the application layer to the MAC layer in smart grids. Using robust security measures at each layer is crucial for protecting the smart grid infrastructure against cyber threats, assuring the reliability and resilience of the entire system. As cyber threats evolve, cryptographic functions must also evolve. Integrating security across these layers involves dealing with complex and sometimes conflicting requirements. There is a need for a balanced approach that ensures security without compromising the performance and functionality of the smart grid.

7.6 Cyber security testbed platforms implementation to investigate the vulnerabilities of the power infrastructures

Cyber security testbeds are critical tools for simulating power infrastructure and assessing its vulnerabilities to cyberattacks. These platforms enable researchers and security professionals to conduct controlled experiments, identify weaknesses, and develop mitigation strategies without risking real-world systems. Testbeds are classified into three types. Hardware-in-the-loop (HIL) testbeds replicate physical components like smart meters, control systems, and generators. Offer the most realistic simulations but are expensive and complex. Software-in-the-loop (SIL) testbeds model the behavior of power grid components using software simulations. They are more flexible and cost-effective but may not capture all real-world complexities. Hardware-based testbeds require significant investment, while software-based options are more affordable. Hybrid testbeds combine both hardware and software elements, providing a balance between realism and flexibility.

Smart grid testbeds are specifically designed to prioritize addressing security and privacy issues, which results in a notable emphasis on research in these areas. Cyber security testbeds primarily aim to explore vulnerabilities and conduct research on wide-area situational awareness. Security concerns are thoroughly analyzed regarding various components involved in smart grid data acquisition and control, such as Phasor Measurement Units (PMUs), Remote Terminal Units (RTUs), Programmable Logic Controllers (PLCs), smart meters, and Intelligent Electronic Devices (IEDs). In addition to addressing security concerns related to specific components like RTUs, IEDs, PMUs, PLCs, and smart meters, challenges related to network appliance security are also addressed in smart grid systems. These challenges encompass various aspects such as encryption methods, firewalls, potential attack scenarios, intrusion analysis techniques, routers, deployment of countermeasures, and forensic analysis methodologies. Addressing these challenges is crucial for safeguarding the integrity and resilience of the smart grid infrastructure against cyber threats and attacks [29–31].

The cyber security of the smart grid is of critical importance and is currently one of the highest research priorities. Current operational systems cannot execute various types of cyber security experiments aimed at identifying vulnerabilities and implementing solutions. Smart grid security testbeds serve as controlled environments specifically designed for conducting a range of cyberattack and defense experiments. These testbeds utilize a hybrid approach, combining real, simulated, and emulated components to closely replicate real-world conditions. Smart grid security testbeds are essential due to the inherent limitations of traditional simulation-based tools, which fail to adequately capture the complex dynamics of control, communication, and physical systems within a unified environment [32].

Smart grid testbeds serve as platforms for exploring various research areas, including cyber security, wide-area situational awareness, network communications, demand response, and distribution grid management. These testbeds are particularly valuable for conducting cyber security experiments and training exercises, aiding in improving preparedness and response capabilities against cyber threats to critical infrastructures. While we aimed to provide comprehensive information on existing security-oriented smart grid testbeds, it's important to note that our coverage may not encompass all available platforms. Additionally, the capabilities of these testbeds may extend beyond those highlighted in this section.

Table 7.4 provides a comprehensive overview of various smart grid testbeds, highlighting their key attributes and capabilities in terms of network tools, smart grid domains, testing type, communication protocols,

Table 7.4 Security-oriented testbeds for smart grid with main features [33].

Testbed	Tools	Domains	Type	Protocols	Tech.	Type	Security testing capabilities
Virtual power system testbed	Rinse simulator	Operation, transmission	Simulation	SCADA-specific protocols, DNP3	Ethernet	WAN LAN	DoS and IP-based attacks
TASSCS	Opnet	Operation, transmission	Simulation	Modbus, DNP3	Ethernet	LAN WAN	ARP spoofing, unauthorized access, DoS attacks, man-in-the-middle
University College Dublin Testbed	–	Operation, transmission	Hybrid	ICCP, DNP3, IEC 61850, OPC UA	Ethernet	LAN	DoS attacks, anomaly detection
ScadaSim	Omnet++	Operation, service provider	Simulation	Modbus, DNP3	Ethernet	LAN WAN	DoS attacks, ARP spoofing, eavesdropping, man-in-the-middle

Continued

Table 7.4 Security-oriented testbeds for smart grid with main features [33].—cont'd

Testbed	Tools	Domains	Type	Protocols	Tech.	Type	Security testing capabilities
Deterlab	Emulab	–	Hardware	–	Ethernet	WAN LAN	Worm/botnet attacks, pattern detection, DDoS attacks, encryption
SCADA Security Lab	–	Transmission, operations	Real-time simulation	Modbus, C37.118, IEC 61850, DNP3	Cloud, RS232, Ethernet	WAN LAN	–
NIDS	–	Operation	Hybrid	Modbus	Ethernet	LAN WAN	DoS and flood attacks
Cyber security testbed for IEC61850	–	Operation	Real-time simulation	IEC 61850, C37.118	Ethernet	LAN	Man-in-the-middle, database attack, ARP spoofing, DoS ttacks
Cyber physical testbed	–	Operation, transmission	Real-time simulation	IEC 61850, C37.118, DNP3	Ethernet	WAN	–

Vscada	Nmap, Ns-2, Opnet, Wireshark	Operation, transmission	Simulation	OPC DA/ UA	Ethernet	WAN	DoS attacks, cryptography
State University of Newyork	–	Transmission, operations	Hybrid	C37.118, OPC DA	Ethernet Wi-Fi Cloud	WAN	Privacy attacks
Sandia national lab	Wireshark, Opnet	Transmission, operations	Simulation	DNP3, Modbus	Ethernet RS232	WAN LAN	Man-in-the-middle, precision insider, rogue software
FIU testbed	LabVIEW	Customer, operations, transmission	Hardware	IEC 61850, DNP3, Modbus, OPC UA, C37.118	Cloud, RS232, Ethernet	LAN WAN	Privacy attacks

network technologies, network type, and main cyber security testing capabilities. (−) indicates that there is no information related to that feature.

The Virtual Power System Testbed, TASSCS (Telecommunications and Software Systems Cyber Security), DeterLab, ScadaSim, and University College Dublin testbeds primarily focus on cyber security. Meanwhile, the SCADA Security Lab, Network Intrusion Detection System (NIDS), and Cyber Security testbed for IEC61850 are centered around cyber security and network communications. VSCADA (Virtual SCADA), State University of New York, Cyber Physical Testbed, and Sandia National Laboratories (VCSE) are designed for wide area situational awareness and cyber security purposes. Finally, the FIU (Florida International University Smart Grid Testbed) encompasses objectives related to wide area situational awareness, cyber security, network communications, and distribution grid management [33].

7.7 Related case study

Securing the increasingly interconnected smart grid is crucial for ensuring reliable and resilient power delivery. Several frameworks were developed to manage and mitigate cyber security risks in this complex environment. One notable case study regarding frameworks for smart grid cyber security analysis involves the implementation of the National Institute of Standards and Technology (NIST) Cyber security Framework (CSF) within the context of smart grid infrastructure. The NIST-CSF provides comprehensive guidelines, standards, and best practices to manage and mitigate cyber security risks across various sectors, including energy. The NIST profile adapts the broader NIST-CSF to the specific needs of the smart grid. It provides a voluntary, flexible, and risk-based approach to identifying, prioritizing, and implementing cyber security controls across the five core functions of detect, protect, respond, identify, and recover [34].

Utility companies and energy regulators have increasingly embraced the NIST-CSF to assess and strengthen the cyber security stance of smart grid systems. This framework provides a systematic approach for identifying and prioritizing cyber security activities, covering risk assessment, threat detection, incident response, and continuous improvement. By adopting the NIST-CSF, stakeholders in the smart grid ecosystem establish a unified framework and language for evaluating cyber security risks and implementing effective countermeasures. This alignment fosters collaboration among utilities, technology providers, government agencies, and other stakeholders to collectively address emerging vulnerabilities. Additionally, integrating the NIST-CSF with existing regulatory requirements, like the North American Electric Reliability Corporation Critical Infrastructure Protection (NERC-CIP) standards, ensures compliance with regulations while addressing the specific challenges of the smart grid environment. Frameworks like the NIST-CSF reflect a proactive approach to tackling cyber security issues in the smart grid sector, enabling organizations to enhance their defenses and safeguard critical energy infrastructure [35].

According to NIST, the smart grid consists of seven interconnected domains, each playing a vital role in the efficient functioning [36]. Fig. 7.2 depicts the different domains within the smart grid system and their interconnections, along with brief descriptions.

1. Generation: This domain focuses on the generation of energy from various energy sources such as fossil fuels (coal, natural gas, oil), renewable sources (solar, wind, hydroelectric, geothermal), and nuclear energy.
2. Transmission: Transmission systems are responsible for transporting energy from generation facilities to distribution networks over long distances.
3. Distribution: The networks in distribution domain serve as the link between transmission systems and end-users, facilitating the delivery of energy to residential, commercial, and industrial customers. Furthermore, these networks incorporate distributed energy resources and storage systems to improve grid resilience and flexibility.

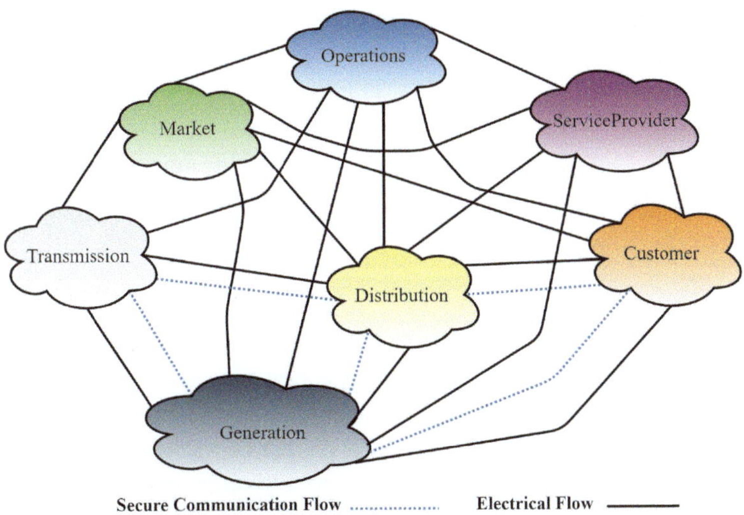

Figure 7.2 NIST smart grid conceptual model [34].

4. Customer: The customer domain encompasses the end-users of energy, including residential, industrial, and commercial consumers. They have the capability to generate, store, and manage their energy usage, contributing to grid stability and efficiency.
5. Market: The energy industry operates within two main markets. Transmission market facilitates the trading of transmission rights. Energy market enables the buying and selling of energy and related products.
6. Service Provider: Service providers offer various services to both customers and utilities, including the installation of smart devices, customer management, and smart building solutions. These services often include features such as demand response capabilities and advanced billing systems.
7. Operation: The operation domain focuses on managing electricity flow within the grid to ensure reliability, safety, and cost-effectiveness. This involves activities such as power system protection, monitoring, analysis, and control.

Although each domain performs unique functions and holds specific responsibilities, they are interconnected and dependent on shared necessities such as communication protocols, networking infrastructure, security measures, and communication mediums. These shared components enable smooth interaction and cooperation among diverse domains, enhancing the overall efficiency and resilience of the smart grid system.

The NIST has identified nine priority research areas within the realm of smart grid development [34]:

1. Demand Response and Consumer Energy Efficiency: Research in this area focuses on strategies and technologies to enable consumers to actively participate in managing their energy consumption and promoting energy efficiency.
2. Advanced Metering Infrastructure (AMI): This research area explores the deployment and optimization of advanced metering systems, including smart meters, to enable real-time monitoring and management of electricity usage.
3. Distributed Energy Resources (DER): DER research investigates the integration and optimization of distributed energy resources such as solar panels, wind turbines, and energy storage systems into the grid to enhance reliability and resilience.
4. Wide-Area Situational Awareness: This area involves the development of technologies and methodologies to enhance grid monitoring and

situational awareness over large geographical areas, enabling proactive response to grid disturbances and emergencies.

5. Distribution Grid Management: Research in distribution grid management focuses on improving the efficiency, reliability, and resilience of distribution networks through advanced monitoring, control, and optimization techniques.
6. Network Communications: This research area addresses the development of robust and secure communication infrastructures to support the reliable exchange of data and control signals within the smart grid ecosystem.
7. Electric Transportation: Electric transportation research explores the integration of electric vehicles and charging infrastructure into the grid, including grid-to-vehicle and vehicle-to-grid technologies to support grid balancing and optimization.
8. Energy Storage: Research in energy storage aims to advance the deployment and utilization of energy storage technologies such as batteries, flywheels, and pumped hydro to enhance grid stability, reliability, and renewable energy integration.
9. Cyber security: This priority area focuses on developing robust cyber security solutions to protect smart grid infrastructure, data, and communication networks from cyber threats and vulnerabilities.

Each of these research areas is critical to the advancement and deployment of smart grid technologies and plays a key role in improving grid efficiency, reliability, and sustainability. Collaboration among researchers, industry stakeholders, and policymakers is essential to address the challenges and opportunities within these priority areas effectively.

The field of cyber security research indeed involves strategies aimed at ensuring the CIA triad of both information and communication technology and control systems necessary for managing, operating, and safeguarding the energy, information technology, and telecommunication infrastructures within the smart grid. Additionally, cyber security and network communication case studies are evaluated across all research areas within the smart grid infrastructure.

7.8 Conclusion

The frameworks presented for smart grid cyber security analysis encompass a comprehensive set of measures aimed at ensuring the security and resilience of smart grid systems. By emphasizing authentication and access control mechanisms, communication flows throughout the system are safeguarded against unauthorized access and potential threats. Pervasive attack detection and countermeasures ensure proactive mitigation of cyber threats across the entire smart grid infrastructure. Additionally, node-wise primary and lightweight cryptographic functions enhance the security of transactions and communications at the node level. Thorough scrutiny and fortification of network protocols from the application layer to the MAC layer play a crucial role in identifying and addressing vulnerabilities within the communication infrastructure. Cyber security testbed platforms provide a controlled environment for investigating vulnerabilities within power infrastructures, enabling the simulation of attacks and evaluation of security measures. Incorporation of a related case study demonstrates the practical application of the framework, offering insights into real-world scenarios and highlighting its effectiveness in enhancing smart grid cyber security. The frameworks serve as a valuable resource for researchers, practitioners, and policymakers to develop and manage cyber security measures for smart grid systems.

References

[1] M. Emmanuel, R. Rayudu, Communication technologies for smart grid applications: a survey, Journal of Network and Computer Applications 74 (2016) 133–148, https://doi.org/10.1016/j.jnca.2016.08.012.

[2] M.Z. Gunduz, R. Das, Smart grid security: an effective hybrid CNN-based approach for detecting energy theft using consumption patterns, Sensors 24 (4) (2024), https://doi.org/10.3390/s24041148. Art. no. 4.

[3] M.O. Qays, I. Ahmad, A. Abu-Siada, M.L. Hossain, F. Yasmin, Key communication technologies, applications, protocols and future guides for IoT-assisted smart grid systems: a review, Energy Reports 9 (Dec. 2023) 2440–2452, https://doi.org/10.1016/j.egyr.2023.01.085.

[4] M.H. Cintuglu, O.A. Mohammed, K. Akkaya, A.S. Uluagac, A survey on smart grid cyber-physical system testbeds, IEEE Communications Surveys and Tutorials 19 (1) (2017) 446–464, https://doi.org/10.1109/COMST.2016.2627399.

[5] M.Z. Gunduz, R. Das, Cyber-security on smart grid: threats and potential solutions, Computer Networks 169 (Mar. 2020) 107094, https://doi.org/10.1016/j.comnet.2019.107094.

[6] T. Chen, X. Yin, G. Wang, Securing communications between smart grids and real users; providing a methodology based on user authentication, Energy Reports 7 (Nov. 2021) 8042–8050, https://doi.org/10.1016/j.egyr.2021.08.125.

[7] G. Ali, M.A. Dida, A. Elikana Sam, A secure and efficient multi-factor authentication algorithm for mobile money applications, Future Internet 13 (12) (Dec. 2021), https://doi.org/10.3390/fi13120299. Art. no. 12.

[8] Y. Nakamura, Y. Zhang, M. Sasabe, S. Kasahara, Exploiting smart contracts for capability-based access control in the internet of things, Sensors 20 (6) (Jan. 2020), https://doi.org/10.3390/s20061793. Art. no. 6.

[9] I. Singh, B. Singh, Access management of IoT devices using access control mechanism and decentralized authentication: a review, Measurement: Sensors 25 (Feb. 2023) 100591, https://doi.org/10.1016/j.measen.2022.100591.

[10] R. Leszczyna, A review of standards with cybersecurity requirements for smart grid, Computers and Security 77 (Aug. 2018) 262–276, https://doi.org/10.1016/j.cose.2018.03.011.

[11] M. Palmer, G.J.W. Kathrine, S. Jebapriya, Comprehensive analysis of smart grid security with intelligent machine learning based framework, in: 2021 7th International Conference on Advanced Computing and Communication Systems (ICACCS), 2021, pp. 307–313, https://doi.org/10.1109/ICACCS51430.2021.9442000.

[12] M.Z. Gunduz, R. Das, Analysis of cyber-attacks on smart grid applications, in: 2018 International Conference on Artificial Intelligence and Data Processing (IDAP), 2018, pp. 1–5, https://doi.org/10.1109/IDAP.2018.8620728.

[13] T.D. Le, A. Anwar, S.W. Loke, R. Beuran, Y. Tan, Grid attack sim: A cyber attack simulation framework for smart grids, Electronics 9 (8) (2020), https://doi.org/10.3390/electronics9081218. Art. no. 8.

[14] B.B. Gupta, T. Akhtar, A survey on smart power grid: frameworks, tools, security issues, and solutions, Annales des Telecommunications 72 (9) (Oct. 2017) 517–549, https://doi.org/10.1007/s12243-017-0605-4.

[15] J. Ding, A. Qammar, Z. Zhang, A. Karim, H. Ning, Cyber threats to smart grids: review, taxonomy, potential solutions, and future directions, Energies 15 (18) (Jan. 2022), https://doi.org/10.3390/en15186799. Art. no. 18.

[16] M.K. Hasan, R.A. Abdulkadir, S. Islam, T.R. Gadekallu, N. Safie, A review on machine learning techniques for secured cyber-physical systems in smart grid networks, Energy Reports 11 (Jun. 2024) 1268–1290, https://doi.org/10.1016/j.egyr.2023.12.040.

[17] M.Z. Gündüz, R. Daş, An effective DNN-based approach for detecting energy theft in smart grids through user consumption patterns, Turkish Journal of Nature and Science 12 (4) (2023), https://doi.org/10.46810/tdfd.1383065. Art. no. 4.

[18] M.Z. Gunduz, R. Das, Cyber-attack process and defense strategies in smart grid applications, Journal of Network Communications and Emerging Technology 11 (11) (2021).

[19] H.M. Rouzbahani, H. Karimipour, L. Lei, Multi-layer defense algorithm against deep reinforcement learning-based intruders in smart grids, International Journal of Electrical Power and Energy Systems 146 (Mar. 2023) 108798, https://doi.org/10.1016/j.ijepes.2022.108798.

[20] M. Ghiasi, T. Niknam, Z. Wang, M. Mehrandezh, M. Dehghani, N. Ghadimi, A comprehensive review of cyber-attacks and defense mechanisms for improving security in smart grid energy systems: past, present and future, Electric Power Systems Research 215 (Feb. 2023) 108975, https://doi.org/10.1016/j.epsr.2022.108975.

[21] A.O. Otuoze, M.W. Mustafa, R.M. Larik, Smart grids security challenges: classification by sources of threats, Journal of Electrical Systems and Information Technology 5 (3) (Dec. 2018) 468–483, https://doi.org/10.1016/j.jesit.2018.01.001.

[22] S. Mishra, K. Anderson, B. Miller, K. Boyer, A. Warren, Microgrid resilience: a holistic approach for assessing threats, identifying vulnerabilities, and designing corresponding mitigation strategies, Applied Energy 264 (Apr. 2020) 114726, https://doi.org/10.1016/j.apenergy.2020.114726.

[23] I.B. Sperstad, E.H. Solvang, S.H. Jakobsen, A graph-based modelling framework for vulnerability analysis of critical sequences of events in power systems, International Journal of Electrical Power and Energy Systems 125 (2021) 106408, https://doi.org/10.1016/j.ijepes.2020.106408.

[24] N. Abosata, S. Al-Rubaye, G. Inalhan, Lightweight payload encryption-based authentication scheme for advanced metering infrastructure sensor networks, Sensors 22 (2) (2022), https://doi.org/10.3390/s22020534. Art. no. 2.
[25] R. Leszczyna, Standards on cyber security assessment of smart grid, International Journal of Critical Infrastructure Protection 22 (2018) 70–89, https://doi.org/10.1016/j.ijcip.2018.05.006.
[26] M. El-hajj, H. Mousawi, A. Fadlallah, Analysis of lightweight cryptographic algorithms on IoT hardware platform, Future Internet 15 (2) (2023), https://doi.org/10.3390/fi15020054. Art. no. 2.
[27] D. Sadhukhan, S. Ray, M.S. Obaidat, M. Dasgupta, A secure and privacy preserving lightweight authentication scheme for smart-grid communication using elliptic curve cryptography, Journal of Systems Architecture 114 (2021) 101938, https://doi.org/10.1016/j.sysarc.2020.101938.
[28] S.U. Jan, A. Ghani, A. Alzahrani, S.M. Saqlain, K. Yahya, H. Sajjad, Bandwidth and power efficient lightweight authentication scheme for healthcare system, Journal of King Saud University - Computer and Information Sciences 35 (7) (2023) 101601, https://doi.org/10.1016/j.jksuci.2023.101601.
[29] A.S. Sani, D. Yuan, J. Jin, L. Gao, S. Yu, Z.Y. Dong, Cyber security framework for internet of things-based energy internet, Future Generation Computer Systems 93 (2019) 849–859, https://doi.org/10.1016/j.future.2018.01.029.
[30] M.S. Alkatheiri, M.A. Alqarni, S.H. Chauhdary, Cyber security framework for smart home energy management systems, Sustainable Energy Technologies and Assessments 46 (2021) 101232, https://doi.org/10.1016/j.seta.2021.101232.
[31] A. Akkad, G. Wills, A. Rezazadeh, An information security model for an IoT-enabled smart grid in the Saudi energy sector, Computers and Electrical Engineering 105 (2023) 108491, https://doi.org/10.1016/j.compeleceng.2022.108491.
[32] A.A. Smadi, B.T. Ajao, B.K. Johnson, H. Lei, Y. Chakhchoukh, Q. Abu Al-Haija, A comprehensive survey on cyber-physical smart grid testbed architectures: requirements and challenges, Electronics 10 (9) (2021), https://doi.org/10.3390/electronics10091043. Art. no. 9.
[33] M.Z. Gunduz, R. Das, A comparison of cyber-security oriented testbeds for IoT-based smart grids, in: 6th International Symposium on Digital Forensic and Security (ISDFS), 2018, pp. 1–6, https://doi.org/10.1109/ISDFS.2018.8355329.
[34] NIST Framework and Roadmap for Smart Grid Interoperability Standards 4.0, NIST, 2021 [Online]. Available: https://www.nist.gov/el/smart-grid-menu/wanted-inputs-nist-framework-and-roadmap-smart-grid-interoperability-standards-40. (Accessed 28 December 2023).
[35] M. Hussain, An Effective Cybersecurity Risk Assessment Framework for a Public Sector Gas Production/Distribution Company, National College of Ireland, Dublin, 2023 [Online]. Available: https://norma.ncirl.ie/6520/. (Accessed 9 February 2024).
[36] J.J. Moreno Escobar, O. Morales Matamoros, R. Tejeida Padilla, I. Lina Reyes, H. Quintana Espinosa, A comprehensive review on smart grids: challenges and opportunities, Sensors 21 (21) (2021), https://doi.org/10.3390/s21216978. Art. no. 21.

CHAPTER 8

Cyberattack detection techniques for smart grid

Shyam V. Aradhye, Ved S. Bilaskar, Snehal S. Shinde, Deepak D. Kshirsagar and Pushparaj R. Nimbalkar
Department of Computer Science and Engineering, School of Computational Sciences, COEP Technological University, Pune, Maharashtra, India

8.1 Introduction

The intricate web of networks known as an electric grid, transports energy from places of production, such as power plants, to end users, including homes and businesses. The conventional American electrical grid [1] was established more than a century ago and is based on the one-way movement of power from a source to a destination. But now, energy production, storage, and saving on the grid are being altered by technology, which is also creating the opportunity for a growing smart energy infrastructure. The smart grid [1] is just one component of the new smart energy ecosystem that is being formed by the integration of intelligent Internet of Things (IoT) sensors, microgrids, digitization, distributed renewable energy sources, and automation. Specifically, IoT-powered smart grid technology [1] is an important instrument for the secure and sustainable energy future we require. Smart grids can optimize energy solutions through flexible management of different energy supplies, enhanced operational insights, improved maintenance predictions, and increased security when connected to the conventional energy infrastructure.

The rise of the IoT has totally changed how we see and interact with the world around us. Thought to be futuristic technology, it has remarkably influenced our daily lives and is now a big part of it. This interconnected system, with seamless communication between devices, has found an application in the Industrial Internet of Things (IIoT), where its impact on industrial processes is profound. IIoT has emerged as a focal point, designed to meet the needs of industrial processes. As industries increasingly keep on

Cyber Security Solutions for Protecting and Building the Future Smart Grid
ISBN 978-0-443-14066-2
https://doi.org/10.1016/B978-0-443-14066-2.00001-3

adopting IoT technologies, the integration of IIoT becomes essential revolutionizing industrial processes.

Ensuring robust security in IIoT is crucial due to the high stakes involved. Potential weaknesses could put physical systems in industrial settings at serious risk in addition to data intrusions. To protect against these hazards, comprehensive security measures are essential. These measures include encrypted communication routes and Intrusion Detection Systems (IDS). The dynamic nature of the IoT emphasizes the necessity of a proactive and flexible security strategy that can change to meet new threats. Cyberattacks against industrial systems are becoming more sophisticated and frequent as a result of the integration of IIoT technology. Hackers exploit vulnerabilities in IIoT networks to disrupt operations, steal data, or cause physical harm. Common attack vectors include malware injections, ransomware attacks, and denial-of-service incidents. Vigilance against these attacks is crucial, requiring a proactive cybersecurity strategy that anticipates and counters evolving threats in real-time.

In the context of IIoT, Machine Learning (ML) models play an important role in extracting insights from dynamic datasets. This chapter delves into three categories of ML models: Rules-based, Trees-based, and Bayesian-based models. Rules-based models operate on predefined conditions and are effective at detecting known attack patterns. Trees-based models excel at identifying complex and nonlinear patterns associated with attacks and interpret using decision-making processes. Bayesian-based models utilize probability theory to update beliefs based on new information, making them adaptive to emerging threats.

The goal of this study is to investigate the benefits and impacts of using various ML models on the Edge-IIoT dataset. By delving into the nuances of this dataset, we aim to study ML models for industrial applications. The need to address the particular difficulties posed by IIoT and provide insightful information that can guide the creation of ML models that are best suited for industrial environments is what drives this research. By doing this, we hope to significantly contribute to the advancement of ML, IIoT, and IoT synergy, ultimately influencing the development of intelligent and effective industrial processes in the future.

Our remaining work is organized as follows: Section 8.2 summarizes the relevant studies. The model being proposed is detailed in Section 8.3. Section 8.4 elaborates on the implementation and analysis of the results obtained. Finally, Section 8.5 concludes the entirety of our work.

8.2 Related work

The two-step system proposed in Ref. [2] aimed to work on smart grid security, calling it a high risk application. In the first stage, the main focus was on detecting attacks accurately and efficiently. While the second stage analyzed available data to predict specific attack class. It made use of DNP3 dataset and Recursive Feature Elimination (RFE) to cut down to 12 features from 96. It achieved an accuracy of 99.5% with Random Forest (RF) and 98.82% with Extreme Gradient Boosting (XGB).

The work in Ref. [3] studied various ML techniques from academic databases to implement cybersecurity countermeasures against attacks in smart grid cyber-physical systems (SG-CPS). It concluded that Deep Learning (DL) is the most common technique for attack detection. This encourages the use of classifier models for this chapter.

The Authors of [4] proposed an Intrusion Detection Scheme for Smart Power-Grid. For it, they used the publicly available 15 Mississippi State University and Oak Ridge National Laboratory (MSU-ORNL) datasets. Application of Binary Gray Wolf Optimization based Feature Selection significantly reduced the datasets to contain only relevant features. They also proposed the Optimized Ensemble classification approach which gave the best results for all datasets after feature selection.

In the research article [5], the proposed Intrusion detection method involved use of XGB and long short-term memory (LSTM). It also implemented a proposed Bayesian-based method to optimize XGB parameters. The method outperformed other algorithms and gave an accuracy of 89.21%. After having looked at Smart grid Intrusion Detection Systems, we now switch our attention to IIoT in specific.

A key component of assessing IDS has been the KDD Cup 1999 dataset [6]. Its main advantage is the variety of attacks it covers, which offers a thorough standard by which to measure the effectiveness of detection techniques. KDD99's relevance in the field of cybersecurity research stems from its goal of simulating real-world network traffic and security problems, despite its synthetic production. Its large dimensionality, however, presents difficulties for several ML algorithms and calls for the application of feature selection methods. Furthermore, the dataset's depiction of current cyber-threats may have been limited by its gradual aging.

NSL-KDD [6], an improvement over the KDD Cup 1999 dataset, aimed to solve some of the shortcomings of its forerunner. Especially, NSL-KDD provides a smaller feature set, which helps to address the problem of high dimensionality and makes the analysis more manageable. The representation of both normal and attack instances in NSL-KDD is enhanced by efforts to achieve a balanced class distribution. NSL-KDD's depiction of real-world circumstances is impacted by the artificial aspects it still has. Furthermore, the dataset might not encompass all contemporary cyberthreats, which would restrict its usefulness for some study goals.

The UNSW-NB15 [7] dataset, which was created to simulate actual network traffic conditions, has become well-known for its emphasis on contemporary cybersecurity issues. For supervised learning methods, the dataset's well-labeled examples offer a trustworthy ground truth. UNSW-NB15 is not without limits, though, especially when it comes to the types of attacks it does not address. Though realistic, the dataset might not cover as much ground in terms of attacks as the KDD Cup 1999 dataset, which could limit its applicability. Moreover, UNSW-NB15 focuses mostly on network traffic.

One notable addition to the realm of IDSs in the IoT and IIoT domains is the TON_IoT dataset [8]. Understanding the growing threat landscape around IoT applications and the imperative need for efficient security measures, the authors give a carefully crafted dataset that includes network traffic, operating system logs, and telemetry data from various IoT/IIoT services. This extensive dataset addresses a significant void in the current research by providing labeled data with normal and attack classes, as well as sub-classes for multi-classification problems. It is produced from a genuine medium-scale network at the UNSW Canberra Cyber Range and IoT Labs. Numerous IoT/IIoT scenarios, including simulated appliances like smart refrigerators, garage doors, motion lights, GPS trackers, thermostats, Modbus services, and weather monitoring systems, are added to the collection. The TON_IoT dataset is a useful tool for assessing and improving the precision and effectiveness of IDSs designed especially for IoT and IIoT contexts because of its wide range of sources and emphasis on real-world IoT applications. The freely accessible dataset has the potential to further cybersecurity research in this quickly developing field.

Although the X-IIoTID [9] dataset represents a significant addition to the field of IIoT security research, it is critical to recognize several inherent limitations in its architecture. A significant limitation of the present edition is its emphasis on cyberthreats, which may mean that physical and cyber-physical attacks against field equipment are not fully covered. Furthermore, even with efforts to guarantee connectivity-agnostic features, specific IIoT contexts may have subtleties that need to be taken into account further for improved interoperability. Like any dataset, X-IIoTID is a snapshot in time, and the dynamic nature of growing cyber threats may have an impact on its applicability. Additionally, even while the dataset is excellent at providing a variety of attack types and up-to-date IIoT protocols, periodic updates may be necessary to preserve its representativeness given how quickly cybersecurity is evolving. Although thorough, the empirical evaluation and analysis carried out on X-IIoTID [9] may need continuous review and validation in a variety of settings to guarantee the generalizability of its conclusions. Perspective of these constraints helps to develop a more nuanced perspective of the dataset's usefulness and potential areas for improvement. It also lays the groundwork for future modifications.

The WUSTL-IIoT-2021 dataset [10] is a significant resource with distinct features that is designed for cybersecurity research in the industrial Internet of things (IIoT) area. The dataset, which was created from an IIoT testbed with the goal of accurately simulating industrial systems in the real world, contains 1,194,464 observations altogether. It has 1,107,448 normal samples and 87,016 attack samples, which were purposefully skewed to represent actual industrial conditions. The dataset contains a variety of attack types, with DoS making up over 90% of the attacks. Other attack types include reconnaissance, command injection, and denial of service. Based on statistical data, the percentage of attack traffic is 7.28% and the regular traffic is 92.72%. This intentionally unbalanced distribution is meant to replicate actual industrial control system situations. The features of the dataset include 41 unique characteristics that are typical of network flows and were carefully chosen based on changes during attack stages compared to normal operation periods. These features cover a wide range of topics, including traffic rates, jitters, packet and bytes counts, and protocol details.

By contrasting it with benchmark datasets that fulfilled comparable functions, the Edge-IIoT dataset can be analyzed. Analysis of datasets like X-IIoTID, WUSTL, TON-IoT, NSL-KDD, UNSW-NB15, and KDD Cup 99 is essential for that. We build upon Edge-IIoTset, which addresses the shortcomings of the benchmark datasets while introducing new features to make it a perfect dataset for IIoT IDS research, taking into account all the datasets stated above.

8.3 Proposed architecture

Fig. 8.1 depicts the flow diagram of the architecture pipeline.

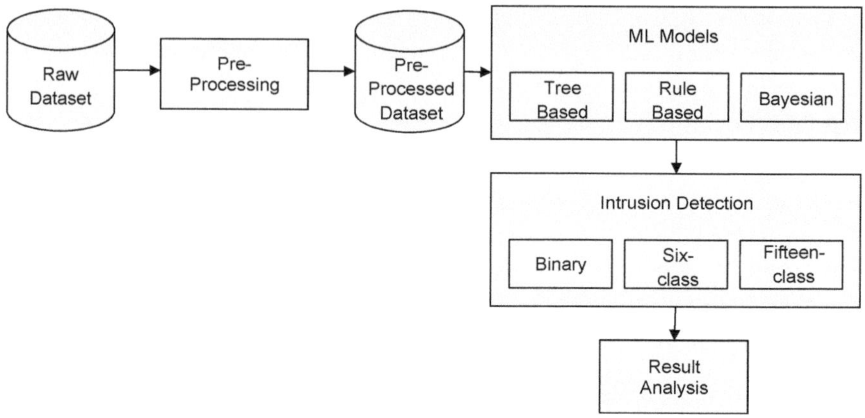

Figure 8.1 The architecture pipeline flow diagram.

Data Pre-processing: Pre-Processing is a crucial step in ML pipelines to ensure data quality for model effectiveness. It involves various tasks, including handling missing data by substituting with median, mean, or mode; identifying and removing duplicate entries to prevent bias, dropping unnecessary features, and encoding categorical variables using one hot or label encoding.

ML Models: The pre-processed dataset, obtained after pre-processing the raw data, undergoes evaluation using various ML models. The results, encompassing accuracy, True Positive (TP) rate, False Positive (FP) rate, mean absolute error, and Model Build-up Time (MBT), are assessed. To outperform others, a model should exhibit high accuracy and TP rate, along with low FP rate, mean absolute error, and MBT. The data is classified into Binary and multiclass, which includes Six-class and Fifteen-class classification. Various models from different model categories are applied and tested. The model category defines the specific algorithm or methodology utilized for evaluation.

The ML model categories and their respective models utilized in our study are:

1. *Rule Based models*: The field of Rule-Based Machine Learning (RBML) includes techniques for identifying, learning, and evolving 'rules' for the storage, processing, and application of information. These interpretable and transparent models are appropriate for jobs that need rationality. In contrast to conventional rule-based systems, RBML uses learning algorithms in place of manual rule creation to automatically find rules from data [11].

- Decision Table: This decision backing tool depicts decision logic using a table of conditions and corresponding actions. The table lists the associated action or result for each row, which represents a distinct set of criteria.
- JRip (Java RIPper): It uses the RIPPER algorithm—Repeated Incremental Pruning to Produce Error Reduction—as its foundation. JRip classifies instances by repeatedly developing rules, which are subsequently pruned to increase precision and simplify the algorithm.
- OneR (One Rule): This method chooses a single rule based on a single data set attribute, usually the one that yields the greatest accuracy when applied alone. Next, new instances are categorized using the chosen rule and the attribute value.
- PART (Partial Rule-Based Classifier): It combines the advantages of decision tree and rule-based methodologies. It functions by first constructing a decision tree, which is subsequently divided into rules that are used to categorization.
- ZeroR (Zero Rule): It functions by forecasting, for each occurrence in the test data, the class that occurs the most frequently in the training data. Although ZeroR is not appropriate for intricate classification tasks, it can function as a reference point to evaluate the efficacy of more sophisticated algorithms.

2. *Tree Based models*: Based on input variables, tree-based models estimate target values using a decision tree structure. In order to build the tree, the models repeatedly divide the input data into subsets and assess the computational efficiency and prediction accuracy of the various branches. It is possible to reduce tree depth and complexity by changing the variable splitting order. Important features are notably elevated to the top of the tree hierarchy while insignificant ones are lowered. Because of their interpretability, scalability to huge datasets, and little need for data preprocessing, tree-based models are preferred [12].
   - Decision Stump: This is a basic decision tree with two leaf nodes and one decision node. It is a mediocre learner that is frequently employed as a foundational element in group learning techniques like boosting. Although decision stumps have a limited ability to convey ideas, they can inevitably identify certain patterns in the data.
   - Hoeffding Tree: This approach uses a decision tree to handle streaming data that can be too big to fit in memory. By dividing nodes according to the Hoeffding bound, which offers theoretical assurances on the split's accuracy, it progressively constructs a decision tree.

- J48: It seeks to optimize the efficacy of the resultant subsets by iteratively splitting the data according to the information gain criterion in order to construct a decision tree. J48 finds widespread application in classification tasks.
- LMT (Logistic Model Trees): This hybrid classification approach blends logistic regression models and decision trees. More adaptable and comprehensible categorization is made possible by fitting logistic regression models to the leaf nodes of a decision tree structure. When working with data that shows nonlinear correlations between characteristics and the goal variable, LMT is especially helpful.
- Random Forest: This ensemble learning technique builds several decision trees and aggregates the predictions from each to arrive at a conclusion. A random subset of characteristics is taken into consideration for splitting at each node, and each tree in the forest is trained using a bootstrap sample of the data.
- Random Tree: This straightforward decision tree technique creates a single tree by choosing characteristics at each node at random for splitting. While it constructs a single tree rather than an ensemble, it is comparable to Random Forest.
- REP Tree (Reduced Error Pruning Tree): Is a decision tree method that divides the data into subsets according to attribute values in a recursive manner before creating the tree. It employs reduced error pruning to eliminate nodes that do not substantially increase the overall accuracy in order to enhance the tree's generalization performance.

3. *Bayesian Models*: Based on the Bayes' Theorem, Bayesian Machine Learning (BML) estimates the posterior distribution ($p(\theta \mid x)$) from the prior distribution ($p(\theta)$) and likelihood ($p(x \mid \theta)$). While classical machine learning aims to maximize the likelihood, Bayesian Machine Learning (BML) aims to optimize the posterior distribution. To do this, a prior distribution ($p(\theta)$) over the parameters of the model must be incorporated, conveying past opinions of their appropriateness [13].
   - Bayes Net: A directed acyclic graph is used to depict a set of variables and their conditional relationships in a probabilistic graphical model. The graph's edges show the probabilistic relationships between the nodes, which each represent a random variable.
   - Naive Bayes: This straightforward probabilistic classifier relies on the independence, or "naïve," between features and is based on Bayes' theorem. Naive Bayes is frequently utilized in a variety of

classification applications and works effectively in practice despite its simplicity and simplistic premise.

- Naive Bayes Multinomial: This classifier variation, which represents counts or frequencies of events, is especially made for multinomially distributed data. It uses a multinomial distribution to describe the likelihood of each feature value and makes the assumption that features are conditionally independent given the class label.
- Naive Bayes Multinomial Text: This variation is designed specifically for text classification applications. It makes the assumption that the characteristics correspond to word frequencies or occurrences in documents and uses a multinomial distribution to estimate the likelihood of each word.
- Naive Bayes Multinomial Updateable: An improvement on the Naive Bayes Multinomial classifier, this permits incremental learning so that novel information may be added to the model without requiring it to be retrained. This is especially helpful in situations where the model needs to adjust to shifting distributions or patterns as new data becomes available over time.
- Naive Bayes Updateable: This classifier is capable of being updated with recent data and supports incremental learning. For different types of data, several forms of updateable Naive Bayes classifiers may be utilized, depending on the particular assumed distributions and feature representations.

Binary Classification: All machine learning models are employed to categorize the dataset as either attack or normal, also known as Binary classification. This basic detection method solely determines whether traffic is malicious. However, it lacks insights into specific attack types, vulnerabilities, and repercussions. Subsequently, further classification is pursued to identify specific attack types. One best-fit model within each category is selected, aiding in identifying the most effective model for each specific ML model category Consequently, only these selected best ML models from each category should be applied for multiclass classification, given their proven enhancement.

Six-class and Fifteen-class Classification: In contrast to Binary classification, further classifying attacks into specific categories provides deeper insights into network attributes, compromised areas, and aids in developing a more secure Intrusion Detection System (IDS). This approach also highlights the system's vulnerability to specific attack types, as demonstrated in the Six-class classification. The Fifteen-class classification offers a detailed analysis of incoming traffic, extending beyond Binary classification. It involves interpreting features more deeply, enhancing the analysis process. The Six-class and Fifteen-class classifications are detailed in Table 8.1. Only the selected ML models from the 'Selecting the best-fit model from each category' step are to be utilized for this purpose.

Result Analysis: Following the Binary, Six-class, and Fifteen-class classifications, the model exhibiting superior performance across all classifications is selected. This model surpasses others in accuracy, TP rate, FP rate, mean absolute error and MBT, establishing itself as the optimal choice for our dataset.

8.4 Implementation and result analysis

For implementation, the system utilized macOS as the operating system, equipped with a 1.8 GHz Intel i5 dual-core processor, 8 GB RAM, and 1600 MHz DDR3 memory. Experimentation was conducted using Weka-3.8.6, Google Colab, and Python 3.

Dataset: In this study, Edge-IIoT dataset [14] is being used. The data collection process, tools, devices and technologies used for dataset preparation is as follows. The data collection environment consisted of interconnected layers such as the Blockchain layer, Software Defined Networking (SDN) layer, perception layer, etc. Advanced technologies like

Table 8.1 Various classes for classification.

Binary classification	Six-class classification	Fifteen-class classification
Normal	Normal	Normal
Attack	Injection	SQL injection
		Uploading attack
		Cross-site scripting (XSS)
	MITM	MITM
	DoS/DDoS	TCP flood
		UDP flood
		ICMP flood
		HTTP flood
	Malware	Password cracking
		Backdoor attack
		Ransomware
	Information gathering	Vulnerability scanning
		Port scanning
		OS fingerprinting

the Things Board IoT platform, Modbus TCP/IP, Mosquitto MQTT brokers, etc. were employed to generate heterogeneous networks. The network traffic was analyzed using the Tshark and Zeek tools. These tools aided in selecting 61 out of the original 1176 features for the final dataset. A total of 14 distinct attacks are detected in the dataset, which can be categorized into five groups. Thus, the data can be classified into Binary (attack or normal), Six-class (five attacks and one normal), and Fifteen-class (fourteen attacks and one normal), as illustrated in Table 8.1.

The major five attack categories and their description, as given in Table 8.1:

1. *Injection Attacks*: When malicious code targets sensors, protocols, or controllers in IIoT environments, it is inserted into data or commands to be input to alter processes, interfere with operations, or cause harm.
2. *Man-in-the-Middle (MITM) Attacks*: Attacks affect device communication and industrial processes by intercepting communications to compromising data security and integrity. This allows attackers to eavesdrop, alter, or impersonate devices.
3. *Denial-of-Service (DoS) and Distributed Denial-of-Service (DDoS) Attacks*: Involves flooding target systems with excessive traffic in an attempt to disrupt services. This can cause production delays, system outages, and monetary losses in the, which can impact IoT/IIoT devices, networks, and control systems.
4. *Malware Attacks*: Assaults involve the use of malicious software to breach systems, devices, or network infrastructure, causing harm or unauthorized access. This poses a danger of sabotage, data theft, and other attacks.
5. *Information Gathering Attacks*: Include gathering information about target networks or infrastructure in order to find possible points of exploitation and vulnerabilities. This information gives attackers the ability to design and carry out complex operations that jeopardize IIoT infrastructure, assets, and environments.

Data Preprocessing: Initially, the raw dataset comprised 157,800 instances and 61 features, with two label columns designated for Binary and Fifteen-class classification. Additionally, a third label column, Six-class, was derived by grouping the Fifteen-class into respective attacks, as outlined in Table 8.1 Although no missing values were detected, a few duplicate entries were identified and subsequently removed. Unnecessary features such as timestamps and ports as specified in Ref. [14], were also eliminated. Features with singular values were dropped. Ordinal categorical features were label encoded, while nominal were one hot encoded. Consequently, the pre-processed dataset comprised 152,028 instances and three label columns. The preprocessed dataset obtained is then used for ML model application.

Binary Classification: The preprocessed data was evaluated using all available Rule, Tree based, and Bayesian models within Weka-3.8.6, employing 10-fold cross-validation to optimize data utilization, ensure stable performance, and facilitate hyper parameter tuning. Performance metrics including accuracy, TP rate, FP rate, mean absolute error, and MBT were recorded for each model, as summarized in Table 8.2. The Binary classification classifies each instance as attack or normal.

After analyzing Table 8.2, PART from Rule-based models with an accuracy of 97.5162%, J48 from Tree-based models with an accuracy of 97.5176%, and Bayes Net from Bayesian models with an accuracy of 97.4235% were chosen. These models exhibited the highest accuracy, TP rate, and lowest FP rate, along with low mean absolute error and model build-up time, compared to other models in their respective categories. Thus, these three models are deemed the most effective within their categories. Moving forward, only these models were utilized for Six-class and Fifteen-class classification due to their proven efficiency.

Six-class Classification: The best-fit models selected from each category were employed for Six-class classification, and their performance is detailed in Table 8.3. This classification offers detailed insights into each attack category, namely injection, MITM, DoS/DDoS, malware, and information gathering. It provides a more comprehensive analysis compared to Binary classification, enhancing the understanding of the dataset's intricacies. Among the three models, J48 stands out with the highest accuracy of 90.9477%, TP rate 0.909, and lowest FP rate 0.22, mean absolute error 0.0385 and moderate MBT of 45.11 s.

Fifteen-class Classification: The same best-fit models selected from each category were utilized for Fifteen-class classification, and their performance is summarized in Table 8.4. This classification provides a detailed view of the exact attack, unlike the attack categories in the Six-class classification and the Binary classification. Here, J48 achieves the highest accuracy of 89.1106%, TP rate 0.891, and low FP rate 0.013, mean absolute error 0.0175 and moderate MBT 42.36 s, compared to others.

Result Analysis: Upon reviewing the results of the Binary, Six-class, and Fifteen-class classification experiments, the J48 ML model from the Tree based category consistently demonstrates superior performance. It achieves the highest accuracy of 97.5176%, 90.9477%, and 89.1106% in the

Table 8.2 Binary classification results.

ML model category	ML model	Accuracy	TP rate	FP rate	Mean absolute error	MBT (s)
Rules based	Decision table	97.5136	0.975	0.132	0.0423	52.66
	JRip	95.8238	0.958	0.222	0.0796	26.6
	OneR	94.0643	0.941	0.309	0.0594	2.01
	PART	97.5162	0.975	0.132	0.0390	63.73
	ZeroR	84.1503	0.842	0.842	0.2668	0.03
Tree based	Decision stump	87.4806	0.875	0.665	0.2180	3.57
	Hoeffding Tree	97.3715	0.974	0.137	0.0414	4.75
	J48	97.5176	0.975	0.131	0.0408	28.72
	LMT	91.8068	0.918	0.405	0.1278	711.41
	RandomForest	96.6460	0.966	0.118	0.0387	311.6
	Random Tree	96.2020	0.962	0.106	0.0384	5.36
	REPTree	97.4939	0.975	0.130	0.0382	17.54
Bayesian	BayesNet	97.4235	0.974	0.135	0.0367	8.02
	NaiveBayes	41.1062	0.411	0.111	0.5890	1.46
	NaiveBayes Multinomial	44.0932	0.441	0.254	0.5591	0.19
	NaiveBayes MultinomialText	84.1503	0.842	0.842	0.2668	0.08
	NaiveBayes Multinomial Updateable	44.0932	0.441	0.254	0.5591	0.08
	NaiveBayes Updateable	41.1062	0.411	0.111	0.5890	1.39

Table 8.3 Six-class classification results.

ML model category	ML model	Accuracy	TP rate	FP rate	Mean absolute error	MBT (s)
Rule based	PART	90.9247	0.909	0.22	0.0378	161.97
Tree based	J48	90.9477	0.909	0.22	0.0385	45.11
Bayesian	Bayes Net	79.5334	0.795	0.66	0.0672	10.73

Table 8.4 Fifteen-class classification results.

ML model category	ML model name	Accuracy	TP rate	FP rate	Mean absolute error	MBT (s)
Rule based	PART	89.0691	89.1	1.3	0.0175	197.2
Tree based	J48	89.1106	89.1	1.3	0.0175	42.36
Bayesian	Bayes Net	84.1042	84.1	1.1	0.0207	8.49

Binary, Six-class, and Fifteen-class classifications, respectively. Additionally, J48 exhibits the high TP rate and the low FP rate, mean absolute error and moderate MBT compared to the other ML models utilized, as indicated in Tables 8.2–8.4. Therefore, it can be concluded that the J48 Tree based ML model is the optimal choice for the Edge-IIoT dataset in this study, proving effective for both Binary and multiclass classification tasks.

8.5 Conclusion

IoT is crucial to the smart grid since it enhances resource management, operational insights, and other areas. IIoT is its next evolution, and IDS development is required for it. The chapter examines which machine learning model (ML) suits the Edge-IIoT dataset the research is using. The raw data is preprocessed first, followed by Binary classification. Among the 3 ML model categories, Rule based, Tree based, and Bayesian, models with the highest accuracy, TP rate, and low FP rate, mean absolute error, MBT were selected. PART from Rule-based models achieved an accuracy of 97.5162%, J48 from Tree-based models achieved 97.5176% accuracy, and Bayes Net from Bayesian models attained 97.4235% accuracy. These best models were then applied for multiclass classification into six and fifteen classes due to their proven efficiency. J48 achieved the highest accuracies of 90.9477%, and 89.1106% in Six-class, and Fifteen-class classifications, respectively. Thus, upon analyzing Binary, Six-class, and Fifteen-class classifications, the J48 model from the Tree based category emerged as the optimal choice based on the mentioned evaluation parameters.

References

[1] Particleio, How IoT Enables the Smart Grid — Applications, Benefits, and Use Cases, 2021. https://www.particle.io/iot-guides-and-resources/iot-smart-grid-applications-benefits-and-use-cases/.

[2] M.M. Alani, L. Mauri, E. Damiani, A two-stage cyber attack detection and classification system for smart grids, Internet of Things 24 (2023) 100926, https://doi.org/10.1016/j.iot.2023.100926. https://www.sciencedirect.com/science/article/pii/S2542660523002494.

[3] M.K. Hasan, R.A. Abdulkadir, S. Islam, T.R. Gadekallu, N. Safie, A review on machine learning techniques for secured cyber-physical systems in smart grid networks, Energy Reports 11 (2024) 1268—1290, https://doi.org/10.1016/j.egyr.2023.12.040. https://www.sciencedirect.com/science/article/pii/S2352484723016323.

[4] M. Panthi, T.K. Das, Intelligent intrusion detection scheme for smart power-grid using optimized ensemble learning on selected features, International Journal of Critical Infrastructure Protection 39 (2022) 100567, https://doi.org/10.1016/j.ijcip.2022.100567. https://www.sciencedirect.com/science/article/pii/S1874548222000518.

[5] C. Song, Y. Sun, G. Han, J.J.P.C. Rodrigues, Intrusion detection based on hybrid classifiers for smart grid, Computers & Electrical Engineering 93 (2021) 107212, https://doi.org/10.1016/j.compeleceng.2021.107212. https://www.sciencedirect.com/science/article/pii/S0045790621002056.

[6] S. Revathi, A. Malathi, A detailed analysis on NSL-KDD dataset using various machine learning techniques for intrusion detection, International Journal of Engineering Research and Technology 2 (12) (2013).

[7] N. Moustafa, J. Slay, The evaluation of network anomaly detection systems: statistical analysis of the UNSW-NB15 data set and the comparison with the KDD99 data set, Information Security Journal: A Global Perspective 25 (1—3) (2016) 18—31.

[8] A. Alsaedi, N. Moustafa, Z. Tari, A. Mahmood, A. Anwar, TON_IoT telemetry dataset: a new generation dataset of IoT and IIoT for data-driven intrusion detection systems, IEEE Access (2020). ieeexplore.ieee.org.

[9] M. Al-Hawawreh, E. Sitnikova, N. Aboutorab, X-IIoTID: A Connectivity- and Device-Agnostic Intrusion Dataset for Industrial Internet of Things, 2021.

[10] M. Zolanvari, M.A. Teixeira, L. Gupta, K.M. Khan, R. Jain, WUSTL-IIOT-2021 Dataset for IIoT Cybersecurity Research, Washington University in St. Louis, USA, 2021. http://www.cse.wustl.edu/~jain/iiot2/index.html.

[11] Wikipedia Contributors, Rule-based machine learning, in: Wikipedia, The Free Encyclopedia, Wikipedia, The Free Encyclopedia, 2023.

[12] C3ai, Tree-Based Models, 2017. https://c3.ai/glossary/data-science/tree-based-models.
[13] DataRobot, Bayesian Machine Learning, September 3, 2020. https://www.datarobot.com/blog/bayesian-machine-learning/.
[14] M.A. Ferrag, O. Friha, D. Hamouda, L. Maglaras, H. Janicke, Edge-IioTset: a new comprehensive realistic cyber security dataset of IoT and IIoT applications for centralized and federated learning, TechRxiv (2022), https://doi.org/10.36227/techrxiv.18857336.v1.

Further readings

[1] M. Zolanvari, M.A. Teixeira, L. Gupta, K.M. Khan, R. Jain, Machine learning-based network vulnerability analysis of Industrial Internet of Things, IEEE Internet of Things Journal 6 (2019) 6822–6834.
[2] N.K. Singh, M.A. Majeed, V. Mahajan, Statistical machine learning defensive mechanism against cyber intrusion in smart grid cyber-physical network, Computers & Security 123 (2022) 102941, https://doi.org/10.1016/j.cose.2022.102941. https://www.sciencedirect.com/science/article/pii/S0167404822003339.

CHAPTER 9

Anomaly and intrusion detection systems for smart grids

Ipek Abasıkeleş-Turgut[1] and Resul Daş[2]
[1]Computer Engineering, Faculty of Engineering and Life Sciences, Iskenderun Technical University, Hatay, Türkiye; [2]Software Engineering, Technology Faculty, Firat University, Elazığ, Türkiye

9.1 Introduction

Traditional electricity networks based on one-way interaction from generation to the end user have been replaced with the bidirectional smart grid (SG) system, using communication devices such as sensors and meters and providing blackout-free, self-healing service with the help of information technology [1]. Besides the advantages, SG systems have a complex structure, which causes different types of problems. Smart components of SG usually use Internet-based algorithms and protocols for advanced functioning [2]. Hence, as components of an SG system, Advanced Metering Infrastructure (AMI), Substations (especially Intelligent electronic devices (IEDs)), Phasor Measurement Units (PMUs) and Supervisory Control and Data Acquisition (SCADA) are vulnerable to several types of attacks [3]. Therefore, detecting, preventing and/or mitigating these intrusions are necessity.

Intrusion detection systems (IDS) are commonly used in SG, as other communication network structures for years. Recent researches focus on designing an intrusion detection system either with or without a mitigation and/or prevention approach. In order to shed light on future studies on the security of SG systems, which technique is useable for which types of attacks in an SG component basis should be investigated. Although there are various review papers on both SG security and IDS systems, as far as we know, the studies in literature about a detailed evaluation and comparison of recent research papers for all types of attacks classifying on an SG component basis are insufficient.

Cyber Security Solutions for Protecting and Building the Future Smart Grid
ISBN 978-0-443-14066-2
https://doi.org/10.1016/B978-0-443-14066-2.00005-0

In this paper, a total number of 58 recent articles published in SCI indexed journals are reviewed in detail. The approaches are compared in terms of target SG component, attack category, attack type, detection category, used technique, evaluation platform, evaluation metric, action to intrusion, used dataset, used standard, used protocols, performance of the proposed approaches and the main objective of the papers in a table. SG system has been handled on a component basis and all attacks with all techniques on four major components have been examined. The reviewed articles cover the year range of 2017 and 2023. Therefore, the contributions of this study, differently from its equivalents in literature, can be listed as follows.

- Including articles in SCI indexed journals published between 2017 and 2023
- No technique, attack type, component or subject restrictions
- Comparing articles for various parameters on a detailed table, rather than simply summarizing or briefly mentioning them
- Performing a separate analysis for each SG component so that the development in literature can be monitored on a component basis
- Investigating a total number of 58 recent publications

The remainder of the paper is organized as follows. Section 9.2 provides the literature review of survey articles about security solutions in SG, whereas Section3 describes the SG architecture and components. Section 9.4 presents general structure, types and properties of IDS. Section 9.5 details recent IDS approaches in SG components, including AMI, Substation, PMU and SCADA. Finally, Section 9.6 concludes the paper and discusses the future work.

9.2 Literature review

Since SG is a popular topic in literature in recent years, it has been the subject of both research and review articles. Some of the review articles [3–5] have discussed SG security in general and have presented IDS among other safety issues in a limited approach. There are also studies that examine intrusion detection methods in detail in industrial control systems, including SG [2,6,7]. However, the unique components and structure of SG systems require the investigation of intrusion detection techniques specific to these systems. There are some reviews published for this purpose in recent years [8–14].

Peng et al. [12] have focused on SG systems, especially on the effect of DoS attacks and protection methods against these attacks. Ferrag et al. [13] have examined IDS approaches in SG SCADA systems under nine topics. Quincoze et al. [14] have investigated 11 IDS studies for digital substations published between 2010 and 2019. Although these reviews can be regarded as detail researches about IDS in SG, their scope is narrow since they focus on a single attack or a single SG component.

Liu et al. [8], have reviewed rule-based IDS methods in SG. They have grouped SG under four topics, including SCADA, AMI, transmission system and substation. Rule-learning based IDSs are divided into 17 groups according to learning techniques. A total of number 21 articles, which was published between 2002 and 2020, are reviewed. They presented the results on a table showing the target system, used technique, used protocol, the purpose of the attack, the type of attack, used dataset and the performance of the reviewed studies.

Nguyen et al. [9] have classified the topic of SG protection from cyber-attacks under the following titles: FDIA detection, PMU protection and Machine Learning detection. The reviewed studies cover the year range of 2010 and 2019. Neither the content of the studies are adequately mentioned nor they are compared with each other over different parameters.

Ghiasi et al. [10] has categorized the detection mechanisms on the basis of attacks. Eight studies about Denial of Service (DoS) attacks that have been published between 2016 and 2019 and numerous conference and journal papers for the false data injection attacks publishing since 2010 have been reviewed. Apart from these studies, a number of 17 studies are examined for replay attack, covert attack, resonance attack, time-delay switch attack and man-in-the-middle attacks. However, the studies are not compared with each other over different parameters.

Panagiotis and Panagiotis [11] have examined IDS in SG systems on a component basis, as in this study. A total number of 37 conference and journal papers publishing between 2011 and 2018 are reviewed. The studies are classified as AMI, SCADA, substation, synchrophasors and entire SG. The target system, the detection method, used protocol, the attack types, the performance, used software and used datasets of the studies are compared in detail.

As summarized above, some of the security studies on SG are lacking in IDS topic, while in others single component-based or single attack-based studies have been examined. The range of years examined in most of the studies is outdated compared to this article. Besides, the number of studies examined in other reviews is less than that of this article. Although other reviews have investigated both conference and journal papers, only SCI journal papers are examined in this study since the conference papers are narrow-scoped in comparison with the SCI articles. Moreover, we have compared the studies in detail over various parameters. In these respects, it can be said that the contribution of this article to the literature is valuable.

9.3 SG architecture

As is seen in Fig. 9.1 [15], SG can be regarded as a combination of two types of infrastructure, including power and information. Power infrastructure is responsible for electricity distribution, whereas information infrastructure is used to monitor and control various equipment in SG. In this article, IDS systems for SG, which includes multiple components and have a complex structure, are examined on four components called AMI, Substation, PMU and SCADA.

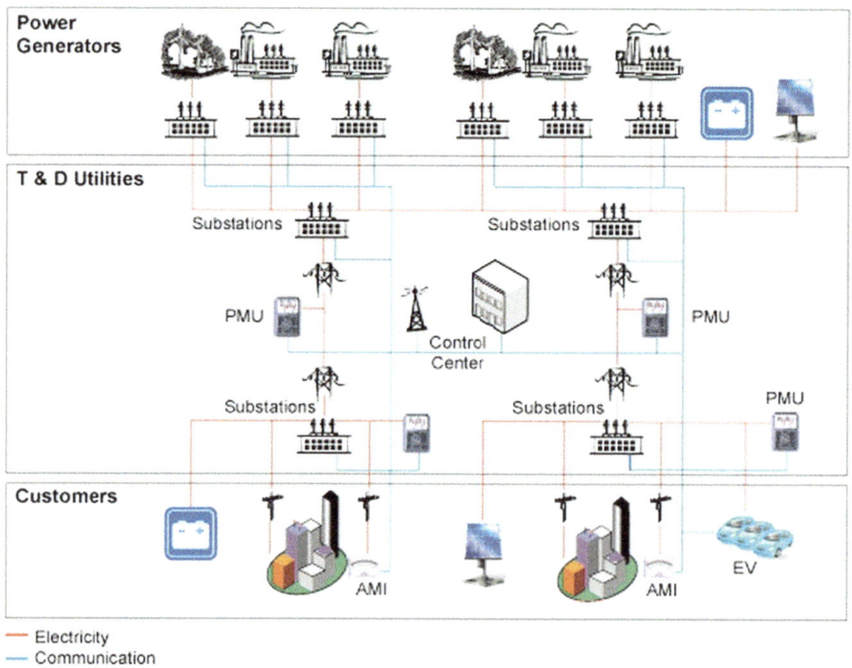

Figure 9.1 General SG architecture [15].

9.3.1 AMI

AMI serves with bidirectional communication with data it collects, manages and analyses from smart meters [16]. The advantages of AMI include power metering service, adaptive charging of power, client-side management and self-healing. The AMI architecture (Fig. 9.2) consists of three networks, Home Area Network (HAN), Neighborhood area network (NAN), and Wide Area Network (WAN). HAN offers communication between home appliances and smart meter, while NAN provides connection between smart meter and data collectors. On the other hand, WAN enables data collectors to connect to a central system called AMI headend in the Utility Network. AMI security is necessary not only for the SG but also for the protection of the communication network [16]. For this reason, there are many studies in the literature to protect AMI [17–34].

9.3.2 Substation

There are two types of substations in SG, including transmission and distribution. In this article, distribution substations are named as substations. According to the IEC-61850 substation architecture seen in Fig. 9.3, there

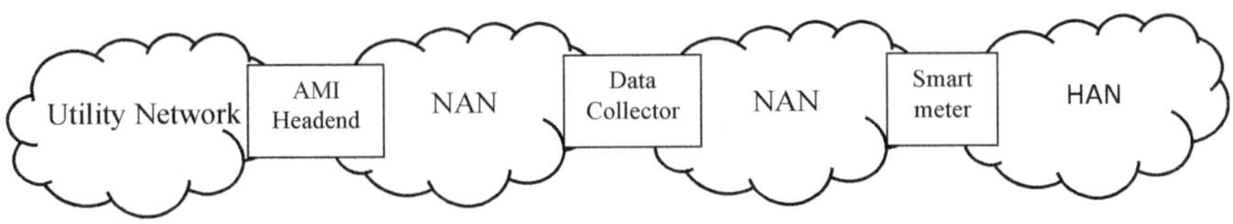

Figure 9.2 AMI communication architecture.

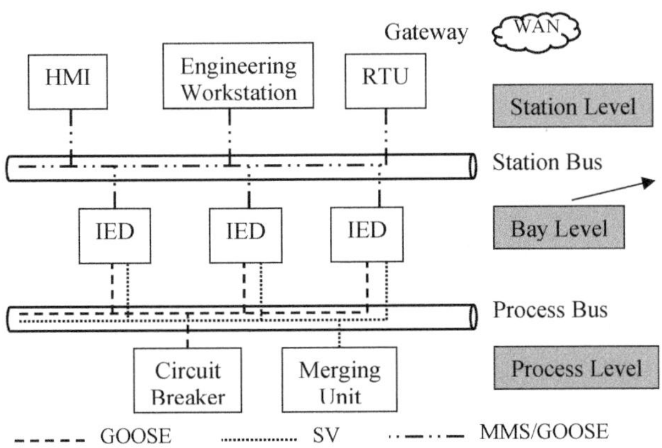

Figure 9.3 IEC-61850 substation architecture.

are three levels as station, bay and process. The station level, which is on the top, includes an engineering workstation; a Remote Terminal Unit (RTU), which provides communication with the external network; and an HMI for managing the station. Bay level IEDs work interoperable without human intervention and perform time synchronization, metering, control and protection functions. Finally, at the process level, there are switchyard equipment such as Circuit Breaker (CB) and Merging Unit (MU). IEC−61850 standard brings about two new communication protocols called Generic Object Oriented Substation Event (GOOSE) and Sampled Values (SV). SV provides communication between MU and IED, whereas GOOSE is used to communicate IEDs with CB. MMS ensures IED and station level communication over Ethernet. Many substations work without any security mechanisms [14]. Along with the traditional attacks, new types of attacks targeting IEC-61850 architecture should be detected. Therefore, there are many studies in literature focusing on this problem [35−50].

9.3.3 PMU

PMU is usually installed in the transmission domain offering real-time measurement of phasor quantities such as voltage and current, using GPS for synchronization. Fig. 9.4 shows the communication architecture of a wide area measurement system in SG [51]. PMUs are usually grouped locally and send their measurements to a phasor data concentrator (PDC). The data is then forwarded by the PDC to a control center in the utility network. Specifications for PMU communication are defined in the IEEE C37.118-2005 standard. Since PMU measurements should be protected against various attacks, many studies have been conducted for this purpose [52−66].

9.3.4 SCADA

As seen in Fig. 9.5, SCADA is used in SG, especially in transmission and distribution domains, to effectively collect, analyze and manage data in the system [67]. Within the distribution substation, the SCADA implementation is made at the station level. Thus, both manual labor operation is reduced and faultless automatic operations minimum disturbance is offered. In addition, there are also studies on PMU integration of SCADA [68]. Since the protection of SCADA systems that is used in many components and in many stages of SG is important, the design of attack detection systems for SG SCADA are popular in literature [69−77].

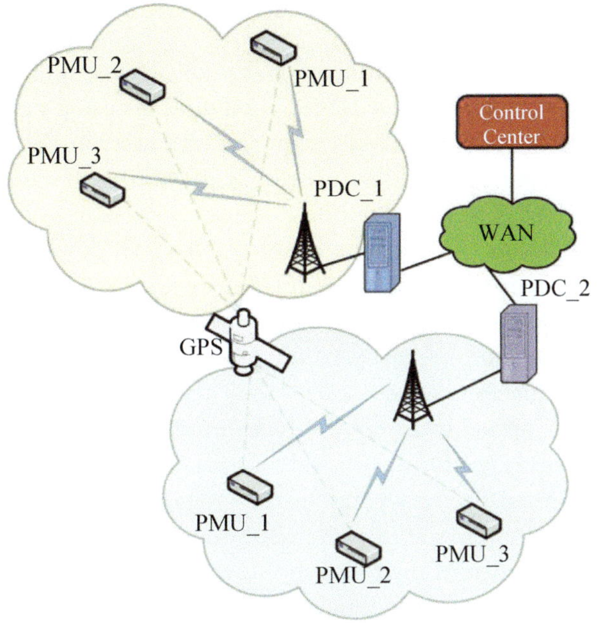

Figure 9.4 Communication architecture of a wide area measurement system in SG [51].

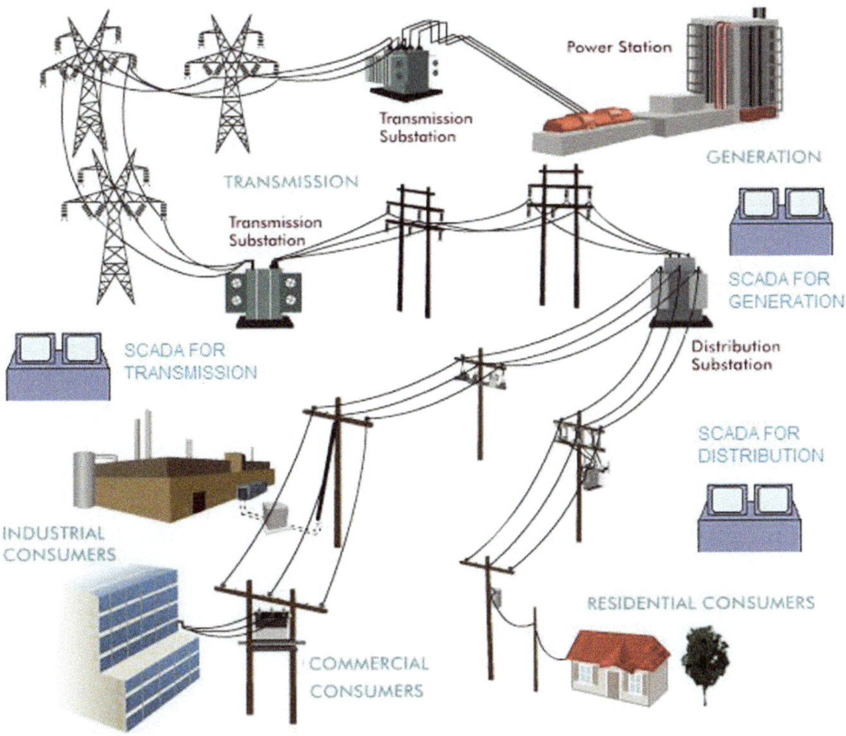

Figure 9.5 Scada for SG [67].

9.4 Overview of IDS

In SG systems as other communication-based systems, the actions against intrusions are divided into three classes, including intrusion detection (IDS), intrusion prevention (IPS) and intrusion response systems (IRS). While the majority of the studies in literature focus on intrusion detection, some studies include prevention or response mechanisms along with the detection. IPS aim to protect the system before the intrusion occurs, whereas IDS focuses on detecting the intrusion targeting the system and usually uses an alarm triggering mechanism to inform the system about the attack. IRS includes the mitigation mechanisms for recovering the system.

Three different approaches can be used to detect an intrusion threatening a system: anomaly-based (or behavior based), signature-based (or misuse-based), and specification-based (or rule-based) [8,11,14]. Anomaly-based detection tracks how much the system deviates from normal behavior, and the system status is labeled abnormal if this deviation is usually greater than a predetermined threshold value. System behavior is also monitored in the specification-based approaches, as in anomaly-based. The difference is that it is checked whether the system complies with the predetermined behavioral rules. Benign system behavior is detected as soon as there is a rule violation. Unlike these two approaches, signature-based detection is based on the behavior of the attacks, not the behavior of the system. It monitors the signatures/behaviors of the attacks on the system and activates the attack warning system if necessary.

Various parameters are used in literature to measure the performance of an IDS system. The most common parameters are detection rate (i.e., sensitivity, recall), accuracy, precision (i.e., true positive rate), false alarm (positive) rate and F measure (F1-score). All of the parameters are actually different equations of four variables (Eqs. 9.1–9.5). These variables are false positive (FP), true positive (TP), false negative (FN), and true negative (TN). TP and TN values refer that the system is working successfully and make the predictions correctly. FN indicates that IDS is not able to catch the attack, and TN means that an alarm is triggered even though there is no attack, indicating that a wrong prediction is made.

$$Detection\ rate = \frac{TP}{TP + FN} \tag{9.1}$$

$$False\ alarm\ rate = \frac{FP}{FP + TN} \tag{9.2}$$

$$Precision = \frac{TP}{TP + FP} \quad (9.3)$$

$$Accuracy = \frac{TP + TN}{TP + TN + FP + FN} \quad (9.4)$$

$$F1 - Score = 2 \times \frac{precision \times recall}{precision + recall} \quad (9.5)$$

9.5 IDS in SG

In our study, 58 SCI Journal papers published between 2017 and 2023 are reviewed. The distribution of the papers according to (a) year, (b) SG component, (c) method of handling the intrusion and (d) IDS type can be seen in Fig. 9.6. As is seen in Fig. 9.6, most of the recent studies focus on anomaly-based IDS regardless of the SG component. About one-fifth of the studies use specification-based IDS, whereas signature-based methods are rarely preferred for SG. Similarly, while IRS has been utilized in addition to IDS, IPS has not been handled much. The attack types, solution techniques and used protocols vary with the SG component. Therefore, detailed analyses should be made on a component basis. The reviewed studies for AMI, substation, PMU and SCADA can be seen in Tables 9.1–9.4, respectively.

9.5.1 IDS for AMI

The reviewed studies about AMI security [17–34] usually focus on a component of AMI instead of entire AMI. The majority of the papers examine smart meter data. The target network is either HAN or NAN. However, a small number of studies are network free or cover entire AMI. Most of the studies use the simulation methodology and MATLAB is mostly preferred simulation tool. Along with general datasets such as KDD Cup '99, CICIDS-2017, NSL-KDD, real data collected from electrical power networks are also used. The performances of the reviewed approaches are analyzed using various metrics such as detection rate, accuracy, false alarm rate, precision and f-score. False data injection attacks for energy consumption with energy theft are popular attack types. Although most studies propose anomaly-based IDS, there are also approaches preferring other methods. Generally, deep learning and machine learning methods are utilized for detecting attacks, but statistical methods and linear programming are also utilized.

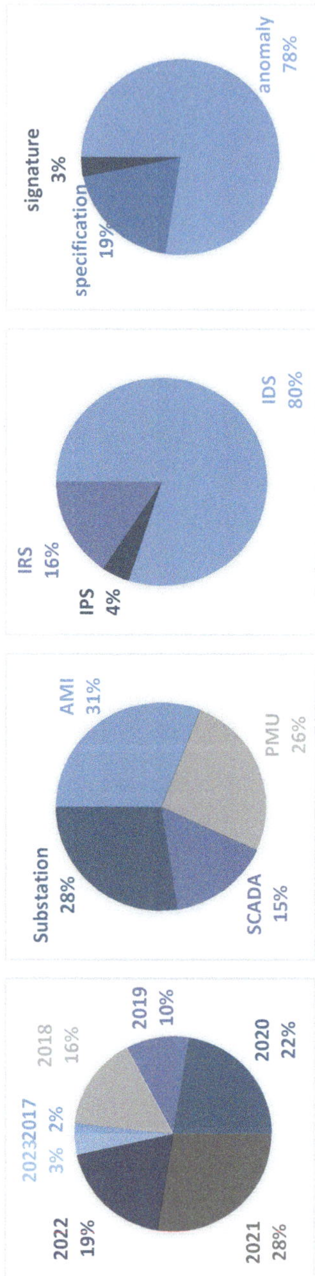

Figure 9.6 The distribution of reviewed papers by (a) year, (b) SG component, (c) method of handling the intrusion, and (d) IDS type.

Table 9.1 IDS for AMI in literature.

Refs.	Year	Target network	Attack category	Attack types	Detection category	Used technique	Evaluation hardware/software	Evaluation metrics	Action type	Protocol/standard/dataset	Performance	Main objective
[17] DeepCoin	2020	BAN and NAN	Network attacks	Brute force, bot, DoS, DDoS, Web, Infiltration, SQL injection, data theft, keylogging, power attack, etc.	Specification-based	Deep learning, recurrent neural networks	Google colaboratory	1. Accuracy 2. Detection rate 3. False positive rate	IDS	1. CICIDS-2017 dataset, 2. A power system Dataset, 3. A Bot-IoT dataset	DeepCoin shows higher detection rate and lower false alarm rate than SVM, Random forest and Naive Bayes.	Detecting network attacks in the blockchain-based energy network by using deep learning
[18] HANIDPS	2018	HAN	IEEE 802.15.4 attacks	Radio jamming, replay protection, steganography, back-off manipulation, DoS	Anomaly-based & specification-based	Q-learning, model-based approach	1. Testbed having TelosB motes as nodes and 4 AMs 2. MATLAB	1. Detection rate 2. Prevention rate 3. False positive rate	IDS & IPS	1. ZigBee for HAN	HANIDPS performs 92.5% detection rate for 0% false positive rate. Moreover, 97% prevention rate is provided.	Detecting and preventing attacks targeting HANs by using model-based and Q-learning approaches
[19] OCSVM	2019	Smart meter	Power quality problems	Power disturbance	Anomaly-based	One class support vector machine	1. MATLAB 2. Simulink toolbox	1. Accuracy 2. F1 score	IDS	1. Simulation generated data	OCSVM yields 93% accuracy for disturbances detection.	Detecting and classifying power disturbances by using support vector machines
[20]	2020	HAN	False Data injection	Forecast shifting, Forecast scaling	Anomaly-based	Game-theoretic approach, statistical methods	Theoretical analysis is performed.	1. Peak-to-average ratio 2. Bill change 3. Revenues change	IDS & IRS	1. U.S. Dept. Of energy: Commercial and residential hourly load profiles for all TMY3 locations in the United States	Although mix approaches are offered, in some cases, no monitoring can be the best strategy.	Offering the most appropriate action for defensing against false data injection attacks by using game-theoretic approach
[21] MIAMI-DIL	2021	Network free	Integrity and availability attacks	False data injection, DoS	Anomaly-based	Statistical methods	Simulation platform is not stated.	1. Detection delay 2. False positive rate	IDS & IRS	1. Simulation generated data	MIAMI-DIL detects the attacks faster than compared approaches.	Detecting and mitigating attacks in a timely manner by using statistical methods

Continued

Table 9.1 IDS for AMI in literature.—cont'd

Refs.	Year	Target network	Attack category	Attack types	Detection category	Used technique	Evaluation hardware/software	Evaluation metrics	Action type	Protocol/standard/dataset	Performance	Main objective
[22] KSVMeans	2020	Network free	Zero-day attacks	DoS, Probe, Root to local attacks (R2L), and user to Root attack (U2R)	Anomaly-based	K-means, support vector machine (SVM)	1. MATLAB	1. Accuracy 2. Recall 3. F-score 4. detection rate 5. False positive	IDS	1. KDD Cup '99 dataset	KSVMeans yields 99.9% accuracy.	Detecting 0-day attacks by using k-means based SVM
[23] CPADF	2020	HAN	Power overloading attacks	Bill reduction, Forming a peak energy load, Increasing load fluctuation	Anomaly-based	Regression decision trees, feature engineering	1. Weka	1. False negative rate 2. False positive rate 3. Accuracy 4. True positive rate 5. True negative rate	IDS	1. Dataset from Irish Social Science Data achieve 2. OCPP protocol	The proposed method yields high detection rate with low false positive rate.	Detecting power overloading attacks by using feature engineering and regression-based learning
[24]	2021	NAN	Integrity and availability attacks	DoS, false data injection, RAM exhaustion, CPU overloading	Anomaly-based & signature-based	Support vector machine (SVM), Temporal Failure Propagation Graph (TFPG)	1. NS-3 2. Single board computers (SBCs)	1. Accuracy 2. Detection rate	IDS	1. Collected from smart city testbed (SCT) having AMI 2. IEEE 802.15.4	The proposed SVM model performs better in terms of training time than neural network approach. SVM model shows high detection rate and accuracy.	Detecting attacks within two-stages by using SVM and TFPG
[25]	2021	Smart meter	Data falsification attacks	False power consumption data	Anomaly-based	Statistical methods	1. Virtual AMI	1. False positive rate 2. Detection rate	IDS	1. Dataset collected from Pecan Street project 2. Dataset collected from Ireland Social Sciences Data Archives	The proposed approach yields good performance for real time detection of attacks and has low false alarm rate	Detecting data falsification attacks targeting AMI by using statistical methods

[26]	2021	Smart meter	Non-technical losses (NTLs)	False data injections	Anomaly-based	Big data-driven methods	1. Oracle VM VirtualBox 2. Apache Hadoop 3. Apache Spark 4. Hortonworks data platform (HDP)	1. Accuracy 2. Specificity 3. Sensitivity 4. Precision 5. False positive rate	IDS	1. Private SG dataset in Turkish distribution network 2. Keras library	The proposed approach shows low false positive rates; high accuracy, specificity and precision.	Detecting false data injections at the distribution level by using big data-driven methods
[27]	2020	Smart meter	Abnormal electricity consumption	Abnormal electricity consumption	Anomaly-based	Feature engineering, unsupervised learning, data mining	1. Python package of tsfresh	1. F1 score 2. False positive rate	IDS	1. Load dataset of an industrial park	The proposed approach yields better performance than other feature strategies.	Detecting abnormality of electricity consumption by using feature engineering approach
[28]	2021	Smart meter	Abnormal electricity consumption	Abnormal electricity consumption	Anomaly-based	Percentage change, K-nearest neighbor, histogram buckets, Principal component Analysis	1. Raspberry Pi controllers 2. Simple measuring and Actuation Profile (sMAP) 3. Weather Underground Web Site	1. Anomaly score	IDS	1. Collected from IIIT-Delhi campus	Multiple abnormal behaviors are found within the collected dataset.	Identifying abnormal energy consumption by using four heuristic approaches
[29]	2022	Smart meter	Energy theft	Meter tampering, false data injections	Anomaly-based	Artificial neural network, recurrent neural network, statistical methods	Simulation platform is not stated.	1. Accuracy 2. Precision 3. Recall 4. F1 score 5. False positive rate	IDS & IRS	1. Dataset from electricity load diagrams 2011–14	The proposed method yields 99.77% accuracy, precision of 0.998, recall of 1.0, F1 score of 0.999 and false positive rate of 0.999.	Detecting and verifying energy theft against AMI by using machine learning and statistical methods

Continued

Table 9.1 IDS for AMI in literature.—cont'd

Refs.	Year	Target network	Attack category	Attack types	Detection category	Used technique	Evaluation hardware/software	Evaluation metrics	Action type	Protocol/standard/ dataset	Performance	Main objective
[30]	2021	NAN	Energy theft	Energy theft	Anomaly-based	Hadoop multi-node cluster model, interpolation search algorithm	1. Intel Core i3 processor with 8 GB RAM 2. Phyton 3. Ubuntu operating system 4. Apache Hadoop	1. Accuracy 2. False negative rate 3. Execution speed 4. Inspection steps	IDS	1. Dataset from UCI machine learning Repository individual household electric power consumption, 2012 2. Synthetic dataset for 100 users	The proposed method yields better performance than other approaches in terms of accuracy and false negative rate	Detecting energy theft against NAN by using machine learning algorithms
[31] Enhanced ADF	2018	NAN	Energy theft	Falsify energy cost	Anomaly-based	Linear programming (LP)	1. MATLAB	1. Detection rate 2. False positive rate	IDS	1. Dataset from Irish smart energy trial	Enhanced ADF yields 100% detection rate with 0% false positive.	Detecting non-technical losses caused by energy theft against AMI by using linear programming
[32]	2018	Entire AMI	Integrity and availability attacks	Price manipulation, DoS	Anomaly-based	Cumulative sum algorithm, statistical methods	1. MATLAB	1. Detection rate 2. False positive rate 3. Accuracy	IDS	1. Simulation generated data	The proposed method yields more than %95 detection rate with maximum 5% false positive rate.	Detecting lethal attacks against whole AMI by using statistical methods

[33]	2023	Smart meter	Attacks involved in used datasets	Attacks involved in used datasets	Anomaly-based	Federated learning (FL)	1. Raspberry Pi	1. Precision 2. Recall 3. Accuracy 4. F1-score	IDS	1. KDD Cup '99 dataset, 2. NSL-KDD dataset, 3. CICIDS-2017 dataset 4. AUSGRID dataset 5. SSL/TLS	The proposed federated model performs better than centralized models for all datasets.	Detecting anomalies of smart meters by using federated learning
[34]	2022	Not provided	Attacks involved in used datasets	Attacks involved in used datasets	Anomaly-based	Convolutional neural network	1. Intel i7-10700 with Windows 10 2. DL library TensorFlow 2.3-GPU of Python	1. Accuracy	IDS	1. KDD Cup '99 dataset, 2. NSL-KDD dataset, 3. CICIDS-2017 dataset	The proposed method yields above 91% accuracy for all used datasets.	Detecting attacks against AMI by using deep learning theory

Table 9.2 IDS for Substation in literature.

Refs.	Year	Target SG component	Attack category	Attack types	Detection category	Used technique	Evaluation hardware/ software	Evaluation metrics	Action type	Protocol/Standard/ Dataset	Performance	Main objective
[35] FARIMA	2019	Substation	Substation network cyber-attacks	Cyber storm, DoS, remote-to-user (R2L), user-to-root (U2R) attack, probing attack	Anomaly-based	Fractional ARIMA threshold model and probability calculation algorithm	1. MATLAB 2. Intel Core Processor I5-4460 (4 cores, 3.7 GHz) with 16 GB RAM	1. Error of filtering 2. Prediction performance	IDS	1. IEC 61850 2. Dataset from Tianjin Jintang Road Substation 3. KDD99	FARIMA yields better prediction performance than ARIMA for data traffic.	Detecting traffic anomalies in substation communication network by using a threshold model
[36] DIP	2020	Distribution system	Man-in-the-middle (MiTM)	Flooding, packet falsification/ modification, replay, brute-force password hack	Anomaly-based	Max Consensus Algorithms, statistical methods	1. IEEE 13-Node Test Feeder 2. Volttron 3. Linux-Mint 4. Scapy 5. DNP3 library with SAv5 6. Power Factory DIgSILENT 7. Matrikon OPC server	1. False positive rate 2. False negative rate 3. Prediction success rate 4. Offline computation time 5. Online attack correlation time	IDS & IPS	1. DNP3 2. Secure Authentication v5 (SAv5) 3. TCP/IP	The performance of DIP with inter node communication is validated. DIP is lighter than NIDS-REF; shows higher prediction success rate than CENTRAL-REF.	Detecting and preventing attacks against distribution systems by using a distributed and collaborative approach
[37]	2020	Distribution system	Distribution automation network attacks	Man-in-the-middle (MiTM), remote access to open switch attack, configuration change attack, DoS	Anomaly-based	Authentication algorithms, domain-based analysis, Decision Agreement algorithms	1. MATLAB	1. Bus voltages 2. Open and close commands under attack 3. Captured GOOSE packets	IDS & IRS	1. IEC 61850 2. GOOSE 3. UDP	The proposed methods are validated and tested by simulation.	Detecting and mitigating the attacks targeting distribution automation network
[38] STRAYER	2022	Substation	Remote access attacks	1. Sniffing 2. Brute Force 3. Rootkit	Signature-based	ICMP Snort signature	1. A created prototype 2. OMICRON™ IEDScout© software 3. TCPDump 4. John the Ripper	1. Rate of affected IED 2. Rate of open CBs 3. IOC access elapsed time 4. Real time to access an IED	IDS	1. IEC 61850 2. GOOSE 3. SV 4. MMS	STRAYER reduces the effect of the remote attacks targeting IEDs by 87.5%; that of circuit breakers by 88.9%.	Detecting remote access attacks targeting substation automation systems by conducting modifications to IEC 61850

[39] CIDS	2019	Substation (IEDs)	IEC 61850 related attacks	GOOSE injection attack, SV injection attack	Specification-based	IEC 61850 rules checking	1. Real time digital simulator (RTDS)	1. Time delays	IDS & IRS	1. IEC 61850 2. GOOSE 3. SV	CIDS is able to detect both GOOSE and SV based intrusions for both separately and simultaneously attacks.	Designing a distributed and collaborate IDS for IEDs against IEC 61850-based attacks
[40]	2018	Substation	IEC 61850-based attacks	MAC address-based attacks	Specification-based	Pattern Matching, Shift-and Algorithm	1. Xilinx Zynq-7030 FPGA 2. Xilinx Vivado 2017.17	1. Number of compared regular expressions 2. Maximum operating clock	IDS	1. IEC 61850 2. GOOSE	The proposed method can compare 137 more regular expressions than BP-NFA and maximum operating clock is 133.35 MHz.	Detecting IEC −61850-based intrusions against substations by using Shift-and Algorithm
[41]	2022	Substation	IEC 61850-based attacks	TCP SYN flooding, land, smurf, FTP semi connection, GOOSE replay, GOOSE & high "stNum", MMS malformed packet, MMS req. & MMS conf. Flooding	Anomaly-based	Discrete Wavelet Transform, locally linear Embedding Algorithm, Long Short term Memory Autoencoder	1. IEC 61850-based Substation Test bed 2. Wireshark 3. Acheron testing platform 4. Intel Core i7 with the processor of 2.9 GHz clock Speed 5. LPDDR3 with 16 GB memory and 2133 MHz frequency 6. Weka	1. Precision 2. Recall 3. F1 score 4. Accuracy 5. Matthews Correlation Coefficient	IDS	1. IEC 61850 2. GOOSE 3. TCP 4. MMS 5. Data set collected from Test bed	The proposed LSTM-Autoencoder outperforms RNN and GRU-based approaches.	Detecting IEC 61850-based network attacks against substations by using DWT, LLE and LSTM-Autoencoder

Continued

Table 9.2 IDS for Substation in literature.—cont'd

Refs.	Year	Target SG component	Attack category	Attack types	Detection category	Used technique	Evaluation hardware/software	Evaluation metrics	Action type	Protocol/Standard/Dataset	Performance	Main objective
[42]	2021	IEC 61850-based communication network	IEC 61850 GOOSE message attacks	IEC 61850 GOOSE message attacks	Anomaly-based	Machine learning algorithms (including DT, RF, SVM, k-NN and AdaBoost)	1. IEC 61850 GOOSE emulator 2. Intel Core i7-6700 2.60 GHz with 16 GB RAM 3. Raspberry Pi 3 4. Python	1. Accuracy 2. Detection rate 3. False positive rate	IDS	1. IEC 61850 2. GOOSE 3. Data set from IEC 61850 GOOSE emulator	The proposed method yields above 94% accuracy and acceptable detection rates for all machine learning methods.	Detecting IEC 61850 GOOSE based attacks by using various machine learning methods
[43]	2021	IEC 61850-based communication network	IEC 61850 SV message attacks	1. Symmetric & Asymmetric faults 2. False data injection	Anomaly-based	Machine learning algorithms (including DT, RF, XRT and ANN)	1. IEEE 14-bus having renewable energy generators 2. PSCAD-Python 3. Intel Core i7-6700 2.80 GHz with 32 GB RAM	1. Accuracy 2. Precision 3. F1 score 4. Detection time	IDS	1. IEC 61850 2. SV 3. Data set from IEEE 14-bus system	The proposed method yields above 82% accuracy and catches the fault for all machine learning methods.	Detecting IEC 61850 SV based attacks by using various machine learning methods
[44]	2020	Substation	IEC 61850 GOOSE message attacks	Denial of service (DoS)	Anomaly-based	Statistical methods (including Generalized Likelihood Ratio Test and Cumulative sum)	1. R Programming language version 3.6.2 2. Intel Core i7 2.00 GHz with 32 GB RAM	1. False positive rate 2. False negative rate 3. Detection delay	IDS	1. IEC 61850 2. GOOSE 3. Synthesized Data set from advanced digital Sciences Center (ADSC)	The proposed statistical method yields better performance than two other approaches in terms of false positive rate and detection delay.	Detecting IEC 61850 GOOSE based DoS attacks by using statistical methods
[45] AMS	2018	Distribution substation	Data integrity attacks	False voltage data injection	Anomaly-based & specification-based	Markov games	1. MATLAB/Simulink 2. D-STATCOM	1. Load shedding cost 2. Convergence time	IDS	1. Simulation generated data	AMS shows higher performance than conventional Nash equilibrium (NE) approach	Detecting voltage stability attacks against the distribution substations by using Markov game strategy

[46] CRM-CSq	2022	Distribution system	Attacks aiming to mislead the central controller	Stealthy attacks, false data injection (FDI), changing the accuracy of the CRM	Anomaly-based	Chi-square test (CSq)	1. OPAL-RT (OP5700) (a digital real-time simulator (DRTS)) 2. Cyber-physical power system Resilience (CPSR) testbed	1. Detection score 2. False positive rate 3. True positive rate 4. Detection time	IDS	1. IEC60255	CRM-CSq performs better than neural network, Kalman filter and SVM for attack detection accuracy and time	Detecting the attacks against distribution systems by using a cyber-physical real-time reference model based on physical properties of the system
[47]	2019	Substation	Remote access	False data injection (FDI), configuration change (CC), direct circuit breaker control (DCBC) attacks	Anomaly-based & specification-based	Rule checking, cross correlation, Transient fault signature	1. Real time digital simulator (RTDS)	1. SV current 2. SV frequency	IDS & IRS	1. IEC 61850 2. GOOSE 3. SV 4. MMS	The performance of secured CC function, FDI security and DCBC security are validated.	Detecting and de-escalating the attacks targeting digital substations by using domain based-approaches
[48]	2022	Substation (IEDs)	Man-in-the-middle (MiTM) on DNP3 protocol	False data injection (FDI), false control injection (FCI)	Anomaly-based	Multitagging with RTAC tags	1. RESLab testbed 2. Wireshark 3. PowerWorld Dynamic Studio (PWDS) 4. Common open research emulator (CORE) 5. RTAC 6. MiTM attack library 7. Layer-3 network switch	1. Generator set point Control value 2. Generator real power output measurement 3. RTAC alert logs	IDS & IRS	1. DNP3 2. Cyber-physical dataset of Hardware-in-the-loop cyber-physical power systems testbed under MITM attacks 3. IEC 61131	The proposed IDS creates the alert logs as accurate as Snort and generates less false alarms than Snort	Detecting and mitigating the attacks targeting IEDs by using status information

Continued

Table 9.2 IDS for Substation in literature.—cont'd

Refs.	Year	Target SG component	Attack category	Attack types	Detection category	Used technique	Evaluation hardware/software	Evaluation metrics	Action type	Protocol/Standard/Dataset	Performance	Main objective
[49] FODN4PS	2021	Substation	Foreign objects	Foreign objects	Anomaly-based	Image processing, deep learning, ViBe algorithm	1. Python 3.6 with CUDA version 10.1 2. NVIDIA GeForce GTX TITAN X with 12G memory 3. Intel i7 CPU	1. Accuracy 2. Running speed	IDS	1. GAN (Generative Adversarial networks) 2. Google NET	FODN4PS performs better than Faster RCNN and Mask RCNN in terms of speed and accuracy in object recognition.	Detecting foreign objects in substations by using image processing methods
[50] TSCDNet+	2021	Substation	Abnormal samples of substation equipment	Abnormal samples of substation equipment	Anomaly-based	Siamese neural network	Not provided	1. Accuracy 2. Precision 3. Recall 4. F1 score 5. F2 score	IDS	1. Data set created for substation equipment anomaly detection	The robustness of TSCDNET+ is sufficient for various threshold values and And the model has the F1 scores up to 0.945.	Detecting anomalies from the distance between the normal equipment image and the image to be measured by using supervised learning

Table 9.3 IDS for PMU in literature.

Refs.	Year	Target SG component	Attack category	Attack types	Detection category	Used technique	Evaluation hardware/ software	Evaluation metrics	Action type	Protocol/standard/ dataset	Performance	Main objective
[52] HAT + DDM ADWIN	2018	Transmission system	Command-based attacks	1. Trip command injection attacks 2. 1LG fault replay attack 3. Relay disabled attacks 4. Aurora attack	Specification based	Data mining methods, including Hoeffding adaptive trees, Drift detection, adaptive Windowing	1. WAMS cyber-physical Test bed with 3-bus 2-generator system 2. Snort	1. Classification accuracy 2. Kappa statistics 3. RAM hours 4. Evaluation time	IDS	1. Data set from Test bed and SNORT logs from simulations	The proposed method yields above 94% classification accuracy for multiclass; 98% for binary class.	Detecting attacks against transmission systems by using data mining methods
[53] NNGE + STEM	2018	Transmission system	Command-based attacks	1. Trip command injection attacks 2. 1LG fault replay attack 3. Relay disabled attacks 4. Aurora attack	Specification based	Data mining methods, including Non-nested generalized exemplars and state Extraction method	1. A HIL testbed with 3-bus 2-generator system 2. Snort	1. Accuracy 2. Precision 3. Recall 4. F-measure	IDS	1. Data set from Test bed and SNORT logs from simulations	The proposed method yields above 94% classification accuracy for multiclass; 97% for binary class.	Detecting attacks against transmission systems by using data mining methods
[54] CACDM	2021	PMU	False data injection	False data injection	Specification-based	Voting mechanism, trust-based approach	1. MATLAB2017b 2. PowerWorld simulator 3. IEEE 39-bus system simulation	1. Accuracy 2. Detection rate	IDS	1. Data generated by PowerWorld simulator	CACDM yields better performance than DHCD for single rule false data injection.	Detecting false data injection attacks against PMU data by using trust and voting based collaborative approach

Continued

Table 9.3 IDS for PMU in literature.—cont'd

Refs.	Year	Target SG component	Attack category	Attack types	Detection category	Used technique	Evaluation hardware/ software	Evaluation metrics	Action type	Protocol/standard/ dataset	Performance	Main objective
[55] hybrid NIDS & PADC	2022	PMU and MMS	Man-in-the-middle (MiTM)	1. Attacks against IEEE C37.118 2. Attacks against the IEC 61850-8-1	Specification based & anomaly-based	Model based traffic prediction, rule checking	1. A real-time simulation engine: HYPERSIM (OPAL-RT)	1. F1 score 2. Average processing time	IDS	1. IEEE C37.118 (PMU) 2. IEC 61850-8-1 (MMS) 3. IEC 60870-5-104	The performance of the proposed methods are verified with simulation.	Designing a network intrusion and PMU anomaly detection system for the attacks against TCP/IP-based communication protocols
[56]	2020	PMU	Attacks against PMU	1. False data injection 2. Stealthy attack	Anomaly-based	Multi-port equivalent circuit model	1. IEEE 118-bus system simulation 2. 7-bus transmission power system simulation 3. 39-bus new England system simulation	1. Attack indicator value 2. ISI index value	IDS& IRS	1. IEEE C37.118	The performance of the proposed methods are verified with simulation	Detecting and mitigating the attacks against voltage stability of transmission power systems by using Thevenin equivalent (TE) based indicators
[57]	2018	Synchrophasor units	Tampered attacks	False data injection	Anomaly-based	Markovian decision model	1. MATLAB 2. 6-bus system simulation 3. IEEE 24-bus system simulation 4. 64 bit 3.5 GHz Intel Xeon processor with 16 GB RAM	1. Detection efficiency 2. Relative error 3. Expected cost	IDS& IRS	1. Data generated by simulations	Optimal responses are affected from the probabilities of intrusions and should be made according to observations	Detecting tampered attacks against synchrophasor units by using observations and mitigating the attacks by cost benefit actions

[58]	2019	PMU	False data injection	Falsified data streams	Anomaly-based	Convolutional neural network, Nesterov Adam Gradient	**1.** MATLAB 2017 **2.** Intel Core i5-4460 CPU 3.20 GHz and 8 GB RAM **3.** IEEE 118-bus system simulation **4.** IEEE 30-bus system simulation **5.** TSAT Transient security Assessment tool	**1.** Accuracy	IDS	**1.** Data generated by simulations	The proposed method yields the best classification accuracy among the compared approaches.	Classifying false data injection attacks against PMU data by using deep neural network
[59]	2021	PMU	False data injection	**1.** Internal false data injection **2.** Unintentional errors	Anomaly-based	Support vector regression	**1.** IEEE 118-bus system simulation **2.** 246 bus NRPG of India	**1.** Magnitude error **2.** Phase Engle Error **3.** Total vector error	IDS & IRS	**1.** Data set from NRPG of India **2.** Data generated by simulations	The proposed method yields low error rate when correcting false data and total vector error is less than 1%.	Detecting and correcting false data in PMU data by using support vector regression
[60]	2022	PMU	False data injection	False data injection	No detection	Shortest Path Computing techniques, i.e., FloydWarshall algorithm	**1.** MATLAB R2018a **2.** IEEE 14-bus system simulation **3.** New England 39-bus system simulation **4.** HP ProDesk 600 G4 SFF	**1.** Giant-component Size **2.** Number of components	IPS	**1.** Data generated by simulations	The proposed method yields 46% and 57% secured measurements in the tested environments.	Developing a PMU deployment framework, which is resistant to false data injection attacks

Continued

Table 9.3 IDS for PMU in literature.—cont'd

Refs.	Year	Target SG component	Attack category	Attack types	Detection category	Used technique	Evaluation hardware/software	Evaluation metrics	Action type	Protocol/standard/dataset	Performance	Main objective
[61]	2020	PMU	Reconnaissance attacks	Reconnaissance attacks	Anomaly-based	Model-based approach	1. Bro network monitoring tool	1. Voltage 2. Time delay 3. Bandwidth	IDS	1. Data generated by simulations 2. Real data from 2 OLTC transformers	The performance of the proposed method is verified by simulation and utility data.	Detecting anomalies in PMU data by monitoring and passively learning the regulation equipment
[62]	2020	PMU	Attacks against power systems	Attacks against power systems	Anomaly-based	Recurrent neural networks, including SimpleRNN, Long Short-Term Memory and Gated recurrent Unit	1. TensorFlow 2. Python with Keras API 3. Intel Xeon E5-2630 v4 CPU, 64 GB RAM, an NVIDIA RTX 2070 card 4. Ubuntu 16.04.6	1. Accuracy 2. Precision 3. Recall 4. F1-score	IDS	1. Dataset 1: Power system datasets	The proposed method yields 99.99% classification accuracy for multiclass; 100% for binary class.	Classifying cyber-attacks against power systems by using three different RNN-based methods
[63]	2021	PMU	PMU data anomalies	PMU data anomalies	Anomaly-based	Principal component analysis, K-means, Gaussian Mixture model	1. MATLAB 2018b	1. F1-score 2. Recall 3. Precision	IDS	1. PMU Data set from EPFL institutions	F1 score of the proposed method varies between 0.91 and 1 based on the number of features.	Detecting anomalies of multivariate PMU data by using unsupervised learning methods
[64]	2021	PMU	GPS spoofing attack (GSA)	1. Step attack 2. Ramp attack	Anomaly-based	Statistical methods, data-driven methods	Not provided	1. True positive 2. False positive 3. True negative 4. False negative	IDS	1. Data set from 2 transmission lines from China 2. Data generated by simulations	The proposed method can effectively detect and identify GSA.	Detecting GSA against PMU data by using data-driven methods

[65]	2023	PMU	PMU data faults	1. Symmetrical 2. Asymmetrical	Anomaly-based	Statistical methods	1. MATLAB/SIMULINK 2. IEEE 73-bus system simulation	1. Fault detection time 2. Number of required PMUs	IDS	1. IEEE C37.118.1 2. IEC/IEEE 60,255−118−1 3. Data generated by simulations	The performance of the proposed method is verified with simulation.	Detecting faults in a transmission system by using PMU-based analytical methods
[66]	2022	μPMU	μPMU data faults	μPMU data faults	Anomaly-based	Machine learning, i.e., Random forest Classifier	1. MATLAB/SIMULINK 2. IEEE 13 node system simulation 3. Lenovo ThinkPad i7 laptop	1. Accuracy 2. Precision 3. Detection time	IDS	1. Data generated by simulations	The performance of the proposed method has 98.91% accuracy, 99.89% precision and exhibits a detection time of 8.5 ms.	Detecting faults in a renewable distributed generation system by using μPMU-based machine learning methods

Table 9.4 IDS for SCADA in literature.

Refs.	Year	Target SG component	Attack category	Attack types	Detection category	Used technique	Evaluation hardware/ software	Evaluation metrics	Action type	Protocol/standard/ dataset	Performance	Main objective
[69]	2021	SCADA	Malicious events on SCADA traffic	Malicious events on SCADA traffic	Anomaly-based	Gradient Boosting Algorithm, Decision-tree-based algorithms	1. WEKA 2. Intel Core i5-8300H 2.30 GHz processor, 16 GB RAM and Nvidia Geforce GTX 1060 6go GPU 3. Python on a Jupyter Notebook	1. Accuracy 2. Detection rate 3. Precision 4. F1 score 5. Miss rate	IDS	1. Three Data set from Oak Ridge National Laboratories (ORNL)	The proposed method yields higher performance than other approaches.	Detecting malicious activities on SCADA traffic by using machine learning approaches
[70]	2020	SCADA	Inside attacks against SCADA	1. Remote tripping command injection attack 2. Relay setting change 3. Data injection	Anomaly-based	K-nearest-neighbor instance —based learning, correlation-based feature selection	The evaluation platform is not provided.	1. F-measure 2. Accuracy 3. Precision 4. Recall	IDS	1. 15 public Data set from an SCADA system	The proposed approach yields the highest detection rate among the corresponding approaches using 17 out of 128 features.	Detecting attacks against SCADA by using machine learning techniques
[71] PC-IDS	2021	SCADA	Attacks against SCADA	Attacks against SCADA	Anomaly-based	Particle swarm optimization, neural network	1. Python on Windows 10 2. i5 CPU & 8 GB RAM	1. False positive rate 2. Detection rate 3. Accuracy	IDS	1. Power system Data set 2. UNSW-NB15 Data set	PC-IDS yields 96.03% and 95.91% detection rate; 0.18% and 0.14% false positive rate for used Data sets.	Detecting attacks against SCADA by using machine learning techniques
[72]	2019	SCADA	Attacks against DNP3 in SCADA	1. Modification 2. Interception 3. Interruption 4. Fabrication	Anomaly-based	Machine learning algorithms, i.e., K-nearest-neighbor, Decision tree, support vector machine, Naïve Bayes, Random forest	1. PCAP 2. Pepper tool 3. Wireshark	1. Classification of attacks	IDS	1. DNP3 2. Stuxnet malware	The proposed method successfully detects attacks.	Detecting attacks against DNP3/ SCADA by using machine learning techniques

[73]	2020	IEEE 1815.1-based SCADA	IEEE 1815.1-based attacks	**1.** CPS network traffic **2.** CPS malware behavior **3.** False data injection **4.** Disabling reassembly	Anomaly-based	Bidirectional recurrent neural network	**1.** Triangle Microworks' distributed test manager (DTM) **2.** Intel Core i7-8700 CPU, 16 GB, GeForce GTX 1070 Ti **3.** TensorFlow library	**1.** Detection rate	IDS	**1.** IEEE 1815.1 **2.** DNP3 **3.** Data set from Korean Substation	The proposed method successfully detects five types of CPS malware behavior, three types of false data injection and disabling reassembly attacks	Detecting IEEE 1815.1-based malicious activities by using bidirectional recurrent neural network (a deep learning method)
[74]	2022	SCADA	Attacks against SCADA	**1.** Reconnaissance **2.** Dos **3.** Command injection **4.** Response injection	Anomaly-based	Convolutional neural network, Gated recurrent Unit	**1.** Python 3.6 on Windows 10 **2.** i5 4.2 GHz & 8 GB RAM	**1.** Accuracy **2.** Precision **3.** False positive rate	IDS	**1.** NSL-KDD Data set	The proposed approach yields 98.84% detection accuracy.	Detecting attacks against SCADA by using deep learning techniques
[75] ZESO-DRKFC	2022	SCADA	Attacks against SCADA	Attacks against SCADA	Anomaly-based	Zaire Ebola search optimization, deep Random Kernel forest classification	The evaluation platform is not provided.	**1.** Accuracy **2.** Precision **3.** F1-score **4.** Detection rate	IDS	**1.** 7 Data sets including NSL-KDD, CAN intrusion, CIC-IDS 2017, Bot-Iot, DS2OS, UNSW-NB 15 and CIRA−CIC−DoHBrw-2020	ZESO-DRKFC outperforms other approaches and yields 99%accuracy.	Detecting attacks against SCADA by using meta-heuristic based approach

Continued

Table 9.4 IDS for SCADA in literature.—cont'd

Refs.	Year	Target SG component	Attack category	Attack types	Detection category	Used technique	Evaluation hardware/ software	Evaluation metrics	Action type	Protocol/standard/ dataset	Performance	Main objective
[76]	2017	IEC 61850-based SCADA	IEC 61850-based SCADA attacks	1. External malicious attacks 2. Internal unintended misuse	Specification-based & anomaly-based	Definition of normal behavior of the system by traffic and file analyses	1. Ubuntu 12.04 2. ITACA 3. Test bed in the state grid Key Laboratory of Substation Intelligent equipment testing Technology in China 4. Wireshark 5. PCAP	1. Accuracy 2. Process time	IDS	1. IEC 61850 2. Substation Configuration Description Language (SCL) 3. GOOSE 4. SV	The proposed method, yielding 100% accuracy and process time lower than 0.3 ms, is validated in a real Test bed.	Detecting internal and external attacks against IEC 61850 by designing SCADA-specific multi-layered IDS
[77]	2018	SCADA	Control-related attacks	1. False data injection 2. Malicious feedback control algorithms	Specification-based	Semantic analyses	1. Ubuntu 12.04 2. The Bro IDS 3. MATLAB 4. Intel i3 (3.07 GHz), 8 GB 5. IEEE 24-bus system 6. IEEE 30-bus system 7. IEEE 39-bus system 8. 2736-Bus system	1. False positive rate 2. False negative rate 3. Detection latency	IDS & IRS	1. DNP3	The proposed system yields 0.78% false positive and 0.01% false negative rate for large scale 2736-bus system; completes the detection in 200 ms.	Detecting control-related attacks against SCADA by a semantic analysis framework

DeepCoin [17] is a specification-based IDS targeting to protect blockchain-based BAN and NAN from network attacks by using deep learning techniques. Three different datasets, i.e., CICIDS-2017, a Power System dataset and a Bot-IoT dataset, are used for evaluating the performance of the DeepCoin, which shows higher detection rate and lower false alarm rate than SVM, Random forest and Naive Bayes.

HANIDPS [18] aims to detect IEEE 802.15.4 attacks against HAN and protect the network by using both anomaly and specification based methods. Q-learning and model-based approaches are used on Zigbee-based HAN. The simulation results shows that proposed approach yields 97% prevention rate; 92.5% detection rate and 0% false positive rate.

OCSVM [19] detects and classifies power disturbances by using one class support vector machine. OCSVM yields 93% accuracy for disturbances detection.

Detecting and mitigating false data injection attacks against HAN by using game-theoretical approach and statistical methods are proposed in Ref. [20]. The results show that no monitoring can be the best strategy for some cases. MIAMI-DIL [21], which is a network-free approach, also uses statistical methods for detecting and mitigating false data injection and DoS attacks in a timely manner. KSVMeans [22] is another network-free anomaly-based approach. Support Vector Machine and K-Means are used for detecting 0-day attacks targeting AMI. The result of the simulations on KDD Cup '99 dataset yields 99.9% accuracy.

CPADF [23] is an anomaly-based IDS, which aims to detect power overloading attacks against HAN by using feature engineering and regression-based learning. The proposed method yields high detection rate with low false positive rate.

Anomaly and signature based IDSs are used together for detecting integrity and availability attacks against NANs by using Support Vector Machine and Temporal Failure Propagation Graph [24]. The simulations are performed on NS-3 and data is collected from smart city testbed having AMI. The proposed IDS shows better training time than Neural Network; high accuracy and detection rate.

Data falsification [25], false data injection [26], abnormal electricity consumption [27,28] attacks against smart meters are detected by using anomaly-based detection and various techniques including statistical [25], big data-driven [26], feature engineering [27] and heuristic [28] approaches. All studies utilize real datasets collected from a campus [28], an industrial park [27], a distribution network [26] and various projects [25].

Energy theft is another popular attack against AMI. Different machine learning methods [29,30], statistical methods [29] and linear programming [31] are used to detect and mitigate this attack. The simulations are conducted on real datasets including electricity load diagrams [29], individual household power consumption [30] and smart energy trial [31].

Although KDD Cup '99 dataset is popular for evaluating the performance of the proposed approaches, some studies use simulation generated data. For instance Ref. [32], has proposed an anomaly-based IDS against DoS and price manipulation attacks for entire AMI by using Cumulative Sum Algorithm with Statistical methods. The results show that the proposed method yields more than %95 detection rate with maximum 5% false positive rate.

Other anomaly-based IDSs, using KDD Cup '99, NSL-KDD, and CICIDS-2017 datasets, utilize federated learning [33] and convolutional neural network [34] approaches.

9.5.2 IDS for substation

As can be seen in Table 9.2, the majority (50%) of the reviewed studies in this section [35—50] aim to protect the substation in a general approach. Four studies are about distribution systems; four studies specifically focus on IED and IEC 61850-based communication network. Unlike AMI, with the MATLAB tool, a wide variety of evaluations including the use of simulation, real time simulators, RESLab testbed, FPGA, prototype, IEC 61850-based Testbed, emulator has been conducted.

FARIMA [35] aims to detect traffic anomalies in substation communication network by using fractional ARIMA threshold model. Simulations are conducted by using MATLAB and it is demonstrated that the proposed method yields better performance than ARIMA for data traffic.

For detecting and preventing attacks against distribution systems, DIP [36] proposes a distributed and collaborative anomaly-based approach by using Max Consensus Algorithms and Statistical Methods. Another study [37] aims to protect the distribution system from Distribution Automation Network Attacks by proposing a detection and mitigation algorithm based on Authentication Algorithms, Domain-based Analysis, Decision Agreement Algorithms.

STRAYER [38] uses ICMP Snort signatures and modifies IEC 61850 for detecting remote access attacks against substations. A prototype is created for evaluations and it is stated that STRAYER reduces the effect of the remote attacks targeting IEDs by 87.5%; that of circuit breakers by 88.9%.

Studies about IEC 61850 related attacks are popular in literature. For instance, CIDS [39] has proposed a specification-based IDS and IRS for GOOSE and SV injection attacks. In the evaluations conducted by the Real Time Digital Simulator, it is stated that both types of attacks are successfully detected. In another study [40], specification-based detection is performed using Pattern Matching and Shift-And Algorithm to detect IEC 61850-based MAC address attacks. In their evaluations with using Xilinx Zynq-7030 FPGA, it has been proven that the proposed method compares 137 regular expressions more than its equivalents. A study [41] that performs anomaly-based detection for a very large set of attacks among IEC 61850-based attacks such as TCP SYN flooding, land smurf, GOOSE replay, MMS attacks [41] has shown that the proposed method outperforms the equivalents in the evaluations made on the substation testbed. While some studies focus on GOOSE message attacks [42,44], there are also studies suggesting attack detection for SV attacks [43]. In these studies, anomaly-based detection is generally designed by using machine learning [42,43] or statistical [44] methods. Performance evaluation is made on parameters such as accuracy and detection rate.

False data injection is another popular attack in substations. There are studies that evaluate both simulation [45−47] and testbed [48] in detection processes using different methods such as multitagging [48], Markov games [45], chi-square test [46], cross correlation [47]. In addition to anomaly-based methods, specification-based [45,47] approaches are also used in the detection of false data injection attacks.

Detecting foreign objects or abnormal samples in substation are studied by FODN4PS [49] and TSCDNet + [50]. FODN4PS [49] has proposed an anomaly-based IDS in order to detecting foreign objects by using image processing and deep learning methods, whereas TSCDNet + [50] aims to detect abnormal samples with a neural network based IDS.

9.5.3 Intrusion detection systems for phasor measurement unit

The majority of reviewed attacks targeted PMU [52−66] are false data injection (FDI) attacks. Apart from FDI, faults for PMU, spoofing attacks, anomalies, reconnaissance attacks, MITM attacks and command based attacks for transmission system are also studied. For evaluating the performance of the proposed approaches both synthetic data generated by simulation and datasets produced from various resources including testbeds and real-time systems, have been utilized. Although half of the studies use MATLAB simulation tool, there are also studies evaluated on real-time testbeds or real time simulator.

Refs. [52,53] have proposed a specification-based IDS against command-based attacks targeting transmission systems by using various data mining techniques. They have realized their proposed method on testbed and achieves the performance up to 98% accuracy. Another specification-based IDS against PMU FDI is CACDM [54] utilizing voting mechanism with a trust-based approach. In the simulations designed by MATLAB and PowerWorld, the proposed method has been tested over the parameters of accuracy and detection rate. [55] uses both specification and anomaly-based approach for designing a network intrusion and PMU anomaly detection system for the attacks targeting TCP/IP-based communication protocols.

FDI against PMU are usually detected with anomaly-based methods [ ]. For instance, for detecting and mitigating the attacks against voltage stability of transmission power systems [56], uses Thevenin Equivalent (TE) based indicators. [57] aims to detect tampered attacks against synchrophasors by using observations and to mitigate the attacks by cost benefit actions. Convolutional Neural Network and Nesterov Adam Gradient are also used to classify FDI against PMU in Ref. [58]. The authors [59] detect and correct false data in PMU data by using support vector regression. Apart from IDS, one study [60] performed only IPS. FloydWarshall algorithm is used for developing a PMU deployment framework, which is resistant to FDI. The proposed IPS yields 46% and 57% secured measurements in the tested environments.

Ref. [61] proposes an anomaly-based IDS against reconnaissance attacks targeting PMUs. The authors aim to detect anomalies in PMU data by monitoring and passively learning the regulation equipment. The performance of the proposed method is verified by simulation and utility data.

Ref. [62] classifies cyber-attacks against power systems by using three different recurrent neural networks (RNN)-based methods and their proposed method yields 99.99% classification accuracy for multiclass; 100% for binary class.

Another study [63] aims detecting anomalies of multivariate PMU data by using various unsupervised learning methods. The simulations are conducted by using MATLAB and it is shown that F1-score of the proposed method varies between 0.91 and 1 based on the number of features.

GPS spoofing attacks against PMU is detected by Ref. [64] by using statistical and data-driven methods. Both simulation generated data and Dataset from two transmission lines from China are used for evaluating the performance of the proposed approach.

Finally, both PMU data faults [65] and μPMU data faults [66] are detected with anomaly-based methods [65]. uses statistical methods, whereas [66] benefits from machine learning approaches. Both studies uses MATLAB/SIMULINK and simulation generated data.

9.5.4 Intrusion detection system for SCADA

Most of the recent studies aiming to protect SCADA [69–77] use machine learning and deep learning methods regardless from the type of the attacks. For instance Ref. [69], uses Gradient Boosting Algorithm and Decision-Tree-based Algorithms for anomaly-based detection of malicious activities on SCADA traffic. Three Data Set from Oak Ridge National Laboratories are utilized to evaluate the performance of the proposed method. Other machine learning methods including K-nearest-neighbor and Correlation-based feature selection are used to detect insider attacks against SCADA [70]. PC-IDS [71] is an anomaly-based IDS that tries to detect the attacks against SCADA with the help of particle swarm optimization and Neural network. It is reported that PC-IDS yields 96.03% and 95.91% detection rate; 0.18% and 0.14% false positive rate for used Data Sets, including Power System Data Set and UNSW-NB15 Data Set. Five different machine learning algorithms (i.e., K-nearest-neighbor, Decision tree, Support vector machine, Naïve Bayes, Random forest) are used to design an anomaly-based detection attacks against DNP3/SCADA [72].

Two different deep learning methods, including bidirectional neural network [73] and convolutional neural network [74] are used to detect IEEE 1815.1-based malicious activities and the attacks against SCADA, respectively. Both approaches are anomaly-based [73]. uses data Set from Korean Substation and their method successfully detects five types of malware behavior, three types of false data injection and disabling reassembly attacks, whereas [74] uses NSL-KDD Data Set and their method yields 98.84% detection accuracy.

ZESO-DRKFC [75] is a meta-heuristic based anomaly detection algorithm for securing SCADA by utilizing Zaire Ebola Search Optimization and Deep Random Kernel Forest Classification. Seven datasets are used for evaluating the performance and the results show that ZESO-DRKFC yields 99%accuracy.

Differently from anomaly-based methods, two specification-based attack detection [76,77] are proposed [76]. focus on IEC 61850-Based SCADA attacks. They design an SCADA-specific multi-layered IDS by defining normal behavior of the system with traffic and file analyses. Their proposed method, yielding 100% accuracy and process time lower than 0.3 ms, is validated in a real Test Bed [77]. designs a semantic analysis framework for control-related attacks against SCADA and their system shows 0.78% false positive and 0.01% false negative rate for large scale 2736-bus system; completes the detection in 200 ms.

9.6 Conclusion

Nowadays, proposing an IDS for SG is a popular topic in literature. In this study 58 SCI journal papers published between 2017 and 2023 are investigated in detail. Four components of SG, including AMI, substation, PMU and SCADA, which are prone to various types of attacks, are taken into account. Both general and component-based inferences are reported.

The following general conclusions have been made as a result of the investigations.

- The studies on SG have intensified in last 3 years. The most focused component is AMI.
- IDS dominates over IPS and IRS; whereas anomaly-based detection is the most common IDS approach.
- MATLAB is a popular simulation tool and both general and SG component specific datasets are used in simulations. Testbeds are less frequently used.
- Machine learning and deep learning approaches are popular in AMI and SCADA security; false data injection attacks are widespread in AMI and PMU.

In this study, besides the general evaluation, the comments on the basis of components are also made. Component-based attack and detection system analyses are important since most of the attacks take form according to the physical properties, used standards and used protocols of the components.

It can clearly be said that signature-based approaches are not suitable and hence, rarely used for SG. However, specification-based methods are an open issue. Moreover, real-time applications are quite limited. It is more appropriate to use SG specific dataset instead of general dataset.

References

[1] M.A. Judge, A. Khan, A. Manzoor, H.A. Khattak, Overview of smart grid implementation: frameworks, impact, performance and challenges, Journal of Energy Storage 49 (2022) 104056, https://doi.org/10.1016/j.est.2022.104056.

[2] A. Ayodeji, Y. Liu, N. Chao, L. Yang, A new perspective towards the development of robust data-driven intrusion detection for industrial control systems, Nuclear Engineering and Technology 52 (2020) 2687–2698, https://doi.org/10.1016/j.net.2020.05.012.
[3] M.K. Hasan, A.A. Habib, Z. Shukur, F. Ibrahim, S. Islam, M.A. Razzaque, Review on cyber-physical and cyber-security system in smart grid: standards, protocols, constraints, and recommendations, Journal of Network and Computer Applications 209 (2023) 103540, https://doi.org/10.1016/j.jnca.2022.103540.
[4] M.Z. Gunduz, R. Das, Cyber-security on smart grid: threats and potential solutions, Computer Networks 169 (2020) 107094, https://doi.org/10.1016/j.comnet.2019.107094.
[5] H. Zhang, B. Liu, H. Wu, Smart grid cyber-physical attack and defense: a review, IEEE Access 9 (2021) 29641–29659, https://doi.org/10.1109/ACCESS.2021.3058628.
[6] L. Rosa, T. Cruz, M.B. de Freitas, P. Quitério, J. Henriques, F. Caldeira, E. Monteiro, P. Simões, Intrusion and anomaly detection for the next-generation of industrial automation and control systems, Future Generation Computer Systems 119 (2021) 50–67, https://doi.org/10.1016/j.future.2021.01.033.
[7] M.A. Umer, K.N. Junejo, M.T. Jilani, A.P. Mathur, Machine learning for intrusion detection in industrial control systems: applications, challenges, and recommendations, International Journal of Critical Infrastructure Protection 38 (2022) 100516, https://doi.org/10.1016/j.ijcip.2022.100516.
[8] Q. Liu, V. Hagenmeyer, H.B. Keller, A review of rule learning-based intrusion detection systems and their prospects in smart grids, IEEE Access 9 (2021) 57542–57564, https://doi.org/10.1109/ACCESS.2021.3071263.
[9] T. Nguyen, S. Wang, M. Alhazmi, M. Nazemi, A. Estebsari, P. Dehghanian, Electric power grid resilience to cyber adversaries: state of the art, IEEE Access 8 (2020) 87592–87608, https://doi.org/10.1109/ACCESS.2020.2993233.
[10] M. Ghiasi, T. Niknam, Z. Wang, M. Mehrandezh, M. Dehghani, N. Ghadimi, A comprehensive review of cyber-attacks and defense mechanisms for improving security in smart grid energy systems: past, present and future, Electric Power Systems Research 215 (2023) 108975, https://doi.org/10.1016/j.epsr.2022.108975.
[11] P.I. Radoglou-Grammatikis, P.G. Sarigiannidis, Securing the smart grid: a comprehensive compilation of intrusion detection and prevention systems, IEEE Access 7 (2019) 46595–46620, https://doi.org/10.1109/ACCESS.2019.2909807.
[12] C. Peng, H. Sun, M. Yang, Y.-L. Wang, A survey on security communication and control for smart grids under malicious cyber attacks, IEEE Transactions on Systems, Man, and Cybernetics: Systems 49 (2019) 1554–1569, https://doi.org/10.1109/TSMC.2018.2884952.
[13] M.A. Ferrag, M. Babaghayou, M.A. Yazici, Cyber security for fog-based smart grid SCADA systems: solutions and challenges, Journal of Information Security and Applications 52 (2020) 102500, https://doi.org/10.1016/j.jisa.2020.102500.
[14] S.E. Quincozes, C. Albuquerque, D. Passos, D. Mossé, A survey on intrusion detection and prevention systems in digital substations, Computer Networks 184 (2021) 107679, https://doi.org/10.1016/j.comnet.2020.107679.
[15] Y. Kabalci, A survey on smart metering and smart grid communication, Renewable and Sustainable Energy Reviews 57 (2016) 302–318, https://doi.org/10.1016/j.rser.2015.12.114.
[16] T.N. Le, W.-L. Chin, D.K. Truong, T.H. Nguyen, Advanced metering infrastructure based on smart meters in smart grid, in: M. Eissa (Ed.), Smart Metering Technology and Services - Inspirations for Energy Utilities, InTech, 2016, https://doi.org/10.5772/63631.

[17] M.A. Ferrag, L. Maglaras, DeepCoin: a novel deep learning and blockchain-based energy exchange framework for smart grids, IEEE Transactions on Engineering Management 67 (2020) 1285–1297, https://doi.org/10.1109/TEM.2019.2922936.
[18] P. Jokar, V. Leung, Intrusion detection and prevention for ZigBee-based home area networks in smart grids, IEEE Transactions on Smart Grid 9 (3) (2016) 1800–1811, https://doi.org/10.1109/TSG.2016.2600585, 1–1.
[19] I. Parvez, M. Aghili, A.I. Sarwat, S. Rahman, F. Alam, Online power quality disturbance detection by support vector machine in smart meter, Journal of Modern Power System and Clean Energy 7 (5) (2019) 1328–1339, https://doi.org/10.1007/s40565-018-0488-z.
[20] M. Pilz, F.B. Naeini, K. Grammont, C. Smagghe, M. Davis, J.-C. Nebel, L. Al-Fagih, E. Pfluegel, Security attacks on smart grid scheduling and their defences: a game-theoretic approach, International Journal of Information Security 19 (2020) 427–443, https://doi.org/10.1007/s10207-019-00460-z.
[21] Y. Yılmaz, S. Uludag, Timely detection and mitigation of IoT-based cyberattacks in the smart grid, Journal of the Franklin Institute 358 (2021) 172–192, https://doi.org/10.1016/j.jfranklin.2019.02.011.
[22] T. Rose, K. Kifayat, S. Abbas, M. Asim, A hybrid anomaly-based intrusion detection system to improve time complexity in the Internet of Energy environment, Journal of Parallel and Distributed Computing 145 (2020) 124–139, https://doi.org/10.1016/j.jpdc.2020.06.012.
[23] A. Amara Korba, N. Tamani, Y. Ghamri-Doudane, N.E.I. karabadji, Anomaly-based framework for detecting power overloading cyberattacks in smart grid AMI, Computers & Security 96 (2020) 101896, https://doi.org/10.1016/j.cose.2020.101896.
[24] C.-C. Sun, D.J. Sebastian Cardenas, A. Hahn, C.-C. Liu, Intrusion detection for cybersecurity of smart meters, IEEE Transactions on Smart Grid 12 (2021) 612–622, https://doi.org/10.1109/TSG.2020.3010230.
[25] S. Bhattacharjee, S.K. Das, Detection and forensics against stealthy data falsification in smart metering infrastructure, IEEE Transactions on Dependable and Secure Computing 18 (2021) 356–371, https://doi.org/10.1109/TDSC.2018.2889729.
[26] F. Unal, A. Almalaq, S. Ekici, P. Glauner, Big data-driven detection of false data injection attacks in smart meters, IEEE Access 9 (2021) 144313–144326, https://doi.org/10.1109/ACCESS.2021.3122009.
[27] W. Zhang, X. Dong, H. Li, J. Xu, D. Wang, Unsupervised detection of abnormal electricity consumption behavior based on feature engineering, IEEE Access 8 (2020) 55483–55500, https://doi.org/10.1109/ACCESS.2020.2980079.
[28] A. Sial, A. Singh, A. Mahanti, Detecting anomalous energy consumption using contextual analysis of smart meter data, Wireless Network 27 (2021) 4275–4292, https://doi.org/10.1007/s11276-019-02074-8.
[29] H. Jain, M. Kumar, A.M. Joshi, Intelligent energy cyber physical systems (iECPS) for reliable smart grid against energy theft and false data injection, Electrical Engineering 104 (2022) 331–346, https://doi.org/10.1007/s00202-021-01380-9.
[30] G. Yogarajan, J.A. Vinosh, S.K.A. Prakash, S.M.M. Kumar, Interpolation search-based malicious user detection in smart grids, Electrical Engineering (2021) 1–11, https://doi.org/10.1007/s00202-020-01196-z.
[31] S.-C. Yip, W.-N. Tan, C. Tan, M.-T. Gan, K. Wong, An anomaly detection framework for identifying energy theft and defective meters in smart grids, International Journal of Electrical Power & Energy Systems 101 (2018) 189–203, https://doi.org/10.1016/j.ijepes.2018.03.025.
[32] M. Attia, S.M. Senouci, H. Sedjelmaci, E.-H. Aglzim, D. Chrenko, An efficient Intrusion Detection System against cyber-physical attacks in the smart grid, Computers

& Electrical Engineering 68 (2018) 499–512, https://doi.org/10.1016/j.compeleceng.2018.05.006.

[33] J. Jithish, B. Alangot, N. Mahalingam, K.S. Yeo, Distributed anomaly detection in smart grids: a federated learning-based approach, IEEE Access 11 (2023) 7157–7179, https://doi.org/10.1109/ACCESS.2023.3237554.

[34] R. Yao, N. Wang, P. Chen, D. Ma, X. Sheng, A CNN-transformer hybrid approach for an intrusion detection system in advanced metering infrastructure, Multimedia Tools and Applications 82 (13) (2022) 19463–19486, https://doi.org/10.1007/s11042-022-14121-2.

[35] Q. Yang, W. Hao, L. Ge, W. Ruan, F. Chi, FARIMA model-based communication traffic anomaly detection in intelligent electric power substations, IET Cyber-Physical Systems 4 (2019) 22–29, https://doi.org/10.1049/iet-cps.2018.5052.

[36] J. Appiah-Kubi, C.-C. Liu, Decentralized intrusion prevention (DIP) against Coordinated cyberattacks on distribution automation systems, IEEE Open Access Journal of Power and Energyis 7 (2020) 389–402, https://doi.org/10.1109/OAJPE.2020.3029805.

[37] I.-S. Choi, J. Hong, T.-W. Kim, Multi-agent based cyber attack detection and mitigation for distribution automation system, IEEE Access 8 (2020) 183495–183504, https://doi.org/10.1109/ACCESS.2020.3029765.

[38] A. de O. Paula, R.I. Meneguette, F.T. Giuntini, M.L.M. Peixoto, V.P. Gonçalves, G.P. Rocha Filho, STRAYER: a Smart Grid adapted automation architecture against cyberattacks, Journal of Information Security and Applications 67 (2022) 103195, https://doi.org/10.1016/j.jisa.2022.103195.

[39] J. Hong, C.-C. Liu, Intelligent electronic devices with collaborative intrusion detection systems, IEEE Transactions on Smart Grid 10 (2019) 271–281, https://doi.org/10.1109/TSG.2017.2737826.

[40] J. Kim, J. Park, FPGA-based network intrusion detection for IEC 61850-based industrial network, ICT Express 4 (2018) 1–5, https://doi.org/10.1016/j.icte.2018.01.002.

[41] L. Yang, Y. Zhai, Y. Zhang, Y. Zhao, Z. Li, T. Xu, A new methodology for anomaly detection of attacks in IEC 61850-based substation system, Journal of Information Security and Applications 68 (2022) 103262, https://doi.org/10.1016/j.jisa.2022.103262.

[42] T.S. Ustun, S.M.S. Hussain, A. Ulutas, A. Onen, M.M. Roomi, D. Mashima, Machine learning-based intrusion detection for achieving cybersecurity in smart grids using IEC 61850 GOOSE messages, Symmetry 13 (2021a) 826, https://doi.org/10.3390/sym13050826.

[43] T.S. Ustun, S.M.S. Hussain, L. Yavuz, A. Onen, Artificial intelligence based intrusion detection system for IEC 61850 sampled values under symmetric and asymmetric faults, IEEE Access 9 (2021b) 56486–56495, https://doi.org/10.1109/ACCESS.2021.3071141.

[44] G. Elbez, H.B. Keller, A. Bohara, K. Nahrstedt, V. Hagenmeyer, Detection of DoS attacks using ARFIMA modeling of GOOSE communication in IEC 61850 substations, Energies 13 (2020) 5176, https://doi.org/10.3390/en13195176.

[45] J. Hao, E. Kang, J. Sun, Z. Wang, Z. Meng, X. Li, Z. Ming, An adaptive Markov strategy for defending smart grid false data injection from malicious attackers, IEEE Transactions on Smart Grid 9 (2018) 2398–2408, https://doi.org/10.1109/TSG.2016.2610582.

[46] M.M.S. Khan, J.A. Giraldo, M. Parvania, Attack detection in power distribution systems using a cyber-physical real-time reference model, IEEE Transactions on Smart Grid 13 (2022) 1490–1499, https://doi.org/10.1109/TSG.2021.3128034.

[47] J. Hong, R.F. Nuqui, A. Kondabathini, D. Ishchenko, A. Martin, Cyber attack resilient distance protection and circuit breaker control for digital substations, IEEE Transactions on Industrial Informatics 15 (2019) 4332–4341, https://doi.org/10.1109/TII.2018.2884728.
[48] H. Huang, P. Wlazlo, Z. Mao, A. Sahu, K. Davis, A. Goulart, S. Zonouz, C.M. Davis, Cyberattack defense with cyber-physical alert and control logic in industrial controllers, IEEE Transactions on Industry Applications. 58 (2022) 5921–5934, https://doi.org/10.1109/TIA.2022.3186660.
[49] L. Xu, Y. Song, W. Zhang, Y. An, Y. Wang, H. Ning, An efficient foreign objects detection network for power substation, Image and Vision Computing 109 (2021) 104159, https://doi.org/10.1016/j.imavis.2021.104159.
[50] Z. Huang, W. Xie, W. Liu, C. Song, Y. Zhang, J. Jing, TSCDNet+: a highly robust substation anomaly detection method, Optik 246 (2021) 167808, https://doi.org/10.1016/j.ijleo.2021.167808.
[51] R.H. Khan, J.Y. Khan, A comprehensive review of the application characteristics and traffic requirements of a smart grid communications network, Computer Networks 57 (2013) 825–845, https://doi.org/10.1016/j.comnet.2012.11.002.
[52] U. Adhikari, T.H. Morris, S. Pan, Applying Hoeffding adaptive trees for real-time cyber-power event and intrusion classification, IEEE Transactions on Smart Grid 9 (2018b) 4049–4060, https://doi.org/10.1109/TSG.2017.2647778.
[53] U. Adhikari, T.H. Morris, S. Pan, Applying non-nested generalized exemplars classification for cyber-power event and intrusion detection, IEEE Transactions on Smart Grid 9 (2018a) 3928–3941, https://doi.org/10.1109/TSG.2016.2642787.
[54] Z. Xia, G. Long, B. Yin, Confidence-aware collaborative detection mechanism for false data attacks in smart grids, Soft Computing 25 (2021) 5607–5618, https://doi.org/10.1007/s00500-020-05557-5.
[55] A. Kummerow, K. Schäfer, P. Gupta, S. Nicolai, P. Bretschneider, Combined network intrusion and phasor data anomaly detection for secure dynamic control centers, Energies 15 (2022) 3455, https://doi.org/10.3390/en15093455.
[56] M. Ghafouri, M. Au, M. Kassouf, M. Debbabi, C. Assi, J. Yan, Detection and mitigation of cyber attacks on voltage stability monitoring of smart grids, IEEE Transactions on Smart Grid 11 (2020) 5227–5238, https://doi.org/10.1109/TSG.2020.3004303.
[57] C. Wang, C.-W. Ten, Y. Hou, Inference of compromised synchrophasor units within substation control networks, IEEE Transactions on Smart Grid 9 (2018) 5831–5842, https://doi.org/10.1109/TSG.2017.2697449.
[58] S. Basumallik, R. Ma, S. Eftekharnejad, Packet-data anomaly detection in PMU-based state estimator using convolutional neural network, International Journal of Electrical Power & Energy Systems 107 (2019) 690–702, https://doi.org/10.1016/j.ijepes.2018.11.013.
[59] G. Khare, A. Mohapatra, S.N. Singh, A real-time approach for detection and correction of false data in PMU measurements, Electric Power Systems Research 191 (2021) 106866, https://doi.org/10.1016/j.epsr.2020.106866.
[60] S. De, R. Sodhi, A PMU assisted cyber attack resilient framework against power systems structural vulnerabilities, Electric Power Systems Research 206 (2022) 107805, https://doi.org/10.1016/j.epsr.2022.107805.
[61] C. Roberts, A. Scaglione, M. Jamei, R. Gentz, S. Peisert, E.M. Stewart, C. McParland, A. McEachern, D. Arnold, Learning behavior of distribution system discrete control devices for cyber-physical security, IEEE Transactions on Smart Grid 11 (2020) 749–761, https://doi.org/10.1109/TSG.2019.2936016.
[62] W.-C. Hong, D.-R. Huang, C.-L. Chen, J.-S. Lee, Towards accurate and efficient classification of power system contingencies and cyber-attacks using recurrent neural

networks, IEEE Access 8 (2020) 123297–123309, https://doi.org/10.1109/ACCESS.2020.3007609.

[63] A.L. Amutha, R. Annie Uthra, J. Preetha Roselyn, R. Golda Brunet, Anomaly detection in multivariate streaming PMU data using density estimation technique in wide area monitoring system, Expert Systems with Applications 175 (2021) 114865, https://doi.org/10.1016/j.eswa.2021.114865.

[64] A. Xue, F. Xu, J.H. Chow, S. Leng, H. Kong, J. Xu, T. Bi, Data-driven detection for GPS spoofing attack using phasor measurements in smart grid, International Journal of Electrical Power & Energy Systems 129 (2021) 106883, https://doi.org/10.1016/j.ijepes.2021.106883.

[65] S. Shazdeh, H. Golpîra, H. Bevrani, A PMU-based back-up protection scheme for fault detection considering uncertainties, International Journal of Electrical Power & Energy Systems 145 (2023) 108592, https://doi.org/10.1016/j.ijepes.2022.108592.

[66] S. Dutta, Leveraging a micro synchrophasor for fault detection in a renewable based smart grid—a machine learned sustainable solution with cyber-attack resiliency, Advances in Electrical Engineering 2 (2022) 100090.

[67] K. Sayed, H.A. Gabbar, SCADA and smart energy grid control automation, in: Smart Energy Grid Engineering, Elsevier, 2017, pp. 481–514, https://doi.org/10.1016/B978-0-12-805343-0.00018-8.

[68] H. Bentarzi, M. Tsebia, A. Abdelmoumene, PMU based SCADA enhancement in smart power grid, in: 2018 IEEE 12th International Conference on Compatibility, Power Electronics and Power Engineering (CPE-POWERENG 2018). Presented at the 2018 IEEE 12th International Conference on Compatibility, Power Electronics and Power Engineering (CPE-POWERENG), IEEE, Doha, 2018, pp. 1–6, https://doi.org/10.1109/CPE.2018.8372580.

[69] D. Upadhyay, J. Manero, M. Zaman, S. Sampalli, Gradient boosting feature selection with machine learning classifiers for intrusion detection on power grids, IEEE Transactions on Network and Service Management 18 (2021) 1104–1116, https://doi.org/10.1109/TNSM.2020.3032618.

[70] A. Gumaei, M.M. Hassan, S. Huda, MdR. Hassan, D. Camacho, J. Del Ser, G. Fortino, A robust cyberattack detection approach using optimal features of SCADA power systems in smart grids, Applied Soft Computing 96 (2020) 106658, https://doi.org/10.1016/j.asoc.2020.106658.

[71] I.A. Khan, D. Pi, N. Khan, Z.U. Khan, Y. Hussain, A. Nawaz, F. Ali, A privacy-conserving framework based intrusion detection method for detecting and recognizing malicious behaviours in cyber-physical power networks, Applied Intelligence 51 (2021) 7306–7321, https://doi.org/10.1007/s10489-021-02222-8.

[72] X.C. Yin, Z.G. Liu, L. Nkenyereye, B. Ndibanje, Toward an applied cyber security solution in IoT-based smart grids: an intrusion detection system approach, Sensors 19 (2019) 4952, https://doi.org/10.3390/s19224952.

[73] S. Kwon, H. Yoo, T. Shon, IEEE 1815.1-based power system security with bidirectional RNN-based network anomalous attack detection for cyber-physical system, IEEE Access 8 (2020) 77572–77586, https://doi.org/10.1109/ACCESS.2020.2989770.

[74] S.Y. Diaba, M. Shafie-khah, M. Elmusrati, On the performance metrics for cyber-physical attack detection in smart grid, Soft Computing 26 (2022) 13109–13118, https://doi.org/10.1007/s00500-022-06761-1.

[75] O.B.J. Rabie, P.K. Balachandran, M. Khojah, S. Selvarajan, A proficient ZESO-DRKFC model for smart grid SCADA security, Electronics 11 (2022) 4144, https://doi.org/10.3390/electronics11244144.

[76] Y. Yang, H.-Q. Xu, L. Gao, Y.-B. Yuan, K. McLaughlin, S. Sezer, Multidimensional intrusion detection system for IEC 61850-based SCADA networks, Transactions on

Power Delivery 32 (2017) 1068–1078, https://doi.org/10.1109/TPWRD.2016.2603339.

[77] H. Lin, A. Slagell, Z.T. Kalbarczyk, P.W. Sauer, R.K. Iyer, Runtime semantic security analysis to detect and mitigate control-related attacks in power grids, IEEE Transactions on Smart Grid 9 (2018) 163–178, https://doi.org/10.1109/TSG.2016.2547742.

CHAPTER 10

Artificial intelligence and machine learning in cyber security applications

Ruchika Gupta[1] and Priyank Srivastava[2]
[1]Mangalmay Institute of Management & Technology, Greater Noida, Uttar Pradesh, India; [2]Galgotias University, Greater Noida, Uttar Pradesh, India

10.1 Introduction

Recent years have seen a sharp rise in the complexity and frequency of cyber threats due to the exponential proliferation of digital technology and the widespread usage of the internet [1]. As cybercriminals become more sophisticated, traditional cybersecurity measures alone are no longer sufficient to protect critical systems and sensitive information [2]. The use of AI and ML technologies together has shown to be a game-changer in cybersecurity as a means of countering the constantly changing cyber threat scenario.

Cyber threats rank highly among the world's most significant hazards in terms of both likelihood and effect, according to the World Economic Forum's Global Risks Report [3] (Refer Table 10.1). With the proliferation of interconnected devices and the increasing sophistication of cyber adversaries, organizations face unprecedented challenges in securing their digital assets. The traditional signature-based approaches to cybersecurity are proving inadequate in detecting and mitigating advanced and targeted attacks [4].

The integration of AI and ML technologies represents a seismic shift in cybersecurity, empowering organizations to adopt proactive defense mechanisms. AI, with its ability to simulate human intelligence and adapt to changing patterns, acts as a digital sentry, constantly monitoring and analyzing vast datasets in real-time [5,6]. On the other hand, ML, a subset of AI, excels in learning from historical data, enabling predictive analytics and automation in cybersecurity operations [7,8].

Cyber Security Solutions for Protecting and Building the Future Smart Grid
ISBN 978-0-443-14066-2
https://doi.org/10.1016/B978-0-443-14066-2.00004-9

Table 10.1 Evolving risks landscape.

Year	Rank of cyber threats in top global risks by likelihood
2021	6
2020	6
2019	5
2018	3
2017	5

Source: World Economic Forum's Global Risks Report.

The objective of this chapter is to examine the diverse role that AI and ML play in the field of cybersecurity. It will delve into how these advanced technologies contribute to improving the identification of threats, the management of incidents, and the overall state of security. This investigation makes a valuable contribution to the expanding field of research at the junction of AI, ML, and cybersecurity by effectively bridging the divide between theoretical comprehension and practical application.

The chapter is organized in a manner that offers a thorough examination of the fundamental principles of AI and ML within the context of cybersecurity. It places particular emphasis on the many uses of these technologies in different aspects of security operations. The system traverses several domains like threat detection, incident response, security operations, endpoint security, network security, threat hunting, and cloud security. Furthermore, this study delves into the issues and constraints that arise in the context of AI and ML in the field of cybersecurity. It offers valuable insights into the ethical implications, biases, and possible adversarial assaults that intelligent systems may encounter. The chapter concludes by providing a prospective analysis of forthcoming developments, encompassing the utilization of reinforcement learning, the integration of AI-ML hybrid models, and the convergence of quantum computing and AI-ML in the field of cybersecurity.

10.2 Foundations of AI and ML in cyber security

The incorporation of AI and ML in the field of cybersecurity signifies a fundamental change in the strategy employed to protect digital resources. Artificial Intelligence is a multidisciplinary domain within the area of computer science that focuses on the creation and advancement of intelligent systems with the ability to execute activities that conventionally necessitate human intelligence [9]. Within the realm of cybersecurity, artificial intelligence functions as a digital guardian, dynamically adjusting itself to identify and counteract evolving vulnerabilities promptly and effectively [10]. The capacity of artificial intelligence to adapt allows it to surpass conventional rule-based methods, hence offering a flexible protection mechanism against the always changing realm of cyber threats [11].

Machine Learning, which is a subfield of Artificial Intelligence, enhances the fundamental principles of cybersecurity by incorporating the notion of autonomous learning and pattern recognition. Machine learning algorithms has the capacity to examine extensive datasets, acquire knowledge from past patterns, and adjust to novel information, rendering them crucial in augmenting security procedures [12,13]. The applications of ML in cybersecurity extend to various domains, including threat detection, anomaly identification, and predictive analysis. This foundation allows cybersecurity systems to evolve from static, signature-based approaches to more proactive and adaptive strategies [14].

It is important to establish a clear distinction between AI and ML within the realm of cybersecurity in order to foster a comprehensive comprehension of their respective implementations. AI spans a wider range of intelligent system development, whereas ML especially concentrates on the creation of systems that possess the ability to learn and enhance their performance without the need for explicit programming. The interaction between these two ideas is the fundamental basis of intelligent cybersecurity solutions, wherein AI serves as the overall framework, while ML gives the skills for learning and adaptation [15]. This distinction establishes the foundation for a thorough examination of how various technologies work together to strengthen digital security.

The actual use of AI and ML in the field of cybersecurity is not only a theoretical notion but rather a pragmatic approach to address the increasing magnitude of cyber threats encountered by various businesses. The rise in sophisticated attacks and the need for adaptive defense mechanisms underscore the significance of these technologies in fortifying the digital landscape [16]. Fig. 10.1. Shows the potential application of AI and ML in cybersecurity.

As organizations increasingly recognize the limitations of traditional cybersecurity measures, the foundations of AI and ML become crucial pillars in the ongoing battle against cyber adversaries.

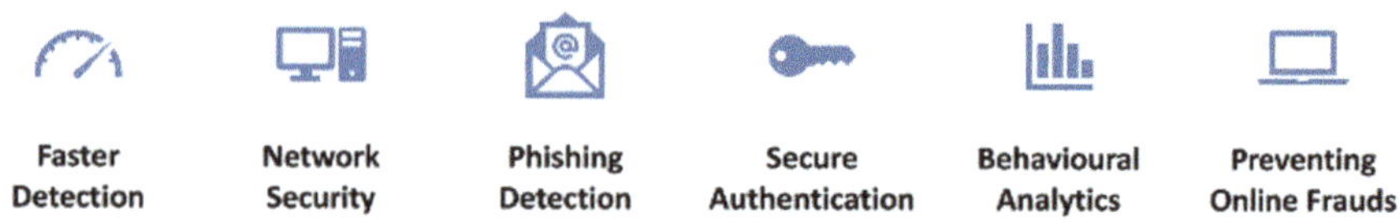

Figure 10.1 Application of AI and ML.

10.3 AI and ML-driven threat detection

The emergence of AI and ML technologies has significantly transformed the domain of threat identification within the field of cybersecurity. By using the capabilities of intelligent algorithms, companies are now able to proactively detect and address possible risks, beyond the constraints associated with conventional, signature-based methods. The merging of artificial intelligence's flexibility and machine learning's pattern recognition skills has led to a paradigm change in threat identification. This integration offers a dynamic defense mechanism that effectively counters the constantly developing tactics employed by cyber attackers.

10.3.1 Utilizing AI for anomaly detection

Anomaly detection is a crucial aspect of threat detection, and the integration of AI significantly enhances its effectiveness. AI utilizes its inherent flexibility and advanced real-time analytic skills to effectively identify and differentiate regular behavioral patterns within a given network. The intrinsic dynamic and responsive nature of AI algorithms enables them to grow in tandem with evolving cyber threats, hence enhancing their flexibility in comparison to static rule-based systems [10]. Through the continual monitoring of network operations, AI can detect deviations from predefined baselines, therefore indicating possible security vulnerabilities. The expeditious detection of anomalies allows businesses to swiftly address new risks, hence minimizing possible harm and decreasing the duration in which adversaries remain undetected within their networks. Fig. 10.2 shows different areas of application of anomaly detection using AI.

Figure 10.2 Anomaly detection applications.

10.3.2 Behavior-based threat detection with machine learning

Behavior-based threat detection, bolstered by the capabilities of Machine Learning, introduces a proactive dimension to cybersecurity. ML models analyze historical data to establish baseline behaviors, facilitating the recognition of deviations indicative of potential security threats [12]. This method transcends the limitations of signature-based approaches, allowing for the identification of novel and previously unseen threats. The dynamic nature of ML enables continuous learning and adaptation to the evolving attack patterns. By extending behavioral analysis to entire networks, ML provides a comprehensive understanding of potential security risks, allowing organizations to stay one step ahead of adversaries. Fig. 10.3 shows the workflow of behavior-based threat detection using ML.

10.3.3 Predictive analysis and threat intelligence

The utilization of machine learning's capacity to evaluate past data enables cybersecurity experts to anticipate and take proactive measures against future risks, thereby enhancing the effectiveness of predictive analysis [17]. Machine learning models, when combined with threat intelligence feeds, contribute to the improvement of situational awareness by consistently updating their understanding of dynamic cyber threats. The integration of predictive analytics and threat information enables the implementation of a proactive defensive plan, therefore mitigating the reactive approach commonly observed in conventional cybersecurity practices. Organizations that possess predictive analysis skills can foresee possible dangers, enabling them to proactively execute preventative measures aimed at mitigating the effect of cyber assaults.

10.4 AI and ML-driven incident response

The foundation of incident response practices is preparedness, which includes both the establishment of an incident response capacity and the prevention of events via the adequate security of systems, networks, and applications (refer Fig. 10.4).

The utilization of AI and ML in incident response is revolutionizing the field of cybersecurity by providing enhanced functionalities in promptly identifying and assessing incidents, automating responses to cyber threats, and employing AI-driven techniques for forensic analysis and investigation. Artificial Intelligence's adaptive learning and continuous

Figure 10.3 Workflow.

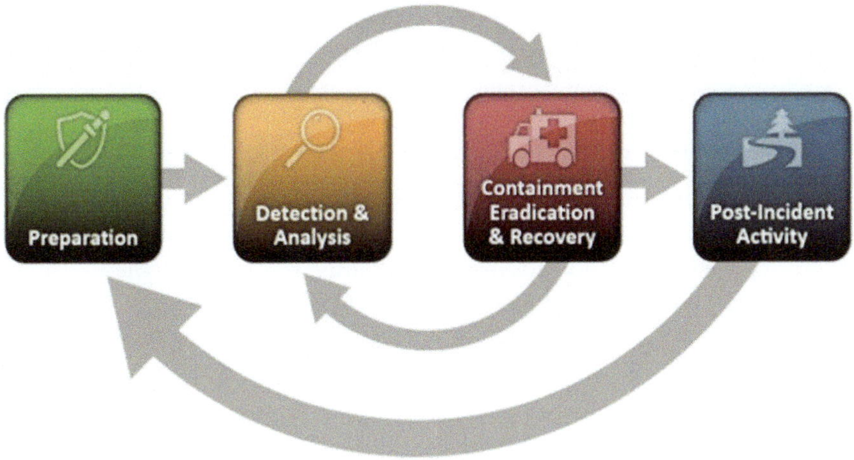

Figure 10.4 Incident response. *(Source: https://www.nist.gov/)*

monitoring capabilities facilitate the prompt detection of possible hazards in real-time event identification and triage scenarios [18]. According to Ref. [10]; the utilization of AI enables the expeditious identification of incidents through the examination of patterns and anomalies in real time. This capability empowers security teams to immediately prioritize these incidents and take immediate action to reduce any damages. The use of this dynamic strategy guarantees that security experts possess up-to-date knowledge, hence promoting a proactive posture in addressing emerging cyber risks.

The implementation of automated responses to cyber-attacks is a crucial aspect of incident response driven by AI and ML. This advancement has significantly transformed the efficiency and swiftness with which threats are mitigated. The utilization of Machine Learning algorithms facilitates the implementation of automated responses that are customized to address distinct categories of cyber threats. This implementation results in the reduction of reaction times and the mitigation of potential human errors [12]. The implementation of automation not only facilitates the efficient handling of routine replies but also enables human resources to allocate their attention toward more intricate threat analysis and decision-making processes. In addition, the use of AI-enabled forensics and investigative techniques increases the incident response procedure by utilizing ML algorithms to examine extensive collections of data, detect recurring trends, and recreate the sequence of events related to the incident. According to

Ref. [17]; this process expedites the determination of the underlying factor, hence enabling comprehensive examinations and the collection of practical knowledge for subsequent preventive measures.

10.4.1 Real-time incident identification and triage

The recognition and prioritization of incidents in real-time have become of utmost importance in the continuously changing field of cybersecurity. The incorporation of AI and ML has significantly transformed these essential procedures. The capacity of AI to dynamically adjust and acquire knowledge in real time, along with its ongoing surveillance capabilities, enables companies to promptly detect possible security issues as they occur. According to Ref. [10]; AI systems can automatically triage and prioritize events by examining trends, abnormalities, and deviations from typical behavior. This enables security teams to receive timely information about potential risks. The use of this dynamic method guarantees a proactive and timely reaction to security issues, hence reducing the adverse effects and length of possible breaches.

ML algorithms play a crucial role in the domain of real-time event detection and triage since they are capable of efficiently processing large volumes of data with high velocity. Machine learning models can examine network activity, detect irregular patterns, and classify occurrences according to predetermined criteria in an automated manner. According to Ref. [12]; the utilization of this approach enhances the expeditious identification of significant occurrences, enabling cybersecurity teams to efficiently deploy resources and prioritize their response efforts. The combination of artificial intelligence's flexibility and machine learning's ability to identify patterns enables enterprises to effectively traverse the ever-changing cybersecurity environment with agility and accuracy.

10.4.2 Automated response to cyber attacks

This aspect of incident response signifies a deviation from conventional manual methodologies, incorporating a dynamic and adaptable component to safeguard against the continuously growing risk environment. According to Ref. [12]; the utilization of Machine Learning algorithms in an automated response framework allows for the autonomous handling of targeted cyber threats, resulting in decreased reaction durations and mitigated potential for human mistakes. This automation not only speeds up regular responses but also allows human resources to focus on more complicated threat analysis and decision-making.

The effectiveness of automated response systems is notably demonstrated in their capacity to swiftly mitigate established hazards. Automated systems can utilize past threat data and predetermined reaction procedures to detect and mitigate well-known attack vectors. The capacity to effectively handle typical cyber events is of utmost importance, as it enables enterprises to sustain a proactive security posture [10]. The increasing variety and severity of threats necessitate the use of automated responses, which play a crucial role in mitigating the impact of cyber assaults and bolstering the overall resilience of cybersecurity measures.

The incorporation of automatic reaction mechanisms is in line with the notion of threat hunting, wherein AI and ML systems consistently analyze trends that may suggest possible dangers. Automated response technologies can promptly detect and mitigate these threats, therefore establishing a proactive defense stance. The adoption of a proactive strategy is crucial for enterprises to maintain a competitive edge against cyber adversaries, particularly in light of the rapid and complex evolution of current cyber threats [17]. The integration of automated reaction capabilities and ongoing threat-hunting activities establishes a resilient security system that effectively adjusts to the ever-changing dynamics of the cyber threat landscape.

10.4.3 AI-powered forensics and investigation

This aspect utilizes the functionalities of AI and ML to optimize the forensic analysis procedure, assisting in the determination of the underlying factors behind occurrences and the assignment of responsibility to malicious individuals.

An essential element of AI-driven forensics is in its capacity to effectively and rapidly process extensive information. Machine learning algorithms can examine substantial amounts of data, such as logs, network traffic, and system activity, to detect patterns and abnormalities that may indicate the occurrence of a security issue. This capability allows cybersecurity experts to reconstruct the chronological sequence of events related to a security issue, therefore offering a full perspective on the advancement of the assault. The expeditious processing of information by AI plays a pivotal role in mitigating the consequences of cyber events and expediting the implementation of countermeasures [17].

Furthermore, the utilization of artificial intelligence in the field of forensics demonstrates exceptional proficiency in the identification of correlations and the revelation of concealed associations inside intricate datasets. Machine learning algorithms can discern intricate connections among apparently disparate occurrences, hence facilitating the identification of complex forms of cyberattacks, such as advanced persistent threats (APTs). The automation of the correlation process using artificial intelligence has been shown to decrease the amount of time and effort needed for investigators to establish connections between various pieces of information. This, in turn, enables a more comprehensive and efficient analysis, as demonstrated by Ref. [12]. As the landscape of cyber threats undergoes constant evolution, artificial intelligence algorithms consistently upgrade their knowledge base. This process ensures that investigative tools retain their effectiveness in countering emerging attack vectors. The capacity to adapt is of utmost importance to maintain an advantage against cyber adversaries that continuously enhance their techniques. The use of AI-enabled forensic techniques not only expedites the process of inquiry but also enhances the proactive defensive stance of companies.

10.5 AI and ML in security operations

The use of AI and ML in Security Operations signifies a comprehensive methodology toward contemporary cybersecurity. The advancements discussed encompass various aspects of cybersecurity, including the analysis of security events in AI-enabled security Information and Event Management (SIEM), the automation and optimization of incident response workflows in Advancements in Security Orchestration and Automation (SOAR), and the collaborative intelligence of Cognitive Security Operations Centers (SOCs). These advancements collectively contribute to a cybersecurity posture that is characterized by proactivity, adaptability, and resilience. The importance of the collaboration between human intuition and AI-driven insights is growing as cyber threats continue to develop and get more complex [19].

10.5.1 AI-enabled Security Information and Event Management (SIEM)

The use of AI in Security Information and Event Management (SIEM) represents a notable progression within the field of cybersecurity. Conventional Security Information and Event Management (SIEM) systems are designed to gather and evaluate log data to identify and address security events. However, the increasing magnitude and intricacy of contemporary data pose significant difficulties for manual analysis. The utilization of

AI-enabled SIEM systems incorporates the application of ML techniques to augment the analysis of security events. These algorithms possess the capability to identify patterns, abnormalities, and trends that may not be discernible by rule-based methodologies. Through the ongoing analysis of past data, AI-enabled SIEM systems enhance their ability to differentiate between normal and malicious activity. This improvement leads to a reduction in false positives, enabling security professionals to allocate their resources toward addressing true threats. The utilization of this technology not only improves the precision of identifying possible threats but also empowers enterprises to take proactive measures in addressing prospective security issues.

10.5.2 Advancements in Security Orchestration and Automation (SOAR)

SOAR solutions signify a significant transformation in the efficiency and efficacy of security operations. AI and ML have a significant impact on the advancement of SOAR capabilities. These systems can automate and optimize incident response workflows, effectively integrating with a range of security technologies to establish a unified and comprehensive defensive mechanism. Machine learning algorithms integrated into SOAR systems exhibit a perpetual learning capability by using past event data, hence enhancing response methods and decision-making procedures. AI-driven SOAR technologies effectively minimize reaction times and human involvement by automating repetitive operations and coordinating intricate response workflows. This leads to a security posture that is more adaptable and resilient, enabling enterprises to proactively address emerging cyber threats.

10.5.3 Cognitive Security Operations Centers (SOCs)

SOCs represent the integration of human knowledge with artificial intelligence capabilities within the cybersecurity domain. These SOC architectures utilize Artificial Intelligence (AI) techniques to perform analysis of threat intelligence, conduct investigations into security incidents, and provide assistance for decision-making processes. Machine learning algorithms are utilized to analyze extensive datasets to detect intricate patterns, trends, and abnormalities, hence offering security analysts important insights. SOCs augment the productivity of human analysts by automating monotonous activities and providing insightful suggestions. The synergy resulting from the collaboration of artificial intelligence and human expertize plays a significant role in addressing complex and dynamic cyber threats. SOCs not only enhance the efficiency and precision of threat detection but also enable enterprises to take proactive measures in anticipating and addressing future security concerns.

10.6 AI and ML for endpoint security

Protecting endpoints (user devices, servers, smartphones, tablets, and virtual environments) against online and offline threats, malware, spyware, and viruses is an important part of endpoint security (refer Fig. 10.5). Endpoint security, strengthened by the integration of AI and ML, occupies a prominent position in safeguarding enterprises from a wide range of cyber threats. Behavioral analysis plays a crucial role in the field of endpoint security, and AI-based solutions demonstrate exceptional performance in this area. Machine learning algorithms exhibit a perpetual learning process and possess the capability to adjust themselves based on the established patterns of endpoint behavior. Consequently, they become proficient in detecting deviations from the norm, so enabling the identification of abnormalities and potential security breaches. The use of this proactive strategy enables enterprises to promptly identify and address potential risks at the endpoint level, hence

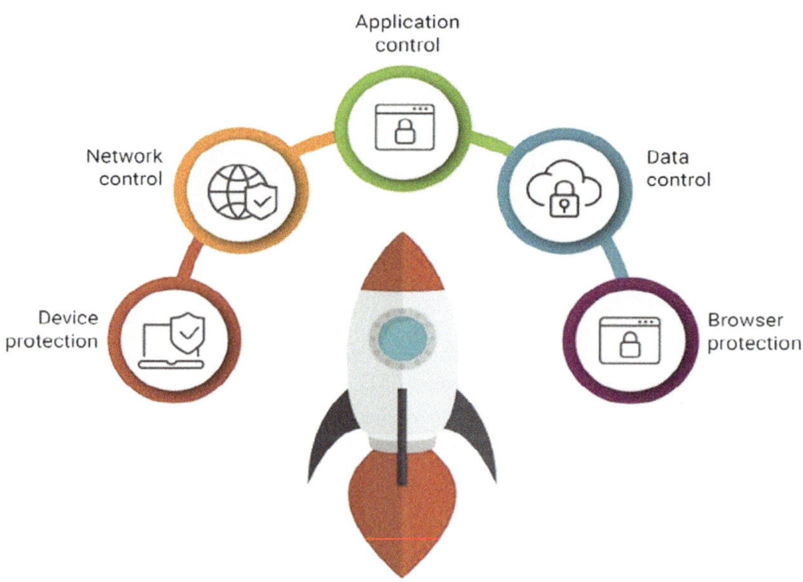

Figure 10.5 Endpoint security component. *(Source: https://www.spiceworks.com/it-security/network-security/articles/what-is-endpoint-security/)*

reducing the likelihood of sophisticated and focused assaults. In addition, AI-enhanced antivirus and anti-malware solutions utilize ML techniques to identify patterns linked to both familiar and unfamiliar malware, so offering a stronger safeguard against constantly growing security risks.

10.6.1 Behavioral analysis for endpoint protection

The practice of behavioral analysis within the realm of endpoint security is the systematic observation and examination of user actions, application behavior, and process operations to detect and pinpoint potentially illicit or questionable activity. AI systems, especially those utilizing unsupervized learning techniques, can identify deviations from predefined behavioral baselines. AI-driven solutions can identify possible security threats by comparing observed actions with established norms for each endpoint. The utilization of this dynamic methodology empowers companies to transcend conventional signature-based detection techniques and proficiently counteract emerging and developing manifestations of malware and cyber threats.

10.6.2 AI-enhanced antivirus and antimalware solutions

Conventional antivirus solutions frequently depend on signature-based detection methods, which might encounter difficulties in effectively adapting to the fast advancements of malware. AI-enhanced antivirus and anti-malware solutions utilize ML techniques to identify and analyze patterns and behaviors that are indicative of dangerous activity. These proposed systems possess the capability to effectively evaluate extensive datasets including pre-identified malware instances, hence acquiring the ability to discern prevalent attributes. By engaging in this practice, individuals may take proactive measures to detect and prevent newly emerging threats, hence enhancing their ability to defend against the always-evolving array of malware variations.

10.6.3 Adaptive and self-learning endpoint protection

The integration of AI and ML technologies inside endpoint security systems facilitates the advancement of adaptive and self-learning mechanisms for defense. These systems undergo continual learning from fresh data and occurrences, leading to their evolution to identify and address emerging risks. Machine learning models provide the capability to autonomously enhance their comprehension of emerging attack vectors, therefore guaranteeing the continued efficacy of endpoint protection measures against contemporary manifestations of cyber threats. The capacity to adapt is of utmost importance in a constantly changing cybersecurity environment, since conventional, unchanging protections may prove inadequate in offering full security.

10.7 AI and ML in network security

The field of network security is being transformed by the emergence of AI and ML, which are bringing about significant changes by enabling the implementation of sophisticated techniques for real-time detection, prevention, and response to cyber-attacks. These technologies enable enterprises to strengthen their network defenses and adjust to the ever-changing nature of cyber dangers.

10.7.1 AI-Based Intrusion Detection and Prevention Systems (IDPS)

With its sophisticated capacities to identify and reduce security risks inside network settings, AI-Based IDPS have emerged as essential elements of contemporary cybersecurity. The accuracy and effectiveness of intrusion detection and prevention are improved by these systems by using the capabilities of AI, specifically ML techniques. Strong network security requires AI-driven solutions as traditional signature-based IDPS techniques are often insufficient to counteract the changing strategies of cybercriminals [20].

Network traffic patterns analysis and the detection of unusual activity that might point to possible security risks are two areas in which AI-based IDPS shine. By using past data to train their algorithms, machine learning systems can identify both typical network activity and departures from the predefined norm. Because of this flexible approach, AI-based IDPS are better able to recognize and identify new threats than they were in the past. These technologies provide businesses with a proactive defensive mechanism that can recognize and neutralize threats instantly by constantly learning from and upgrading their models [21].

Additionally, AI is being used in IDPS in ways that go beyond conventional rule-based methodologies. A more thorough and nuanced knowledge of network activity is provided by AI-based IDPS's essential functions, which include behavioral analysis, anomaly identification, and predictive modeling. This helps security teams concentrate on actual security issues by improving threat detection accuracy and lowering false positives. An increasingly complex digital world can be defended against by companies thanks to the strategic development of network security that the incorporation of AI in IDPS implies.

10.7.2 Network traffic analysis with machine learning

Machine Learning plays a vital role in the study of network traffic, facilitating enterprises in acquiring valuable insights pertaining to the patterns and anomalies present within extensive datasets. Machine learning algorithms provide exceptional performance in the examination of extensive datasets produced by network traffic, facilitating the detection of nuanced anomalies and patterns that may signify possible security breaches. The use of supervised and unsupervized learning methods in Network Traffic Analysis (NTA) facilitates the creation of models capable of distinguishing between regular network activity and potentially harmful or malicious actions. According to Ref. [22]; these models possess the ability to adapt to changing threat environments by using previous data. As a result, enterprises are equipped with a dynamic defensive mechanism that effectively counters developing cyber threats.

Moreover, machine learning-enhanced network traffic analysis surpasses traditional rule-based methods by providing advanced behavioral analysis and predictive modeling capabilities. The use of sophisticated approaches enables businesses to anticipate possible security issues before their escalation, hence facilitating a proactive approach in the identification and response to threats. Using machine learning in network traffic analysis, security teams can effectively analyze extensive quantities of network data, detect recurring patterns, and assign priority to alarms. This approach enhances the efficiency of the investigative process and mitigates the occurrence of false positives. The integration of machine learning with network traffic analysis indicates noteworthy progression in the field of network security, enabling enterprises to proactively combat highly skilled cyber attackers.

10.7.3 AI-driven Network Access Control (NAC)

The use of Artificial Intelligence in conjunction with conventional NAC technologies has resulted in a revolutionary method for effectively controlling and safeguarding network endpoints. Traditional Network Access Control systems are of paramount importance in the maintenance of network security by allowing access only to approved devices and users. The use of artificial intelligence improves this procedure by bringing flexible and responsive decision-making that is informed by real-time contextual elements. The use of Artificial Intelligence in Network Access Control systems involves the application of Machine Learning algorithms to examine user behavior, device well-being, and network circumstances. This enables the implementation of access control choices that are more refined and adaptable [23].

One of the primary advantages of AI-Driven NAC is in its capacity to surpass conventional static access control regulations. Conventional NAC solutions often depend on pre-established regulations that may not sufficiently encompass the intricacies of contemporary network landscapes. The AI-powered NAC system demonstrates the ability to effectively respond to the ever-changing characteristics of networks via its ongoing process of acquiring knowledge and revising its comprehension of typical operations. The system's flexibility allows for the identification of anomalous or suspicious behaviors that might potentially signify security risks. The overall security posture of network environments may be enhanced by AI-driven NAC via the dynamic adjustment of access permissions based on real-time analysis.

Furthermore, the implementation of AI-Driven NAC plays a significant role in mitigating the risks associated with illegal access and possible security breaches. Through the implementation of automated analysis of contextual aspects and the use of predictive modeling techniques, these systems can detect and anticipate possible security issues before their escalation. The proactive strategy described here is in accordance with the changing threat landscape, where the ability to make quick and adaptable decisions is crucial for ensuring the security of a network environment. The use of AI in NAC signifies a notable progression in network security, affording enterprises a heightened level of intelligence and adaptability in regulating network access.

10.8 AI and ML in threat hunting

The practice of threat hunting, which involves actively and repeatedly searching for concealed risks inside the network of an organization, has seen a substantial evolution due to the incorporation of AI and ML technology. These sophisticated technologies enhance the capabilities of cybersecurity experts in identifying and addressing possible attacks with more efficiency, therefore offering a vital level of protection in the always-expanding realm of cyber risks. Fig. 10.6 show the steps involved in the threat-hunting process.

Figure 10.6 Threat hunting process.

10.8.1 Proactive threat hunting using AI

The incorporation of AI-driven technology into the practice of threat hunting enhances its effectiveness by introducing a proactive and dynamic element to the conventional approach. Machine Learning algorithms, particularly those that use supervised and unsupervized learning techniques, are capable of analyzing extensive datasets. Through this analysis, these algorithms can detect patterns, abnormalities, and deviations that might potentially indicate the presence of risks [24]. Through the use of historical data and the continual adaptation to developing tactics and procedures deployed by cyber attackers, the implementation of AI-driven threat hunting enables security teams to proactively anticipate and mitigate emerging risks. The use of a proactive approach is crucial in the identification and mitigation of possible security issues prior to their escalation, hence enhancing the overall resilience of the cybersecurity framework.

10.8.2 Utilizing ML for identifying new and emerging threats

Machine Learning assumes a crucial role in the identification of novel and emerging risks that may not be readily apparent using conventional approaches. Machine learning algorithms demonstrate exceptional performance in the analysis of various data sources, such as network logs, endpoint activity, and threat intelligence feeds, with the purpose of identifying patterns related to new attack methods. According to Ref. [17]; the use of unsupervized learning enables machine learning models to independently detect and classify outliers and abnormalities. This capability empowers security teams to effectively examine and mitigate possible risks that may otherwise remain undetected. The possession of this skill is of utmost importance in order to maintain an advantage against cyber threats that are constantly and swiftly changing.

10.8.3 AI-enhanced threat intelligence gathering

AI and ML significantly contribute to enhancing the process of threat intelligence acquisition, thereby providing security professionals with an ample supply of actionable and context-sensitive data. The collection, analysis, and correlation of threat intelligence data from a variety of sources—including dark web monitoring and open-source feeds—can be automated by these systems. AI-driven threat intelligence solutions provide a holistic perspective on possible threats by including internal data and considering the organization's threat environment. This capability provides threat hunters with vital information, allowing them to make well-informed judgments and effectively prioritize their efforts in the detection and mitigation of concealed dangers.

10.9 The role of AI and ML in cloud security

The emergence of cloud computing has significantly transformed how corporations oversee and implement their information technology infrastructure. The growing reliance on cloud services requires the adoption of robust security measures (see Fig. 10.7) to safeguard sensitive data and preserve the integrity of digital operations. AI and ML are now crucial elements in enhancing cloud security, offering enhanced capabilities in the

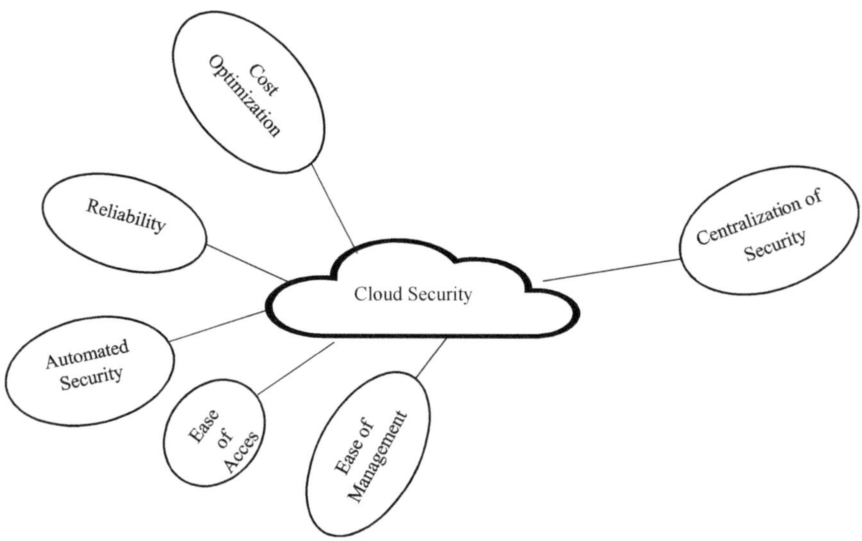

Figure 10.7 Elements of cloud security.

domains of detecting threats, managing incidents, and overseeing holistic risk management.

10.9.1 AI-driven cloud security monitoring

AI, via machine learning techniques, improves the monitoring of cloud security by conducting ongoing analysis of extensive datasets in real-time. These algorithms demonstrate exceptional performance in the task of anomaly detection, effectively recognizing patterns, behaviors, and deviations from the established norm that may indicate possible security problems. AI-driven cloud security monitoring offers enterprises a proactive defensive mechanism that facilitates real-time threat identification and response by using past data and responding to evolving threat landscapes [25].

The use of a proactive strategy to cloud security plays a crucial role in mitigating the adverse effects of security breaches. AI-powered monitoring systems have the capability to identify inconspicuous and previously unobserved signs of compromise, enabling security professionals to promptly respond and address the situation. Furthermore, the ongoing learning skills of artificial intelligence guarantee that the monitoring system progresses in tandem with the ever-changing landscape of risks, modifying its detection algorithms to effectively counter novel and emerging cyber dangers. The outcome is an enhanced and flexible cloud security stance, which is essential for enterprises that depend on cloud services for their digital operations.

The integration of AI-driven cloud security monitoring has become a strategic need as enterprises progressively transition their activities to the cloud. This sophisticated methodology not only improves the ability to identify potential threats but also mitigates the occurrence of inaccurate alerts, enabling security personnel to allocate their attention toward genuine security problems. The scalability of AI-driven monitoring is well-suited to the dynamic characteristics of cloud environments, offering enterprises a resilient and expandable approach to safeguarding their digital assets within an ever-changing threat scenario.

10.9.2 Machine learning for cloud data protection

Machine learning algorithms play a pivotal part in the classification and encryption of data, using content, context, and user behavior to provide a meticulous and contextually sensitive approach to safeguarding data. The aforementioned algorithms have the capability to evaluate extensive datasets, hence detecting patterns that indicate the presence of sensitive information. ML-driven data security systems constantly learn from past data, enabling them to develop and effectively identify emerging forms of sensitive data. This adaptive capability allows these solutions to effectively respond to changes in data kinds and formats. The capacity to adapt is of utmost importance in dynamic cloud systems, as data is in a continual state of flux.

In addition, machine learning-based Data Loss Prevention (DLP) solutions contribute to the overall improvement of security measures by effectively identifying and halting unlawful access or transfer of confidential information. These systems use advanced algorithms to examine user behavior, network activity, and data flows, therefore detecting possible security vulnerabilities in real-time. ML-driven DLP solutions enable enterprises to adopt a proactive approach in protecting their sensitive information inside the cloud by automating the monitoring and enforcement of data protection regulations.

10.9.3 Cloud-native AI security solutions

One of the primary characteristics of cloud-native AI security solutions is their capacity to adapt and develop in reaction to the ever-shifting threat environment. Machine Learning algorithms, which are essential components of these systems, exhibit a continual learning process by assimilating knowledge from previous data and new threat trends. The system's flexibility allows for the identification of emerging and developing threats, therefore equipping businesses with a proactive defensive mechanism that surpasses traditional rule-based methods. Cloud-native AI security solutions provide a dynamic quality that enables them to adapt and grow in tandem with the developing architecture of cloud environments. This adaptability allows these solutions to provide strong protection without affecting the performance of the system [26].

Moreover, these systems demonstrate exceptional performance in the automation of threat detection and response processes, thereby reducing the operational workload on security teams. Cloud-native AI security systems provide the capability to evaluate extensive quantities of data that are produced inside cloud settings, with the ability to promptly detect abnormalities, trends, and possible security events. According to Ref. [24]; the use of automated processes and the coordination of intricate response workflows contribute to the optimization of incident response, resulting in improved efficiency and effectiveness. Consequently, companies can promptly address new risks and mitigate the consequences of security events.

10.10 Challenges and limitations of AI and ML in cyber security

The incorporation of AI and ML in the arena of cyber security has undeniably bolstered the capacities of enterprises to identify and address cyber threats. Nevertheless, the progress in this field presents some obstacles and constraints that need thoughtful deliberation to guarantee the responsible and efficient use of new technologies.

10.10.1 The challenge of explainability and transparency

A notable obstacle encountered in the usage of AI and ML in the arena of cyber security pertains to the dearth of explainability and transparency inside the decision-making mechanism. Numerous machine learning models exhibit intricate "black box" characteristics, hence posing difficulties for cybersecurity experts in comprehending the underlying processes leading to certain decisions or predictions. The absence of transparency poses a hindrance to elucidating the underlying reasoning behind security warnings or actions, a critical aspect in establishing confidence and adhering to legislation that require explainable AI [27].

10.10.2 Bias and ethical concerns

AI and ML systems possess a vulnerability to biases that may exist within the data used for their training, hence giving rise to ethical considerations within the realm of cyber security applications. Biased models have the potential to unintentionally perpetuate discrimination against certain groups or exhibit limitations in identifying risks that are unique to certain demographic categories. The task of addressing equity and minimizing prejudices in AI and ML algorithms is a critical and complex endeavor. It requires meticulous examination of the training data and ongoing vigilance to detect any unintended ramifications [28].

10.10.3 Adversarial attacks on AI and ML systems

The vulnerability of AI and ML models to adversarial attacks poses a significant limitation in cyber security applications. Adversarial attacks involve manipulating input data to deceive the model, leading to incorrect predictions or decisions. In the context of cyber security, attackers can exploit these vulnerabilities to evade detection or manipulate security systems. Developing robust defenses against adversarial attacks is an ongoing challenge, necessitating constant refinement of AI and ML models and the implementation of adversarial training techniques [29].

Addressing these challenges requires a concerted effort from the cybersecurity community, researchers, and organizations leveraging AI and ML in their security strategies. Striking a balance between the power of these advanced technologies and addressing their limitations is essential for building effective and ethical cyber security practices.

10.11 Future perspectives and trends

As we consider the next era, the rise of AI and ML in the field of cyber security has the potential to greatly impact the structure of protecting against constantly evolving cyber threats. The future development of AI and ML in the domain of cyber security is expected to be influenced by several important perspectives and trends.

10.11.1 Reinforcement learning in cyber security

The use of reinforcement learning approaches is a growing trend in the realm of cyber security. Reinforcement learning enables AI systems to acquire knowledge and adjust their behavior by actively engaging with their surroundings. This characteristic renders reinforcement learning particularly suitable for addressing dynamic and developing cyber threat situations. Within the realm of cyber security, the use of reinforcement learning has the potential to augment adaptive threat response mechanisms. This would empower systems to independently modify their defensive methods by using real-time feedback and the ever-changing nature of threat landscapes [30].

10.11.2 AI-ML hybrid models for enhanced cyber resilience

The integration of AI and ML into hybrid models is expected to have a significant impact on the improvement of cyber resilience. By integrating the respective advantages of AI and ML, these hybrid models may effectively use the predictive powers offered by ML in conjunction with the flexibility and decision-making abilities provided by AI. Carlini et al. (2017)

suggest that incorporating these components into cyber security systems might improve their efficiency and sophistication, allowing them to successfully identify, respond to, and mitigate many forms of cyber threats.

10.11.3 Quantum computing and AI-ML in cyber security

The convergence of Quantum Computing and AI-ML within the domain of cyber security signifies a boundary that provides significant potential and complexities. The field of quantum computing has the capacity to fundamentally transform cryptography methodologies, hence presenting novel avenues for safeguarding communication and data. The incorporation of AI and ML into the field of quantum computing has the potential to enhance the efficiency and effectiveness of threat detection and response tactics. Nonetheless, this convergence also gives rise to novel security concerns, including the potential consequences of quantum computing on current cryptographic techniques and the need for quantum-resistant AI-ML models [31].

In light of the intricate and ever-evolving cyber threat environment, it is imperative for enterprises to prioritize investments in research and development pertaining to these domains. The collaboration between human knowledge and the advancing capabilities of AI and ML will play a crucial role in establishing a more secure digital environment in the coming years.

10.12 Conclusion

The integration of AI and ML technologies in the field of cyber security represents a significant advancement in the continuous effort to combat cyber threats. Given the rapid expansion of the digital environment and the more intricate strategies used by cybercriminals, the implementation of AI and ML provide a robust means of safeguarding against potential threats. By using these technologies, companies have the ability to preemptively identify, address, and minimize cyber risks.

The exploration of AI and ML in the field of cyber security has shown the fundamental ideas, practical implementations, and obstacles encountered in this rapidly evolving area. The potential of these technologies to transform cyber security methods has been proved by their skills in augmenting threat detection using anomaly detection and behavior-based analysis, as well as increasing incident response through real-time identification and AI-powered forensics.

Nevertheless, this investigation has also shed light on the obstacles and constraints that come with the use of AI and ML in the field of cyber security. The need of responsible and cautious deployment is underscored by the necessity for explainability and transparency, the ethical considerations related to biases, and the susceptibility to adversarial assaults. It is crucial to tackle these difficulties in order to establish confidence, guarantee ethical use, and optimize the efficacy of AI and ML within the domain of cybersecurity.

When considering the future, it is evident that the viewpoints and patterns in the fields of AI and ML indicate an ongoing process of development and advancement. The incorporation of reinforcement learning, the examination of AI-ML hybrid models, and the convergence of Quantum Computing with AI-ML provide promising opportunities for more progress. In the context of managing the intricate cyber threat environment, the integration of human knowledge and the functionalities of AI and ML will play a crucial role in developing a robust and flexible security strategy.

In summary, the integration of AI and ML in the field of cyber security has initiated a new era of protection against cyber threats. The use of modern technology offers the opportunity to enhance an organization's ability to promptly and precisely identify and address possible threats, hence reinforcing its cyber resilience. Nevertheless, it is important to recognize the obstacles and constraints associated with the integration of AI and ML in the field of cybersecurity and make efforts to tackle them. The continuous evolution of the cyber threat environment necessitates the persistent study and development of AI and ML solutions in order to maintain an advantage over cyber adversaries. It is imperative for organizations to allocate resources toward acquiring suitable AI and ML technologies, while concurrently fostering a culture of openness, accountability, and ethical deliberation in their use. The creation of a more secure digital environment for the future necessitates the integration of human experience with the capabilities of AI and ML.

References

[1] Kaspersky, Cyberthreat Landscape for Financial Institutions in 2021, 2021. Retrieved from: https://www.kaspersky.com/resource-center/threats/cyberthreat-landscape-for-financial-institutions-in-2021.

[2] Verizon, 2021 Data Breach Investigations Report, 2021. Retrieved from: https://enterprise.verizon.com/resources/reports/dbir/.

[3] World Economic Forum, The Global Risks Report 2021, 2021. Retrieved from: http://www3.weforum.org/docs/WEF_The_Global_Risks_Report_2021.pdf.

[4] Cisco, Cisco 2021 Annual Cybersecurity Report, 2021. Retrieved from: https://www.cisco.com/c/dam/m/hu_hu/security/annual-report/cisco-annual-cybersecurity-report-2021.pdf.

[5] S.J. Russell, P. Norvig, Artificial Intelligence: A Modern Approach, third ed., Prentice Hall, 2010.
[6] S. Mohan, A.K. Solanki, H.K. Taluja, Anuradha, A. Singh, Predicting the impact of the third wave of COVID-19 in India using hybrid statistical machine learning models: a time series forecasting and sentiment analysis approach, Computers in Biology and Medicine 144 (2022) 105354, https://doi.org/10.1016/j.compbiomed.2022.105354, 0010-4825.
[7] E. Alpaydin, Introduction to Machine Learning, third ed., MIT Press, 2020.
[8] H. Kumar, Anuradha, A.K. Solanki, K.K. Singh, Progressive machine learning approach with WebAstro for web usage mining, Procedia Computer Science 167 (2020) 1400–1410, https://doi.org/10.1016/j.procs.2020.03.351. ISSN 1877-0509.
[9] De Y. Jiang, H. Zhang, H. Kumar, Q.N. Naveed, C. Takhi, V. Jagota, R. Jain, Automatic control model of power information system access based on artificial intelligence technology, Mathematical Problems in Engineering 2022 (2022) 6, https://doi.org/10.1155/2022/5677634. Article ID 5677634.
[10] G. Dhillon, L. Behera, P. Kaur, Role of artificial intelligence in cyber security: a review, Computers, Materials and Continua 63 (2) (2020) 699–721.
[11] P. Srivastava, et al., Fuzzy methodology approach for prioritizing maintenance 4.0 attributes, in: 2020 International Conference on Computation, Automation and Knowledge Management (ICCAKM), Dubai, United Arab Emirates, 2020, pp. 308–311, https://doi.org/10.1109/ICCAKM46823.2020.9051483.
[12] D. Sculley, G. Holt, D. Golovin, E. Davydov, T. Phillips, Hidden technical debt in machine learning systems, in: Advances in Neural Information Processing Systems, 2015, pp. 2503–2511.
[13] A. Taluja, A.K. Solanki, H. Kumar, A comprehensive approach for assessing the reliability of complex networks using OANN approach, Multidisciplinary Science Journal 5 (2023) 2023ss0104, https://doi.org/10.31893/multiscience.2023ss0104.
[14] Gupta, R. Gupta, S. Kumar, Machine learning assisted early diagnosis of skin cancer, in: 2022 IEEE International Conference on Current Development in Engineering and Technology (CCET), Bhopal, India, 2022, pp. 1–5, https://doi.org/10.1109/CCET56606.2022.10080671.
[15] A.L. Buczak, E. Guven, A survey of data mining and machine learning methods for cybersecurity intrusion detection, IEEE Communications Surveys and Tutorials 18 (2) (2016) 1153–1176.
[16] Ponemon Institute, The Cost of Cyber-Crime Study, 2021. Retrieved from: https://www.ibm.com/security/data-breach.
[17] W. Dai, P. Yin, W. Wu, DeepIoT: compressing deep neural network structures for sensing systems with a compressor-critic framework, in: Proceedings of the 16th ACM Conference on Embedded Networked Sensor Systems, 2018, pp. 1–14.
[18] R. Gupta, R.K. Phanden, S. Sharma, P. Srivastava, P. Chaturvedi, Security in manufacturing systems in the age of industry 4.0: pitfalls and possibilities, in: R.K. Phanden, K. Mathiyazhagan, R. Kumar, J. Paulo Davim (Eds.), Advances in Industrial and Production Engineering. Lecture Notes in Mechanical Engineering, Springer, Singapore, 2021, https://doi.org/10.1007/978-981-33-4320-7_10.
[19] N. Agarwal, R. Gupta, P. Kumar, Role of ICT in imparting quality education and curbing cyber security risks during COVID-19 pandemic, in: R. Agrawal, J. He, E. Shubhakar Pilli, S. Kumar (Eds.), Cyber Security in Intelligent Computing and Communications. Studies in Computational Intelligence, vol. 1007, Springer, Singapore, 2022, https://doi.org/10.1007/978-981-16-8012-0_26.
[20] N. Zhang, R. Zheng, L.T. Yang, A survey on emerging cloud computing technologies, Journal of Computer and System Sciences 82 (2) (2016) 357–373.

[21] M.A. Al-Garadi, A. Mohamed, A.K. Al-Ali, X. Du, I. Ali, M. Guizani, A survey of machine and deep learning methods for internet of things (IoT) security, IEEE Communications Surveys & Tutorials 22 (3) (2020) 1646–1685.
[22] E. Ahmed, I. Yaqoob, I.A.T. Hashem, I. Khan, A.I.A. Ahmed, M. Imran, A. Gani, The role of machine learning in the internet of things, Computers and Electrical Engineering 70 (2018) 605–627.
[23] J. Han, S. Hu, T. Zhang, W. Zhang, C. Zhang, Artificial intelligence-driven network access control: a survey, IEEE Access 7 (2019) 138177–138187.
[24] Gartner, Market Guide for Managed Detection and Response Services, 2020. Retrieved from: https://www.gartner.com/document/3984836.
[25] D. Zissis, D. Lekkas, Addressing cloud computing security issues, Future Generation Computer Systems 28 (3) (2012) 583–592.
[26] S. McCarthy, P. Hennelly, T.M. Chen, M. Ma, Cloud services and security: a research perspective, Journal of Cloud Computing: advances, Systems and Applications 5 (1) (2016) 15.
[27] D. Gunning, Explainable Artificial Intelligence (Xai), Defense Advanced Research Projects Agency (DARPA), 2017. Retrieved from: https://www.darpa.mil/program/explainable-artificial-intelligence.
[28] S. Barocas, M. Hardt, Fairness and abstraction in sociotechnical systems, in: Proceedings of the Conference on Fairness, Accountability, and Transparency (FAT/ML), 2019, pp. 59–68.
[29] C. Szegedy, W. Zaremba, I. Sutskever, J. Bruna, D. Erhan, I. Goodfellow, R. Fergus, Intriguing Properties of Neural Networks, 2013 arXiv preprint arXiv:1312.6199.
[30] R.S. Sutton, A.G. Barto, Reinforcement Learning: An Introduction, second ed., The MIT Press, 2018.
[31] D. Akinwande, N. Petrone, J. Hone, Two-dimensional flexible nanoelectronics, Nature communications 5 (1) (2014) 5678.

Further reading

[1] Cybersecurity Ventures, Cybersecurity Market Report, 2020. Retrieved from: https://cybersecurityventures.com/.
[2] Deloitte, Future of Cyber Survey 2021, 2021. Retrieved from: https://www2.deloitte.com/us/en/insights/topics/cyber-risk.html.
[3] Symantec, Internet Security Threat Report, vol. 26, 2021. Retrieved from: https://www.broadcom.com/company/newsroom/press-releases/2021/symantec-issues-its-26th-internet-security-threat-report.

PART III

Cyber-security standards and best practices

CHAPTER 11

Smart grid: Interoperability and cyber security

Muhammed Zekeriya Gündüz[1] and Resul Daş[2]
[1]Department of Computer Science and Technology, Bingol University, Bingöl, Türkiye; [2]Software Engineering, Technology Faculty, Firat University, Elazığ, Türkiye

11.1 State-of-the-art-interoperability

The dependence of humankind on energy has continued since the first day of its existence. Today, electrical energy is one of the leading types of energy [1]. The importance of effective management of electrical energy at the generation, transmission, distribution, and consumption stages has further increased. The traditional four stages of electrical energy reaching end users are shown in Fig. 11.1.

The smart grid is a sophisticated system of interconnected devices, networks, and services that work together to manage the generation, transmission, distribution, and consumption of energy. Its interoperability refers to the ability of different components of the smart grid system to communicate and exchange data seamlessly, regardless of the manufacturer, technology, or system architecture [2]. In addition, the integration of new communication technologies into power systems has accelerated the evolution of the smart grid. Fig. 11.2 shows a generic communication system for smart grid [3]. Interoperability is crucial to provide that all of these components can communicate and work together seamlessly [4]. Cyber

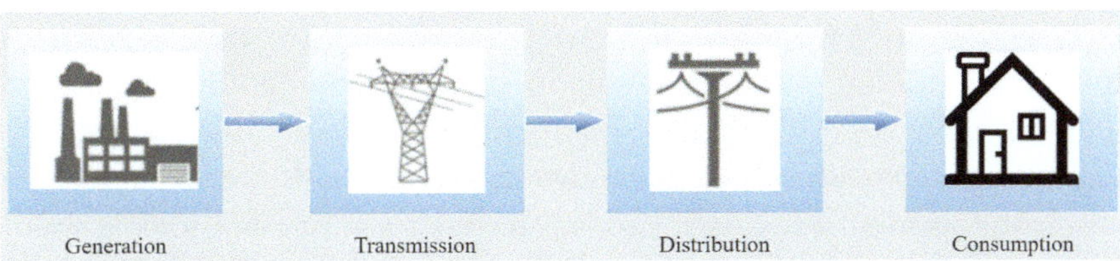

Figure 11.1 Electricity flow stages in a traditional power system.

Cyber Security Solutions for Protecting and Building the Future Smart Grid
ISBN 978-0-443-14066-2
https://doi.org/10.1016/B978-0-443-14066-2.00007-4

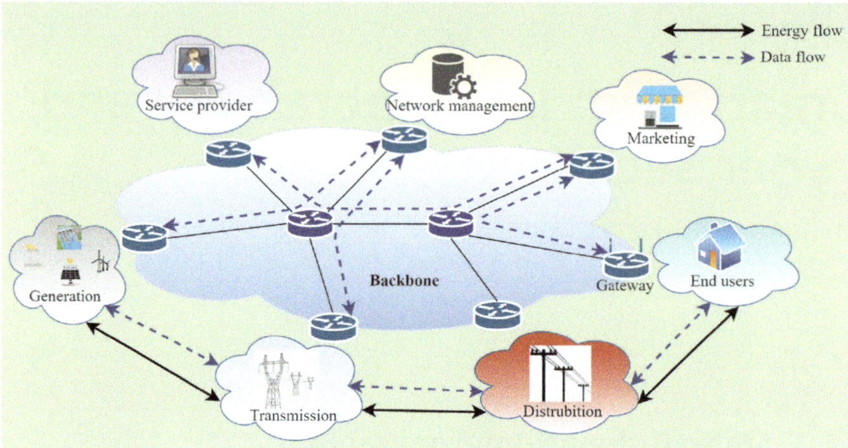

Figure 11.2 A generic communication backbone for smart grid.

security, on the other hand, means to the efforts carried out safe the smart grid from cyber threats such as viruses, worms, malicious activities, DDos attacks, APT attacks, and unauthorized access [5].

The interoperability of different components in the smart grid is critical for efficient and reliable energy delivery. For example, if the communication between the power generation system and the distribution system fails, it can lead to blackouts or other critical system failures. Therefore, the interoperability standards must be designed to ensure that all components can communicate and exchange data seamlessly. The integration of various technologies into the smart grid systems creates a complex and interconnected network that can be security vulnerabilities. A breach in the security of any device or system can potentially compromise the entire smart grid. It is crucial to ensure that all components of the smart grid are interoperable and secure. Therefore, ensuring the cyber security is essential to protect smart grid from cyber threats that can cause significant disruptions and damage [6].

Cyber security measures must be implemented at every stage of the smart grid, from power generation to consumption. Some measures to improve cyber security in the smart grid applications include intrusion detection systems, access controls, encryption, firewalls, network segmentation, and incident response plans [7]. Regular security assessments should also be managed to identify vulnerabilities and mitigate them in a timely manner.

To achieve interoperability, standards and protocols must be established to ensure that devices and systems can communicate with each other seamlessly. Interoperability testing should also be conducted to ensure that all components can work together without any compatibility issues. It is essential to note that interoperability and cyber security are interconnected. If a component is not secure, it may compromise the security of the entire system. Similarly, if a component lacks interoperability, it may face difficulties in establishing communication and coordination with other components [8].

The smart grid interoperability and cyber security are two essential components of a successful smart grid system. They must be considered together when designing and implementing a smart grid system. The interoperability standards and cyber security essentials must be designed to ensure that all components can communicate and exchange data seamlessly, in addition to assuring the reliability, security, and resilience for smart grid applications [9].

Today, with the increasing complexity and diversity of technological systems, achieving state-of-the-art interoperability has become more challenging than ever. The advanced interoperability on the cyber security of smart grid applications includes the followings [1,7–11]:

Standardized security framework: A standardized security framework provides a common set of security practices that all smart grid components must adhere to. It helps in the interoperability of different components, simplifies security management, and reduces the risks associated with cyber security.

Security by design: The design of smart grid components and systems should take into consideration the need for interoperability and cyber security from the outset. The design of the system should incorporate built-in security measures like authentication, encryption, access control, and intrusion detection.

Secure communication protocols: Smart grid components and systems should employ robust communication protocols like TLS, SSL, and IPSec to guarantee the secure transmission of data, preventing interception or unauthorized modification by potential attackers.

Identity and authentication management: A resilient identity and authentication management system is imperative for the smart grid, ensuring that solely authorized users and devices gain access. This system should possess the capability to validate the identity of both users and devices, guaranteeing that only those with proper authorization can enter the system.

Risk assessment: It is an important component of smart grid cyber security, and it should be performed regularly to identify potential vulnerabilities and threats to interoperability. This measure can provide assurance that the system is well-equipped to address and counteract novel and evolving threats as they arise.

Collaboration: Ensuring interoperability and cyber security in the smart grid necessitates collaboration between various stakeholders, such as utilities, vendors, regulators, and cyber security experts. The collaboration could include activities such as sharing best practices, conducting joint assessments, and developing common standards and guidelines.

Microservices architecture: Microservice architecture is a software design pattern that has been gaining popularity in the development of smart grid systems. When the system is subdivided into smaller, autonomous services, it simplifies the process of recognizing and containing possible security risks. Additionally, microservices can be designed with security measures in mind, such as access controls and encryption, to ensure that each service is protected.

Service-Oriented Architecture (SOA): It is a software design pattern that allows different applications to communicate with each other over a network. In the context of smart grid systems, SOA can enhance interoperability and security by allowing different components to exchange data and services. By using standardized interfaces, components can be developed independently and easily integrated into the overall system. Additionally, SOA can help with security by providing a framework for implementing access controls and monitoring services. However, the implementation of SOA requires careful consideration of security issues such as authentication and authorization.

API management platforms: They provide a centralized framework for managing and securing the APIs that allow different components of smart grid systems to communicate with each other. These platforms can enhance interoperability and security by providing tools for managing API access, monitoring usage, and enforcing security policies. Additionally, API management platforms can help with security by providing features like encryption, access control and threat detection. However, the effectiveness of API management platforms depends on careful planning and implementation, including the establishment of secure communication channels and the use of appropriate authentication and authorization mechanisms.

State-of-the-art interoperability for the smart grid involves the use of open standards and protocols that enable seamless communication between various devices and systems. Achieving state-of-the-art interoperability requires a combination of technical expertise, standards adoption, and proper design patterns.

In terms of cyber security, state-of-the-art solutions involve a multi-layered approach that includes network segmentation, access control, encryption, intrusion detection and prevention systems, and analysis of flow data traffic [14]. Advanced threat intelligence and artificial intelligence algorithms are also used to identify and react to cyberattacks in real-time. As technology develops, it is probable that new approaches and tools will appear to support interoperability in even more complex environments [12,13].

11.2 Benefits, control and model for interoperability

The power system's ability to exchange information in a fast and effective manner is interoperability. This capability is crucial, but its development is not yet sufficient. Despite significant grid modernization efforts in recent years, the increase in technology and adoption of new standards have only had a limited impact on improving interoperability. Smart grid interoperability refers to the ability of different devices, systems, and applications within a smart grid to communicate and work together seamlessly [15]. It is essential for the efficient and effective operations of the smart grid. However, with the increasing connectivity and complexity of the smart grid, there is a growing trouble regarding cyber threats. The smart grid faces cyberattacks capable of disturbing its operations, compromising confidential data, and even cause physical damage [16]. Because of the increasing reliance on digital communication and data exchange, cyber security is vital for interoperability.

Interoperability enables the use of standard protocols and interfaces, which can simplify the implementation and management of security measures that can lead to more consistent and effective cyber security practices. Interoperability can enable a rapid response to cyber security threats by allowing different systems and devices to communicate and coordinate in real-time. This can help mitigate the effect of a security incident and prevent it from spreading to the other parts of the grid. Interoperability can provide better monitoring for the smart grid's security by enabling the collection and analysis of data [17]. It may help determine potential vulnerabilities and threats, and enable proactive security measures to be implemented. Interoperability can improve the resilience of the smart grid by enabling the use of distributed security mechanisms. This can ensure the running of the system even in a security incident.

Interoperability gives an exhaustive sight for the entire smart grid system, enabling better monitoring, analysis, and response to security events. Interoperability allows for seamless integration of different systems and devices, reducing the need for manual intervention and enhancing the general efficiency and productivity. Interoperability ensures that the smart grid can counteract security threats minimizing the risk of disruptions and blackouts with improved resilience and reliability. Interoperability eliminates the need for specific interfaces and protocols, reducing development and maintenance costs. Interoperability can enhance cyber security by enabling different devices and systems to share threat intelligence and coordinate responses to cyber threats [18].

Efficient operations of the smart grid are achieved by a combination of cyber-physical operations and economic optimizations. It maintains the system at the highest efficiency level while minimizing potential interruptions or disruptions. As the electrical grid is rapidly affected by policy and market decisions, utility companies need to adapt their architecture and operations according to new and existing devices in a changing environment. Hence, it is crucial to comprehend the association between interoperability and system functions, as it enables the organization of the system in a structured manner.

To secure the dependability and safety of the smart grid, it is crucial to conduct a comprehensive analysis of its constituent parts, uncovering any inherent weaknesses and potential cyber security risks. All parts of the smart grid must meet the principles of the CIA triad. CIA triad is the main cyber security objective. It includes confidentiality, integrity, and availability. This will help safeguard against potential attacks and ensure the practical process of the smart grid infrastructure [19].

The use of secure communication protocols can help to ensure that data transmitted between devices and systems in the smart grid is protected from interception or tampering. Enforcing robust authentication and authorization mechanisms can serve as a deterrent against unauthorized entry into the system. Data encryption may be used to protect sensitive information stored or transmitted into the smart grid. Regular security testing and assessment of smart grid systems could help to determine vulnerabilities and potential weaknesses that could be exploited by cyber attackers. Standards and regulations such as NERC-CIP, NIST Cyber Security Framework,

IEEE 2030.5, and IEC 61850 can provide guidance and requirements for smart grid interoperability and security [20].

Interoperability is a critical requirement for the smart grid, enabling seamless communication and integration among various devices, systems, and technologies. However, interoperability also creates potential vulnerabilities that cyber attackers could exploit to get unauthorized access [21]. Therefore, it is essential to establish controls to ensure interoperability in smart grid cyber security.

One such control is the implementation of strong authentication and access control. This control aims to prevent unauthorized access to the smart grid by requiring users to authenticate themselves using multi-factor authentication, and strong passwords. Also, role-based access control limits access to specific functions relegated the user's role, reducing the likelihood of an attacker gaining access to sensitive systems and data.

Encryption is another means of control that can be used to secure data both in transit and in storage. Encryption stands as a critical security measure, guaranteeing the secure transmission and storage of data by preventing unauthorized access and data theft. Strong encryption algorithms can be utilized to secure sensitive data being transmitted between devices and systems in the grid. It can prevent unauthorized access and interception of data by cyber attackers.

Regular security assessments and penetration testing can identify vulnerabilities in the security controls, allowing them to be addressed proactively. Additionally, regular maintenance and updates to security controls can ensure that they remain effective against the latest threats and attacks.

Implementing network segmentation is another control that can improve interoperability. Network segmentation involves dividing the smart grid network into smaller, isolated parts or zones, with restricted access between them. It helps to control any security breaches and limits the scope of an attack, reducing the risk of the entire smart grid being compromised.

Regular monitoring of the network and systems can detect any anomalies or suspicious activity, enabling real-time action to mitigate any potential threats. A well-executed incident response strategy can also guarantee the swift and efficient handling of security breaches, thus mitigating the consequences of an attack on the smart grid. In conclusion, implementing robust controls for interoperability is vital to ensure the reliable and secure operation of the grid [22].

Interoperability also introduces potential cyber security risks, as it expands the attack surface and exposes the grid to various cyber threats. Therefore, it is a need to develop and implement models for interoperability. Standard protocols and interfaces ensure that different systems and devices can communicate with each other seamlessly and securely [23]. Models play a critical role in interoperability and other concerns related to the smart grid, as they facilitate a common language and communication among stakeholders. However, with the advancements in technology and the changes in smart grid architecture in recent years, these models must be kept up to date. Also, through visual representation, stakeholders can obtain a better understanding of how diverse factors can influence interoperability. Models for interoperability have different approaches for smart grid cyber security.

The smart grid cyber security framework adopts a stratified approach, utilizing a series of security layers to fortify and safeguard the smart grid. Each layer provides a different level of protection, and they run together to ensure holistic security for the system. The SOA model for smart grid cyber security involves breaking down the smart grid into individual services, which can be accessed and controlled through standardized interfaces. This approach simplifies the process of integrating new devices and systems into the smart grid, making it more flexible and adaptable. The reference architecture model involves the use of a standardized architecture that defines the components, interfaces, and protocols used in the smart grid. This method guarantees that various devices and systems can interact effectively by adhering to a shared set of standards, simplifying the process of achieving interoperability [24].

One another model for interoperability is the use of open standards and protocols such as Open Automated Demand Response (OpenADR). The adoption of open standards and protocols make possible seamless integration and communication among different systems and devices, without compromising security. Open standards and architectures encourage innovation and competition, leading to better security solutions and practices.

Another model is the implementation of security oriented design principles. Security oriented design principles aim to embed security considerations throughout the design and development process of smart grid and devices. By considering security as an integral part of the design process, security-by-design can help to identify potential vulnerabilities and threats early on and ensure that security is incorporated into all aspects of the system.

Furthermore, the adoption of risk-based approaches can help to prioritize security measures and controls according to the level of risk they address. Risk-based methodologies encompass a methodical procedure that entails recognizing conceivable threats and vulnerabilities, assessing their probability and potential consequences, and subsequently adopting suitable security measures to diminish these risks. Through the prioritization of risks according to their potential consequences and probability, organizations can optimize resource allocation for addressing the most significant vulnerabilities and reducing the potential harm caused by cyberattacks.

Another model for interoperability is the use of secure APIs. APIs enable different systems and devices to communicate and share data securely, while also ensuring data integrity and confidentiality. Secure APIs can be used to facilitate secure communication between different components of the smart grid, such as data aggregators, sensors, and control systems.

In the following sections, the leading organizations working on interoperability and cyber security for the smart grid are explained in detail.

11.3 Smart grid interoperability standards globally

By implementing the standards, the smart grid becomes more secure, reliable, and resilient, enabling it to meet the growing energy demands. The standards enable different devices and systems to communicate with each other effectively and efficiently, regardless of their origin or manufacturer. Several organizations have developed smart grid interoperability standards globally. Smart grid interoperability standards have been developed and implemented globally to provide a framework for secure communication and interoperability among different devices, systems, and technologies in the smart grid. These standards enable secure data exchange, secure device management, and secure system integration [10].

The lack of global interoperability standards for smart grid cyber security can create fragmentation and complexity, making it challenging to ensure the security and interoperability of different smart grid components. Initiatives such as the International Electrotechnical Commission (IEC) smart grid standardization roadmap, Common Information Model (CIM), Open Automated Demand Response (OpenADR), Institute of Electrical and Electronics Engineers (IEEE), the National Institute of Standards and

Technology (NIST) cyber security framework, The North American Electric Reliability Corporation - Critical Infrastructure Protection (NERC-CIP) and Federal Energy Regulatory Commission (FERC) work toward establishing global smart grid interoperability standards for cyber security.

NERC-CIP, a non-profit regulatory authority, focuses on enhancing power grid reliability and security in North America. It enforces specific cyber security standards aimed at protecting critical assets, ensuring compliance, and managing secure processes. The CIP standards establish a cyber security framework for the identification and protection of critical assets, which have the potential to influence the operational efficiency and reliability of systems. They also oversee the formulation and management of standards, compliance, risk assessments, and the confidential procedures associated with NERC-CIP's computational systems. NERC-CIP standards are mandatory for cyber security standards that apply to bulk power systems in North America and the standards are enforced by the FERC, which is responsible for regulating the interstate transmission of electricity, natural gas, and oil in the United States. Adhering to these standards enables utilities to contribute to the safeguarding of the reliability and security of the extensive power system, concurrently upholding the confidence and trust of both customers and stakeholders [25].

NIST fosters innovation and industrial competitiveness through the progression of measurement science, standards, and technology. NIST develops and publishes standards, guidelines, frameworks, and best practices for a wide range of fields, including cyber security, information technology, and engineering. NIST is also responsible for developing and maintaining the cyber security framework, which provides a voluntary set of guidelines and best practices for organizations to manage and reduce cyber security risk. While the NIST cyber security framework is not mandatory, it is widely used by organizations in the energy sector to enhance their cyber security practices and aligns with NERC-CIP requirements [26].

The FERC is an independent agency of the United States government responsible for regulating and overseeing the interstate transmission of electricity, natural gas, and oil. Its mission is to ensure the reliability, efficiency, and security of the nation's energy infrastructure while also promoting competition and innovation in energy markets. FERC plays a critical role in ensuring the reliability and security of the nation's energy infrastructure, including the electric grid, natural gas pipelines, and oil pipelines. FERC is also responsible for approving and enforcing NERC's

CIP standards, ensuring that utilities comply with them, and taking enforcement actions if necessary. Additionally, FERC oversees the development of energy infrastructure and promotes competition in energy markets [27].

NERC-CIP, FERC, and NIST are all interconnected as they play critical roles in ensuring the cyber security and reliability of the energy infrastructure in North America. NERC-CIP establishes mandatory cyber security standards, FERC enforces compliance with these standards, and NIST provides voluntary guidelines and best practices to enhance cyber security practices and aligns with NERC-CIP requirements.

Detailed information about the leading organizations NERC-CIP, NIST, and FERC is given in the next section. In addition to these organizations, regional and national organizations have also contributed to the development of smart grid interoperability standards. The European Committee for Standardization (CEN) and the European Committee for Electrotechnical Standardization (CENELEC) have developed a series of standards known as the Smart Grid Coordination Group (SGCG) which cover areas such as communication protocols, metering, and cyber security.

DLMS/COSEM is an international standard for communication protocols used in metering and billing systems. It is developed by IEC. It allows utilities to communicate with smart meters to obtain detailed information about energy usage. OpenADR is a standard for automated demand response, which allows utilities to communicate with and control customer loads during peak demand periods. CIM is a standard for power system modeling and simulation, which provides a common language and framework for different applications to exchange data and information.

The IEC is a nonprofit global standards body that formulates and issues international standards for electrical, electronic, and affiliated technologies, facilitating global standardization in these domains. The IEC has developed a number of standards related to smart grid interoperability, including the IEC 61850 communication protocol standard, the IEC 61968 standard for distribution management, and the IEC 62351 for security.

The IEEE is a professional association that develops standards for a wide range of industries, including the power and energy sector. The IEEE has developed a number of standards related to smart grid interoperability, including the IEEE 1547 standard for interconnection and interoperability of distributed energy resources (DERs), and IEEE 2030.5 for the exchange of information between different components in a smart grid system.

Smart grid interoperability standards are essential for ensuring the security and resilience of the energy infrastructure. The lack of global standards can create significant challenges for the industry and increase the risk of cyber-attacks. Standardization efforts such as the NERC, NIST, and FERC initiatives are critical for establishing global smart grid interoperability standards for cyber security. By implementing these standards, smart grid applications can become more secure, reliable, and resilient, enabling them to meet the growing energy demands of societies.

11.4 NERC, NIST, FERC standards

The NERC, NIST, and FERC standards are put in place to ensure that every component of the smart grid, from design to implementation, is prioritized with security and interoperability. By doing so, it strengthens the grid's resilience against cyber threats while simultaneously promoting reliable operation [28]. NERC has developed a set of standards to address the cyber security of the smart grid. These standards, known as the CIP standards, are designed to ensure the security and reliability of the bulk power system in North America. NERC-CIP reliability standards and substandards provide a set of fundamental guidelines. NERC has developed several other standards and guidelines related to smart grid cyber security, including [29]:

Security management controls: These controls require utilities to establish and maintain a cyber security program that includes policies, procedures, and processes to protect critical assets. Moreover, it establishes requirements for security management controls, including risk assessment, personnel training, and security awareness.

Access control: Utilities must implement strong access controls to prevent unauthorized access to critical cyber systems. It includes requiring strong passwords, implementing multi-factor authentication, and limiting access to only authorized personnel.

Personnel and training: These controls require utilities to ensure that personnel who have access to critical cyber assets are properly trained and aware of their cyber security responsibilities.

Physical security: Utilities must implement physical security controls to protect cyber systems from unauthorized access or tampering. It includes securing facilities, restricting access to critical systems, and monitoring for unauthorized access attempts.

Cyber security incident response planning: These controls require utilities to develop and implement plans to respond to cyber security incidents.
Incident response: Utilities must have a robust incident response plan in place to detect and respond to cyber security incidents. The plan should include procedures for reporting incidents, assessing the impact of incidents, and restoring systems to normal operations.
Recovery and restoration planning: The controls require utilities to develop and implement plans to recover from cyber security incidents and restore critical cyber assets to normal operations.
Security monitoring: Utilities must implement security monitoring tools and processes to detect and respond to cyber security threats in real-time. It includes monitoring for unauthorized access attempts, abnormal network activity, and other indicators of potential cyberattacks.
Configuration Change Management and Vulnerability Assessments: This standard establishes requirements for the management of configuration changes and vulnerability assessments for cyber assets within the bulk power system.

NERC-CIP standards mainly include critical infrastructures, asset classification, cyber security policies, access controls, cyber security risks, cyber security incident identification and notification, configuration management and monitoring, recovery plans, and information protection. NERC-CIP standards are given with the number and version in Table 11.1.

Table 11.1 Current NERC-CIP standards [29].

Number and version	Covered main subject
CIP-01-1	Sabotage reporting
CIP-02-1	Asset identification and classification
CIP-03-8	Security management controls (policy and governance)
CIP-04-6	Personnel and training (general security)
CIP-05-6	Electronic security perimeters (network security)
CIP-06-6	Physical security of cyber systems
CIP-07-6	System security management
CIP-08-5	Incident reporting and response planning
CIP-09-6	Recovery plans for cyber assets
CIP-10-3	Configuration change management and vulnerability assessment
CIP-11-2	Information protection
CIP-12-1	Communications between control centers
CIP-13-1	Supply chain risk management
CIP-14-2	Physical security

CIP-01 aims to report incidents suspected or detected to be caused by sabotage to the appropriate systems, government agencies, and regulatory agencies.

CIP-02 identifies and categorizes cyber systems and their associated cyber assets for the application of cyber security requirements. By clearly defining these assets and systems, organizations can better understand their potential risks and vulnerabilities and take appropriate measures to protect them from cyber threats.

CIP-03 aims to establish clear accountability to protect cyber systems. The standard also includes provisions regarding emergencies.

CIP-04 focuses on the training of staff and stakeholders on security awareness. It aims to reduce the exposure to cyber risks from personnel. The training consists of cyber security awareness, training, risk management, and access control management.

CIP-05 focuses on the training of staff and stakeholders on security awareness. It aims to reduce the exposure to cyber risks from personnel. The training consists of cyber security awareness, training, risk management, and access control management. Different from the previous it is about network security.

CIP-06 covers the creation and monitoring of physical security perimeters for physical security plans, visitor control programs, and maintenance and testing programs.

CIP-07 provides technical, operational, and procedural elements required to secure critical and non-critical cyber assets. These elements include ports and services, security patches, prevention of malicious code, security event logging and monitoring, management of shared accounts, password and credential management, and system access controls.

CIP-08 outlines the preparation of assets for cyber incidents and offers directives for responding to such incidents through a comprehensive cyber security incident response plan. This standard aids in the recognition, categorization, response, reporting, and documentation of events linked to critical cyber security assets. It encompasses aspects such as the cyber security incident response plan, its implementation, rigorous testing, periodic evaluation, updating, and communication procedures.

CIP-09 addresses how entities can recover from a cyber security attack that has affected the functioning of the grid cyber security. It is related to backup and restoration. It ensures that a recovery plan is in place and that entities are following established plans for disaster recovery and business continuity. The standard covers recovery specifications, recovery implementation and testing, recovery plan review, update, and communication.

CIP-10 defines the security policy criteria to forestall and identify illicit modifications in cyber systems. Its principal goal is to attain persistent and foundational security by means of system configuration regulations and dynamic assessments of vulnerabilities. The standard encompasses diverse aspects such as managing alterations in configuration, monitoring configuration status, appraising vulnerabilities, and overseeing transient cyber assets.

CIP-11 outlines the conditions for identifying data that has the potential to disrupt the grid's operation if intentionally manipulated or compromised. It additionally establishes guidelines for safeguarding information, repurposing cyber assets, and the proper disposal of digital media.

CIP-12 implements documented plans to mitigate the risks of disclosure and modification of real-time assessment and monitoring data while in transit between control centers.

CIP-13 identifies risks to the grid including the transitioning from one vendor to another. It covers developing, implementing, and periodically reviewing a supply chain risk management plan.

CIP-14 prescribes the criteria for recognizing and securing transmission stations, substations, and their corresponding primary control centers. These protections are necessary because, in the event of a physical attack that renders them inoperable or damaged, it could lead to instability, uncontrolled disconnection, or cascading problems within an interconnection. This standard also details protocols for safeguarding information and managing the reuse and disposal of cyber assets.

By complying with the NERC standards, entities responsible for the bulk power system can ensure that they are implementing efficient cyber security controls to protect the reliability and security of the power grid.

The framework presents a methodology grounded in risk assessment for overseeing cyber security risks within the smart grid. It is structured around five fundamental functions, namely identify, protect, detect, respond, and recover respectively.

The five core functions are designed to be used together in a cyclical process to continually assess and improve an organization's cyber security. The NIST cyber security framework consists of five functions that work together to manage and reduce the security risk to critical infrastructure systems. Identify pertains to the establishment of a comprehensive awareness within the organization concerning cyber security risks related to systems, individuals, assets, data, and functionalities. Protect is centered on formulating and executing suitable measures to secure the continuous provision of vital infrastructure services. Detect involves the formulation and execution of pertinent actions to recognize instances of cyber security incidents. Respond encompasses the development and implementation of suitable procedures for responding to identified cyber security incidents. Lastly, recover encompasses the formulation and execution of appropriate actions for upholding resilience plans and reinstating any impaired capabilities or services resulting from a cyber security incident. The five core functions are designed to be used together in a cyclical process to continually assess and improve an organization's cyber security. Table 11.2 shows five functions and related categories [30].

Table 11.2 Function and categories.

Function	Categories
Identify	Asset management
	Business environment
	Governance
	Risk assessment
	Risk management strategy
	Supply chain risk management
Protect	Identity management and access control
	Awareness and training
	Data security
	Information protection processes and procedures
	Maintenance
	Protective technology
Detect	Anomalies and events
	Security continuous monitoring
	Detection processes
Respond	Response planning
	Communications
	Analysis
	Mitigation
	Improvements
Recover	Recovery planning
	Improvements
	Communications

In addition to the smart grid cyber security framework, NIST has developed a number of other standards and guidelines related to smart grid cyber security, including:

Guidelines for smart grid cyber security(NISTIR 7628): This publication provides guidance on the development of a comprehensive smart grid cyber security strategy [26].

Guide to industrial control systems (ICS) security (NIST SP 800-82): This guide provides guidance to organizations involved in the planning, implementation, and operation of industrial control systems, including those used in the smart grid. The guide covers a range of topics, including risk management, security controls, and incident response planning [31].

Security and privacy controls for information systems and organizations (NIST SP 800-53): This standard provides a catalog of security and privacy controls for information systems and organizations. The catalog includes controls related to smart grid cyber security, such as access controls, incident response planning, and security assessment and authorization [16].

Recommendation for the entropy sources used for random bit generation (NIST SP 800-90B): This standard provides guidance on the sources of randomness used to generate cryptographic keys and other security-related information. By adopting the NIST standards and guidelines, organizations involved in the smart grid can ensure that they are implementing effective cyber security controls to protect the reliability and security of the smart grid.

Academic and industry frameworks have been developed to describe the architecture of the smart grid, with the NIST framework being the most widely used. The NIST framework for the smart grid architecture is divided into seven interconnected domains, as illustrated in Fig. 11.3. The energy ecosystem is divided into customers, distribution, transmission, and generation domains, which oversee the provision, conveyance, and production of energy. On the other hand, the market, operation, and service provider domains are responsible for offering services, managing energy distribution, and overseeing energy market operations. The generation, transmission, and distribution domains encompass energy substations, whereas the operation domain includes control systems like SCADA. The NIST framework provides a comprehensive view of the smart grid architecture and its interconnections, helping to identify potential vulnerabilities and threats to the smart grid's security [32].

The CIP standards are designed to ensure the security and reliability of the power grid by establishing a minimum set of requirements for cyber

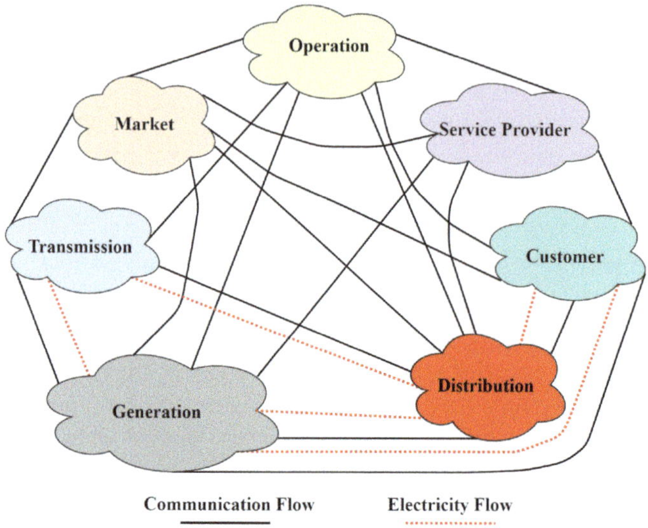

Figure 11.3 NIST smart grid conceptual model.

security controls and practices. NERC is responsible for developing and enforcing the CIP standards for the power system in North America. However, the standards are subject to approval by the FERC, which oversees and regulates the energy industry in the United States. In addition to the CIP reliability standards, FERC has also established a set of guidelines for smart grid cyber security. These guidelines provide recommendations for the implementation of cyber security controls and best practices for the protection of the smart grid.

FERC's guidelines for smart grid cyber security include [25]:

FERC staff report on cyber security in the energy industry: This report provides an overview of the cyber security threats facing the energy industry and provides recommendations for improving cyber security controls.

FERC's guide to cyber security for the electric sector: This guide provides an overview of the cyber security threats facing the electric sector and provides recommendations for implementing effective cyber security controls. It is including:

FERC staff report on cyber security in the bulk electric system: This report provides an overview of the cyber security risks facing the bulk power system and recommends best practices for managing those risks.

FERC Order No. 791: This order establishes requirements for the submission of information related to cyber security incidents and threats to the bulk electric system information sharing and analysis center (BES-ISAC).

FERC Order No. 848: This order directs the NERC to develop new standards for cyber security incident reporting reliability standards.

FERC Order No. 893: Incentives for advanced cyber security investment.

FERC Order No. 887: Internal network security monitoring for high and medium impact bulk electric system cyber systems.

FERC Order No. 850: Supply chain risk management reliability standards.

FERC Orders contains specific solutions for CIP essentials.

Interoperability between NERC, NIST, and FERC standards is also critical to ensure that the electric power system can adapt to new technologies and emerging threats. For example, as new technologies are developed and implemented in the power system, it is important to ensure that they are interoperable with existing systems and can be integrated seamlessly into the power system. Additionally, as cyber threats continue to evolve, it is critical that the electric power system can adapt to these threats by implementing new security measures and ensuring that they are interoperable with existing security systems.

The NERC, NIST, and FERC standards are critical to ensuring the reliable and secure operation of the electric grid. Interoperability between these standards is necessary for effective communication and coordination among grid operators, regulators, and stakeholders. The adoption of common standards and protocols can improve efficiency, reduce costs, and enhance cyber security. The collaboration between these agencies is essential for achieving a more resilient and flexible grid that can support the integration of renewable energy sources and advanced grid technologies.

11.5 Conclusion

The convergence of the smart grid, interoperability, and cyber security underscores a pivotal nexus in the evolution of modern energy systems. The smart grid's integration of advanced technologies and data-driven functionalities promises significant advancements in energy efficiency and sustainability. However, the seamless functionality of this grid hinges upon the concept of interoperability, which necessitates a blended communication framework among various components and systems. Concurrently, the security of these interconnected systems is of paramount concern, particularly in light of evolving cyber threats. In this context, standards and guidelines formulated by regulatory bodies such as the NERC, the NIST, and the FERC emerge as critical instruments. These standards serve as a foundation for establishing a robust security framework within the smart grid, encompassing crucial aspects like access controls, threat detection, and incident response protocols. They are essential in mitigating vulnerabilities and ensuring the resilience of power systems. As the smart grid continues to mature and expand, adherence to these standards becomes not only a compliance requirement but also an academic and practical imperative. This adherence paves the way for realizing the full potential of the smart grid while safeguarding energy infrastructure from the ever-evolving landscape of cyber threats.

References

[1] M.O. Qays, I. Ahmad, A. Abu-Siada, M.L. Hossain, F. Yasmin, Key communication technologies, applications, protocols and future guides for IoT-assisted smart grid systems: a review, Energy Reports 9 (2023) 2440–2452, https://doi.org/10.1016/j.egyr.2023.01.085.

[2] F. Ayadi, I. Colak, R. Bayindir, Interoperability in smart grid, in: 7th International Conference on Smart Grid (icSmartGrid), 2019, pp. 165–169, https://doi.org/10.1109/icSmartGrid48354.2019.8990680.

[3] M.Z. Gündüz, R. Daş, Communication infrastructure and cyber-security in smart grids, Journal of the Institute of Science and Technology 10 (2) (2020). Art. no. 2, https://doi.org/10.21597/jist.655990.

[4] G. Tuna, R. Daş, V.C. Gungor, Communications technologies for smart grid applications: a review of advances and challenges, in: Smart Grid Analytics for Sustainability and Urbanization, IGI Global, 2018, pp. 215–235, https://doi.org/10.4018/978-1-5225-3996-4.ch008.

[5] S. Dong, J. Cao, D. Flynn, Z. Fan, Cybersecurity in smart local energy systems: requirements, challenges, and standards, Energy Informatics 5 (1) (Jun. 2022) 9, https://doi.org/10.1186/s42162-022-00195-7.

[6] M.Z. Gunduz, R. Das, A comparison of cyber-security oriented testbeds for IoT-based smart grids, in: 6th International Symposium on Digital Forensic and Security (ISDFS), 2018, pp. 1–6, https://doi.org/10.1109/ISDFS.2018.8355329.

[7] A.O. Otuoze, M.W. Mustafa, R.M. Larik, Smart grids security challenges: classification by sources of threats, Journal of Electrical Systems and Information Technology 5 (3) (Dec. 2018) 468–483, https://doi.org/10.1016/j.jesit.2018.01.001.

[8] R. Leszczyna, Cybersecurity and privacy in standards for smart grids – a comprehensive survey, Computer Standards and Interfaces 56 (Feb. 2018) 62–73, https://doi.org/10.1016/j.csi.2017.09.005.

[9] S.S. Refaat, O. Ellabban, S. Bayhan, H. Abu-Rub, F. Blaabjerg, M.M. Begovic, Smart grid standards and interoperability, in: Smart Grid and Enabling Technologies, IEEE, 2021, pp. 435–447, https://doi.org/10.1002/9781119422464.ch17.

[10] M. Kuzlu, M. Pipattanasompom, S. Rahman, A comprehensive review of smart grid related standards and protocols, in: 5th International Istanbul Smart Grid and Cities Congress and Fair (ICSG), 2017, pp. 12–16, https://doi.org/10.1109/SGCF.2017.7947600.

[11] A. Akkad, G. Wills, A. Rezazadeh, An information security model for an IoT-enabled smart grid in the Saudi energy sector, Computers & Electrical Engineering 105 (2023) 108491, https://doi.org/10.1016/j.compeleceng.2022.108491.

[12] S. Zahra, W. Gong, H.A. Khattak, M.A. Shah, H. Song, Cross-domain security and interoperability in internet of things, IEEE Internet of Things Journal 9 (14) (Jul. 2022) 11993–12000, https://doi.org/10.1109/JIOT.2021.3132643.

[13] D.K. Panda, S. Das, Smart grid architecture model for control, optimization and data analytics of future power networks with more renewable energy, Journal of Cleaner Production 301 (Jun. 2021) 126877, https://doi.org/10.1016/j.jclepro.2021.126877.

[14] I. Priyadarshini, R. Kumar, R. Sharma, P.K. Singh, S.C. Satapathy, Identifying cyber insecurities in trustworthy space and energy sector for smart grids, Computers and Electrical Engineering 93 (Jul. 2021) 107204, https://doi.org/10.1016/j.compeleceng.2021.107204.

[15] D. Acarali, M. Rajarajan, D. Chema, M. Ginzburg, Modelling DoS attacks and interoperability in the smart grid, in: 2020 29th International Conference on Computer Communications and Networks (ICCCN), 2020, pp. 1–6, https://doi.org/10.1109/ICCCN49398.2020.9209671.

[16] J.T. Force, Security and Privacy Controls for Information Systems and Organizations, National Institute of Standards and Technology, NIST Special Publication (SP) 800-53 Rev. 5, 2020, https://doi.org/10.6028/NIST.SP.800-53r5.

[17] E.Y. Song, G.J. FitzPatrick, K.B. Lee, E. Griffor, A methodology for modeling interoperability of smart sensors in smart grids, IEEE Transactions on Smart Grid 13 (1) (Jan. 2022) 555–563, https://doi.org/10.1109/TSG.2021.3124490.

[18] J. Barata, A. Cardoso, J. Haenisch, M. Chaure, Interoperability standards for circular manufacturing in cyber-physical ecosystems: a survey, Procedia Computer Science 207 (2022) 3320–3329, https://doi.org/10.1016/j.procs.2022.09.390.

[19] M.Z. Gündüz, R. Daş, Smart grids: data transmission with cyber security principles, Bingol University Journal of Technical Science 2 (2) (2021). Art. no. 2.

[20] H.T. Reda, et al., Vulnerability and impact analysis of the IEC 61850 GOOSE protocol in the smart grid, Sensors 21 (4) (2021). Art. no. 4, https://doi.org/10.3390/s21041554.

[21] G.V.S. Raj Kumar, A.V.H. Sai Prasad, B. Padma, B. Raj Koti, Smart Grid communication and information technologies for cyber security, data privacy, and policy issues, in: B.D. Deebak, F. Al-Turjman (Eds.), Sustainable Networks in Smart Grid, Academic Press, 2022, pp. 1–29, https://doi.org/10.1016/B978-0-323-85626-3.00008-9.

[22] R. Leszczyna, Standards with cybersecurity controls for smart grid—a systematic analysis, International Journal of Communication Systems 32 (6) (2019) e3910, https://doi.org/10.1002/dac.3910.

[23] U. Cali, et al., Standardization of smart contracts for energy markets and operation, in: 2022 IEEE Power and Energy Society Innovative Smart Grid Technologies Conference (ISGT), 2022, pp. 1–5, https://doi.org/10.1109/ISGT50606.2022.9817542.

[24] M.K. Hasan, A.A. Habib, Z. Shukur, F. Ibrahim, S. Islam, M.A. Razzaque, Review on cyber-physical and cyber-security system in smart grid: standards, protocols, constraints, and recommendations, Journal of Network and Computer Applications 209 (2023) 103540, https://doi.org/10.1016/j.jnca.2022.103540.

[25] 2023 FERC Orders. https://www.nerc.com/FilingsOrders/us/Pages/2023FERCOrdersRules.aspx (Accessed May 05, 2023).

[26] T. S. G. I. P.-S. G. C. Committee, Guidelines for Smart Grid Cybersecurity, National Institute of Standards and Technology, NIST Internal or Interagency Report (NISTIR) 7628 Rev. 1, Sep. 2014, https://doi.org/10.6028/NIST.IR.7628r1.
[27] "Home Page | Federal Energy Regulatory Commission." https://www.ferc.gov/. (Accessed May 05, 2023).
[28] M.Á. Sánchez Rodríguez, J. Bermejo Higuera, J.R. Bermejo Higuera, J.A. Sicilia Montalvo, R. González Crespo, A systematic approach to analysis for assessing the security level of cyber-physical systems in the electricity sector, Microprocessors and Microsystems 87 (Nov. 2021) 104352, https://doi.org/10.1016/j.micpro.2021.104352.
[29] "Cyber Security (Permanent)." https://www.nerc.com/pa/Stand/Pages/Cyber-Security-Permanent.aspx (Accessed May 05, 2023).
[30] A. Gopstein, C. Nguyen, C. O'Fallon, N. Hastings, D.A. Wollman, NIST Framework and Roadmap for Smart Grid Interoperability Standards, Release 4.0, NIST, 2021 [Online]. Available: https://www.nist.gov/publications/nist-framework-and-roadmap-smart-grid-interoperability-standards-release-40.
[31] K. Stouffer, S. Lightman, V. Pillitteri, M. Abrams, A. Hahn, Guide to Industrial Control Systems (ICS) Security, National Institute of Standards and Technology, NIST Special Publication (SP) 800-82 Rev. 2, 2015, https://doi.org/10.6028/NIST.SP.800-82r2.
[32] V.Y. Pillitteri, T.L. Brewer, Guidelines for Smart Grid Cybersecurity, NIST Interagency/Internal Report (NISTIR) - 7628 Rev 1, 2014, https://doi.org/10.6028/NIST.IR.7628r1.

Further readings

[1] N.K. Kandasamy, et al., An electric power digital twin for cyber security testing, research and education, Computers & Electrical Engineering 101 (2022), https://doi.org/10.1016/j.compeleceng.2022.108061.
[2] A. Chehri, et al., Security risk modeling in smart grid critical infrastructures in the era of big data and artificial intelligence, Sustainability 13 (6) (2021), https://doi.org/10.3390/su13063196.

CHAPTER 12

International standards, regulations, and best practices for cyber security of smart grid

V.S.K.V. Harish[1], Sweta Gupta[1], Jignesh G. Bhatt[2] and Mohit Bansal[3]

[1]Department of Electrical Engineering, Netaji Subhas University of Technology (NSUT), Dwarka, Delhi, India; [2]Department of Instrumentation and Control Engineering, Faculty of Technology, Dharmsinh Desai University, Nadiad, Gujarat, India; [3]Department of Electrical and Electronics Engineering, GL Bajaj Institute of Technology and Management, Noida, Uttar Pradesh, India

12.1 Introduction

Power grids are the lifeblood of modern society, providing electricity for homes, businesses, and essential services. The security of electrical power grids, particularly in the context of modern smart grids, is of paramount importance in our increasingly interconnected world. Smart grids are critical infrastructure systems that enable advanced communication, automation, and electrical grid management [1]. While smart grid offers numerous benefits, such as increased efficiency and reliability, they also introduce new cyber-security challenges and vulnerabilities [2]. Recognizing these challenges, several international standards and regulations have been developed and are being developed, to ensure the security of smart grids against cyber threats.

12.1.1 Significance of cyber security in power grids

The traditional electrical power grid has evolved into a complex and interconnected ecosystem. Smart grids leverage digital technologies, sensors, and communication networks to monitor and control the distribution and consumption of electricity [3]. This transformation has brought significant advantages, including the ability to integrate renewable energy sources, enhance grid stability, and provide consumers with real-time information and control over their energy usage. However, these benefits come with an increased exposure to cyber threats.

Cyber Security Solutions for Protecting and Building the Future Smart Grid
ISBN 978-0-443-14066-2
https://doi.org/10.1016/B978-0-443-14066-2.00010-4

The digital components and interconnectivity of smart grids create new attack surfaces that adversaries can exploit. Cyberattacks on smart grids can lead to power outages, financial losses, and even threats to national security [4]. The potential consequences are significant, motivating the development of international standards and regulations to safeguard these critical systems. As per the U.S. Department of Homeland Security's Industrial Control System Computer Emergency Response Team (ICS-CERT), most Cyber-attacks have been related to the power and energy utilities (Fig. 12.1).

Table 12.1 outlines the factors with their characteristics conveying the significance of cyber security in smart grids.

12.1.2 Significance of understanding the international standards and regulations for cyber security in power grids

Studying, understanding, and analyzing international standards and regulations for the cybersecurity of smart grids is imperative for several compelling reasons. Table 12.2 illustrates the significance in brief. The reliability of electrical supply is a cornerstone of modern society. Any disruption in power distribution, whether caused by cyberattacks or other factors, can have severe consequences, impacting daily life, critical services, and the economy. International standards provide the framework to ensure grid stability, with performance metrics helping gauge its reliability [5].

The economic impact of cyberattacks on smart grids is profound. Downtime, infrastructure repairs, and lost revenue can result in substantial

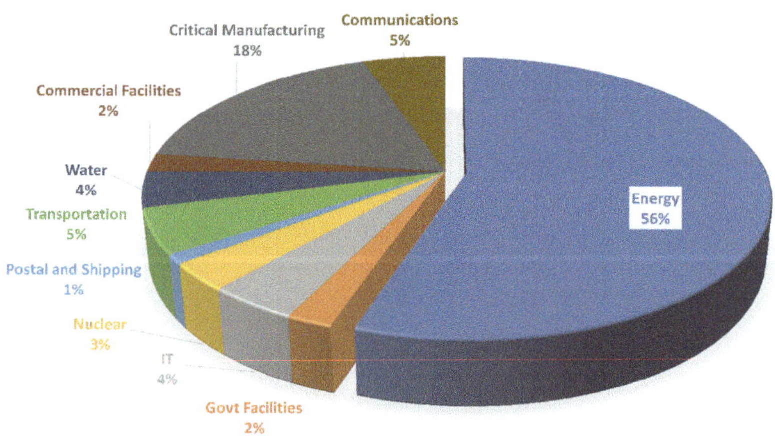

Figure 12.1 Utility as target of cyberattack.

Table 12.1 Factors determining the significance of cyber security of smart grids.

Factors	Characteristics
Resilience and reliability	Ensuring the security of smart grids is essential to maintaining a reliable electricity supply. Cyberattacks can disrupt power distribution and negatively impact businesses and daily life. International standards and regulations aim to enhance the resilience of power grids against such disruptions.
Economic impact	The financial implications of a cyberattack on a power grid can be devastating. The costs associated with downtime, infrastructure repairs, and loss of revenue for businesses and utilities can be astronomical. International guidelines help mitigate these economic risks.
Public safety	Power outages can have life-threatening consequences, especially in critical facilities like hospitals. Ensuring the cybersecurity of smart grids is essential for protecting public safety and preventing unnecessary harm during an attack.
National security	The electrical grid is considered a critical infrastructure sector in many countries. A compromised power grid can have cascading effects on other essential services, making it a potential target for state-sponsored cyberattacks. International standards and regulations play a vital role in national security.

financial losses [6]. Developing and understanding regulations is essential to mitigate economic risks and protect businesses and consumers. Moreover, national security considerations make adherence to international standards and regulations vital. The power grid is an attractive target for state-sponsored cyberattacks, which can have extensive national security implications. Analyzing these standards aids in bolstering national security defenses and preparedness. Public safety is another paramount concern. Power outages in critical facilities like hospitals can be life-threatening. Compliance with cybersecurity regulations is vital to protect public safety during crises, with performance metrics helping assess safety measures. The interconnected nature of smart grids necessitates consistency in cybersecurity standards across borders, while public confidence in the power supply relies on adherence to these regulations. Furthermore, embracing international standards fosters innovation in cybersecurity technologies and practices, ultimately enhancing grid security.

Table 12.2 Factors determining the significance of analyzing the international standards and regulations for cyber security.

Factors	Characteristics
Global collaboration	International standards encourage collaboration among countries and organizations. The shared knowledge and expertize enable collective efforts to combat cybersecurity threats more effectively.
Consistency	Standardized approaches ensure a consistent level of cybersecurity across borders. This consistency is vital in an interconnected world where the failure of one grid can impact neighboring regions.
Innovation	The development of international standards and regulations fosters innovation in cybersecurity technologies and practices. It incentivizes research and development in the field.
Compliance and accountability	Standards and regulations create a framework for compliance, holding organizations and utilities accountable for maintaining cybersecurity measures. Compliance helps prevent lapses in security.
Best practices	These guidelines often encapsulate best practices and recommendations from experts in the field. They serve as a valuable resource for organizations seeking to enhance their cybersecurity posture.
Risk mitigation	By adhering to international standards and regulations, utilities and organizations can effectively mitigate the risks associated with cyberattacks. This proactive approach safeguards against potential financial, operational, and reputational damage.

12.2 International standards for smart grid cybersecurity

Traditional power systems have been transformed into sophisticated, interconnected smart grids due to the integration of Information and Communication Technologies (ICT) into power grids [7]. While there are many advantages to these advancements, they additionally trigger unforeseen challenges, particularly in the domain of cyber security. To ensure the reliability, resilience, and security of the energy infrastructure, smart grids must be safeguarded against malicious cyber assaults.

The smart grid is a modern but complex network of interconnected devices, sensors, and control systems that enable real-time monitoring and management of electric power production, transmission, distribution, and usage. The likelihood of cyber-attacks and security breaches greatly increases with this degree of connectedness. Power outages, financial losses, and a threat to public safety are just a few of the negative effects that could result from a successful cyber-attack on a smart grid.

12.2.1 International electrotechnical commission (IEC) standards for cyber security in power grids

IEC 62351 [8], IEC 61850 [9], and IEC 62443 [10] are essential international standards that play a pivotal role in enhancing cybersecurity in the context of smart grids. Each standard serves distinct purposes and brings unique features to the table to address the specific cybersecurity challenges of modern power distribution systems.

IEC 62351 was first published in 2010, with subsequent amendments to adapt to evolving cybersecurity threats and technologies. This standard series primarily focuses on the security of power systems and associated information exchange. Its salient features include security profiles for various communication protocols used in the power sector. These profiles address encryption, access control, and key management, ensuring the confidentiality and integrity of data. IEC 62351 is known for its wide range of supported protocols, emphasizing encryption mechanisms, and facilitating secure information exchange. IEC 62351 is applicable to utilities, manufacturers, and regulators involved in the power sector. It addresses the security of data exchanged within power systems, making it vital for protecting sensitive information in smart grids.

IEC 61850 was first published in 2004, with amendments over the years to adapt to technological advancements and cybersecurity requirements. IEC 61850 primarily deals with communication protocols for power system automation. Its salient features include defining secure data exchange mechanisms and protection against cyber threats in power systems. This standard plays a critical role in ensuring secure communication between different grid components, contributing to the reliability and resilience of the smart grid. IEC 61850 is relevant to utilities, substation automation vendors, and any organization or entity involved in power system automation. It is essential for securing the communication protocols that underpin modern smart grids.

IEC 62443 was first published in 2010, with subsequent amendments to address evolving industrial automation and control system security needs. IEC 62443 focuses on the security of industrial automation and control systems, which include components used in smart grids. Its salient features include secure development practices, network security, and risk assessment in industrial control systems. This standard aims to protect smart grid components and their associated networks from cyber threats. IEC 62443 is relevant to manufacturers, integrators, operators, and organizations involved in the development and operation of industrial control systems in the context of smart grids. It provides guidelines to secure these systems effectively (Tables 12.3—12.5).

12.2.2 National Institute of Standards and Technology (NIST) standard for cyber security in power grids

NISTIR 7628, published by the National Institute of Standards and Technology (NIST), is a significant document that provides guidelines and best practices for enhancing the cybersecurity of smart grids [11]. It offers a comprehensive framework for addressing the unique security challenges

Table 12.3 IEC 62351 standard for cyber security of power grids.

IEC standard	Description
IEC/TS 62351-1	Introduction and background
IEC/TS 62351-2	Definition of terms and acronyms
IEC 62351-3	Communication security for profiles including TCP/IP
IEC 62351-4	Communication security for profiles including MMS and similar payloads
IEC 62351-5	Security for IEC 60870-5 and derivatives (i.e., DNP 3.0)
IEC 62351-6	Security for IEC 61850 peer-to-peer profiles
IEC 62351-7	Security of the information infrastructure through network and system management (NSM)
IEC 62351-8	Security through role-based access control for the entire pyramid in power system management
IEC 62351-9	Security through handling digital certificates and cryptographic keys to protect digital data and its communication
IEC/TR 62351-10	Security architecture guidelines for power systems
IEC 62351-11	Security requirements for exchanges of XML-based documents
IEC/TR 62351-12	Security for engineering/operational strategies and cyber security techniques that are applied to distributed energy resources (DER) systems
IEC/TR 62351-13	Security aspects — Guidelines for their inclusion into publications
IEC 62351-14	Implementation of security logs: communication, content and semantics

Table 12.4 NERC CIP standards for cyber security of power grids.

Std.	Description	Std.	Description
CIP001	Sabotage reporting	CIP006	Physical security of critical cyber assets
CIP002	Critical cyber asset identification	CIP007	Systems security management
CIP003	Security management controls	CIP008	Incident reporting and response planning
CIP004	Personnel and training	CIP009	Recovery plans for critical cyber assets
CIP005	Electronic security perimeters		

faced by modern electrical power distribution systems. NISTIR 7628 was first published in 2010. Although it is not a standard per se, it serves as a foundational guideline for securing smart grids. NISTIR 7628 is characterized by several key features that make it a valuable resource for enhancing smart grid cybersecurity.

- **Risk Management**: The document underscores the importance of risk management in the context of smart grid cybersecurity. It provides a systematic approach for organizations to assess risks, identify vulnerabilities, and prioritize security measures accordingly.
- **Security Controls**: NISTIR 7628 outlines a set of security controls and measures to protect smart grids from cyber threats. These controls cover areas such as access control, encryption, monitoring, incident response, and physical security, offering a comprehensive approach to security.
- **Smart Grid Components**: The document addresses specific smart grid components, such as advanced metering infrastructure (AMI), grid operations, and demand response systems. It guides securing these components to ensure the reliability and integrity of the grid.
- **Privacy Protection**: NISTIR 7628 emphasizes the protection of customer and grid data privacy. With the proliferation of smart meters and IoT devices, it guides safeguarding sensitive information.
- **Frameworks and Guidelines**: NISTIR 7628 aligns with existing cybersecurity frameworks and standards, such as the NIST Cybersecurity Framework, and offers guidance on their application in the smart grid context.

Table 12.5 Summary of a few significant international standards for cyber security of smart grids.

International standard	Organization	Year	Sub-parts	Focus area/scope	Target audience
IEC 62351	International Electrotechnical Commission (IEC)	2010	IEC/TS 62351-1 - IEC 62351-14	Communication security, access control, key management	Utilities, manufacturers, regulators
IEC 61850		2004	IEC 61850, IEC 60870-5 (101/104 and DNP3), IEC 60870-6 (ICCP), and IEC 61968/61970 (CIM)	Communication protocols for power system automation	Utilities, substation automation vendors
IEC 62443		2010	IEC 62443-2-3, 2-4 and 4-1 security programmes	Industrial automation and control systems security	Manufacturers, integrators, operators
NISTIR 7628	National Institute of Standards and Technology (NIST)	2010/2014	Development strategy, architecture and interfaces of the smart grid, high-level security requirements, cryptography and key management, privacy and the smart grid, vulnerability classes, bottom-up security analysis of the smart grid, research and development themes for cybersecurity in the smart grid, overview of the standards review, key power system use cases for security requirements	Risk assessment, security controls, privacy protection	Grid operators, manufacturers, government

ISO/IEC 27001	International Organization for Standardization (ISO)	2005/2022	Scope of the information security management systems (ISMS), risk assessment, gap analysis, ISMS development, internal audits, certification audit, and certification decision.	Information security management, risk assessment	Organizations implementing information technology based smart grid solutions
ISO 45001		2018	–	Occupational health and safety	Utility personnel, contractors
IEEE 2030	Institute of Electrical and Electronics Engineers (IEEE)	2010	IEEE 2030.1.1 (DC EV charger), IEEE P2030.2 (grid-integrated energy storage systems), IEEE P2030.3 (testing of electric energy storage systems), IEEE P2030.4 (control and automation), IEEE 2030.5 (smart energy profile 2.0 application protocol – Cyber security)	Smart grid interoperability	Utilities, device manufacturers, regulators

Continued

Table 12.5 Summary of a few significant international standards for cyber security of smart grids.—cont'd

International standard	Organization	Year	Sub-parts	Focus area/scope	Target audience
NERC CIP	North American Electric Reliability Corporation (NERC) Critical Infrastructure Protection (CIP)	2006	NERC CIP-002-5.1a (cyber system categorization of the bulk electric systems), NERC CIP-003-8 (security management hierarchy and protocols), NERC CIP-004-6 (capacity building through training), NERC CIP-005-6 (physical security of the cyber physical systems), NERC CIP-008-6 (incidence reporting and response planning), NERC CIP-010-3 (Configuration change management and vulnerability assessments)	Physical security, access control, incident response	North american power utilities
ENISA	The European Union Agency for Cybersecurity (ENISA)	2014	ENISA threat landscape 2023, good practice guide for incident management, baseline security recommendations for IoT, EU cybersecurity certification	Risk assessment, threat intelligence, incident reporting	European grid operators, policymakers

DOE guidelines	United States Department of Energy (US DoE)	2012	Cybersecurity capability maturity model (C2M2), electricity subsector cybersecurity risk management process (RMP), CESER's cybersecurity for energy delivery systems (CEDS) R&D program, cybersecurity training and workshops	Risk management, security practices, incident response	U.S. Energy sector, grid operators

NISTIR 7628 is highly applicable to organizations and utilities operating within the smart grid ecosystem. It is particularly relevant to grid operators, manufacturers, government agencies, and other stakeholders involved in the design, deployment, and operation of smart grid components and systems. NISTIR 7628 is not a regulatory standard but rather a foundational guideline that organizations can use to bolster their cybersecurity posture in the smart grid domain. Its risk-based approach and emphasis on security controls make it a valuable resource for protecting critical infrastructure. Compliance with the principles outlined in NISTIR 7628 is essential for organizations seeking to enhance the cybersecurity and resilience of their smart grid operations.

12.2.3 International Organization for Standardization (ISO) standards for cyber security in power grids

ISO/IEC 27001 and ISO 45001 are essential international standards that contribute to the cybersecurity and safety aspects of smart grids. These standards provide guidelines and best practices for securing sensitive data, ensuring operational integrity, and enhancing the safety of personnel involved in smart grid operations.

ISO/IEC 27001 was initially published in 2005, with an amendment in 2013. It remains a cornerstone in the field of information security management. ISO/IEC 27001 focuses on the establishment, implementation, and continuous improvement of information security management systems (ISMS) [12]. It is designed to safeguard information assets by identifying, assessing, and mitigating security risks. Key features include risk assessment, security controls, and a structured approach to information security management. ISO/IEC 27001 applies to any organization involved in smart grid operations that needs to protect sensitive information and maintain the confidentiality and integrity of data. It provides a globally recognized framework for implementing robust information security measures.

ISO 45001 was published in 2018. It addresses occupational health and safety management systems [13]. While not a cybersecurity standard per se, ISO 45001 is highly relevant to the smart grid sector. It focuses on creating a safe and healthy work environment by preventing work-related injuries and illnesses. Key features include risk assessment, safety controls, and a commitment to ensuring the well-being of personnel involved in grid operations. ISO 45001 is relevant to utilities, grid operators, contractors, and any organization involved in smart grid activities. It is vital for ensuring the safety of personnel working with grid components, which is essential for the overall reliability and continuity of smart grid operations.

These standards, while not exclusive to cybersecurity, complement the broader cybersecurity efforts in the smart grid sector. ISO/IEC 27001 addresses the protection of sensitive information and data integrity, critical for maintaining the confidentiality and security of information in smart grids. ISO 45001, on the other hand, contributes to the safety of personnel, an essential aspect of maintaining the reliable operation of smart grid systems.

12.2.4 Institute of Electrical and Electronics Engineers (IEEE) standard for cyber security in power grids

IEEE 2030.5 is a significant standard that plays a crucial role in enhancing the cybersecurity of smart grids [14]. IEEE 2030.5—2018 was published in 2018, representing the latest advancements in smart grid cybersecurity. This standard, formally known as IEEE 2030.5—2018, is part of the IEEE 2030 family of standards and specifically addresses the interoperability and security of energy management and control systems. It provides a comprehensive framework for securing communication and data exchange within smart grids.

IEEE 2030.5 is characterized by several key features that make it an essential resource for securing smart grids.

- **Interoperability**: The standard places a strong emphasis on interoperability, ensuring that smart grid components and systems can work together seamlessly while maintaining security. This is crucial for the efficient and reliable operation of smart grids.
- **End-to-End Security**: IEEE 2030.5 offers a holistic approach to security by addressing end-to-end protection. It covers aspects such as authentication, encryption, access control, and secure communication protocols to safeguard data from end devices to control centers.
- **Use of Proven Protocols**: The standard leverages established security protocols and best practices, including Transport Layer Security (TLS) for data encryption and Public Key Infrastructure (PKI) for authentication.
- **Security Profiles**: IEEE 2030.5 defines security profiles tailored to the specific needs of smart grids. These profiles include recommendations for securing the various components and systems within the grid.
- **Risk-Based Approach**: The standard promotes a risk-based approach to security, which means that organizations can tailor their security measures based on their unique risk assessments and requirements.

IEEE 2030.5 is highly applicable to organizations, utilities, manufacturers, and stakeholders involved in smart grid operations. It is particularly relevant to the secure communication and data exchange between different smart grid components, ensuring that information flows are protected from cyber threats.

12.2.5 North American Electric Reliability Corporation (NERC) Critical Infrastructure Protection (CIP) for cyber security in power grids

The North American Electric Reliability Corporation (NERC) Critical Infrastructure Protection (CIP) standards, first introduced in 2006 and subsequently revised, are a fundamental set of regulations aimed at ensuring the cybersecurity and resilience of the power grids in North America [15]. These standards apply to entities operating within the Bulk Electric System (BES) and play a vital role in safeguarding the reliability of electrical power distribution.

Key features of the NERC CIP standards include stringent security controls, asset categorization, and measures to protect critical assets. The standards encompass access control, incident response, physical security, personnel training, and more, providing a comprehensive framework to address cybersecurity threats. They emphasize asset categorization based on their significance, applying more rigorous requirements to critical assets. The applicability of NERC CIP standards is mandatory for all entities operating critical assets within the BES, including generation, transmission, and distribution companies, as well as regulatory authorities. These standards undergo regular updates to adapt to emerging cyber threats and technological advancements, ensuring their continued relevance.

Compliance with NERC CIP standards is verified through audits, assessments, and inspections, conducted by regulatory authorities. Noncompliance can result in penalties, including fines, underscoring the importance of adherence to these standards. NERC CIP standards are essential for maintaining the cybersecurity of power grids, reducing vulnerabilities, and ensuring the reliability and resilience of the electric grid in the face of evolving cyber threats.

12.3 Regulations for smart grid cybersecurity

Regulations on cybersecurity for power grids play a vital role in safeguarding the reliability and security of a nation's critical infrastructure. The significance of these regulations lies in their ability to mitigate the ever-growing threat of cyberattacks, which could disrupt the power supply, leading to widespread societal consequences. One of the salient features of these regulations is their focus on critical infrastructure protection [16]. Power grids are considered a cornerstone of national security and prosperity, and as such, regulations are designed to prevent cyber threats from compromising their operation [17]. Furthermore, these regulations emphasize resilience, ensuring that power grids can withstand cyberattacks and continue to provide a consistent power supply even in the face of adversity [18]. By mandating stringent security measures, regulations reduce the risk of extended outages and their associated societal disruptions.

In addition to preventive measures, cybersecurity regulations include provisions for detecting and responding to cyber incidents swiftly. They require robust incident reporting mechanisms to ensure that threats are addressed promptly, limiting their impact [19]. Regulations also hold power grid operators accountable for maintaining cybersecurity through compliance checks, audits, and penalties for non-compliance [20]. Moreover, regulations often necessitate cross-sector coordination, bringing together governmental agencies, power grid operators, and cybersecurity experts to collectively address threats. This collaboration is crucial in confronting the evolving landscape of cyber threats effectively [21].

12.3.1 Cyber security regulations for the Indian power grid

With a target of achieving a trillion-dollar economy, the Indian government is actively engaged in revising its National Cybersecurity strategy to enhance its standing in the digital realm. Although India has made significant strides since the introduction of its policy in 2013, the rapid evolution of technology has brought forth a host of new challenges that necessitate a strategic update [22]. The Indian government has acknowledged the critical role of cybersecurity in the power sector and has been actively promoting initiatives to enhance overall cybersecurity. Notable initiatives include the National Cyber Security Policy, which provides a strategic framework for addressing cybersecurity challenges [23], and the establishment of the National Critical Information Infrastructure Protection Center (NCIIPC) [24] to bolster the protection of critical infrastructure. Various Indian states have also undertaken initiatives to enhance cybersecurity in the power sector. These state-level efforts are tailored to their specific requirements and contribute to the overall security and resilience of the national grid.

The government of India enacted The Information Technology Act, 2000, also known as IT Act 2000 or the ITA-2000, on October 17, 2000, as a significant piece of legislation dealing with e-commerce and cybercrimes. It was passed to provide legal recognition and a regulatory framework for electronic transactions, digital signatures, and electronic records. The Act aimed to promote e-commerce, and e-governance, and facilitate online communication by establishing a legal framework for the use of electronic records and digital signatures. The act was amended in 2018. As per Section 70 of the IT Act, 2000, NCIIPC was established. NCIIPC is a specialized organization responsible for safeguarding India's critical information infrastructure (CII) from cyber threats. Its focus is on identifying and protecting critical assets in various sectors, including energy, finance, telecommunications, and transportation.

Clause 4.6.5 of the Indian Electricity Grid Code pertains to the "Protection of Cyber Systems and Cyber Assets" [25]. This clause is crucial in ensuring the security and reliability of India's electricity grid, especially in the digital age where cyber threats pose a significant risk to critical infrastructure. It is designed to create a comprehensive framework for addressing the growing cyber threats that the electricity grid faces. By emphasizing protection, incident response, and collaboration, it aims to ensure the continued reliability of the grid in the digital age. Compliance with this clause is crucial for entities involved in the operation and management of India's electricity grid to secure their cyber assets and maintain the nation's critical infrastructure [26].

The Central Electricity Regulatory Commission (CERC) Communication System for Inter-State Transmission of Electricity Regulations, 2016 regulations [27] set the framework for the establishment and operation of communication systems that connect various stakeholders in the electricity grid, including transmission utilities, regional load despatch centers, and power exchanges. As per section 13: Cyber Security, CEA shall formulate and notify technical standards, cyber security requirements, and protocol for the communication system for the power sector within the country including the grid integration with the grid of the neighboring countries".

CEA released the Cyber Security in Power Sector Guidelines in 2021, which represent a significant step in safeguarding India's critical power infrastructure from cyber threats [20]. These guidelines were developed to address the evolving cybersecurity challenges facing India's power sector and to ensure the resilience and reliability of the electricity grid. The primary objective of the CEA's guidelines is to establish a robust cybersecurity framework for the power sector. This framework aims to ensure the security and resilience of critical power infrastructure, protecting it from cyber threats and vulnerabilities.

The CEA's initiative to create cybersecurity guidelines for the power sector comes in response to the increasing digitalization and automation of India's electrical grid. As the power sector becomes more reliant on information technology, the need to protect it from cyber threats has become paramount. The urgency was heightened by the escalating global incidence of cyberattacks on critical infrastructure and the ever-increasing digitalization of India's power systems.

The guidelines offer a comprehensive framework for addressing cybersecurity within the power sector, covering various dimensions such as network security, incident response, and regulatory compliance. It emphasized a risk-based approach to cybersecurity, allowing organizations to focus their efforts on the most critical areas based on their specific vulnerabilities and risks. The guidelines classify power sector assets into different categories, enabling the application of security measures commensurate with their significance. It also included provisions for prompt reporting and response to cybersecurity incidents, fostering early detection and mitigation of threats.

The guidelines advocate adherence to relevant international standards, ensuring that cybersecurity measures align with best practices recognized globally. Standardization promotes interoperability and consistency across the sector. In addition, the guidelines highlight the significance of testing procedures to evaluate the effectiveness of security measures and to identify areas for improvement.

12.4 Cyber attack case studies on smart grids

Several incidents have been reported pertaining to cyber-attacks on electrical power grid. Fig. 12.2 describes the different types of attacks which have been identified.

Incidents involving smart grid cyber-attacks around the world. Here are a few case studies.

12.4.1 Ukraine power grid attack (2015 and 2016)

The Ukraine power grid hack refers to a series of cyberattacks that targeted Ukraine's electrical infrastructure, leading to widespread power outages. In December 2015 and December 2016, the Ukrainian power grid was

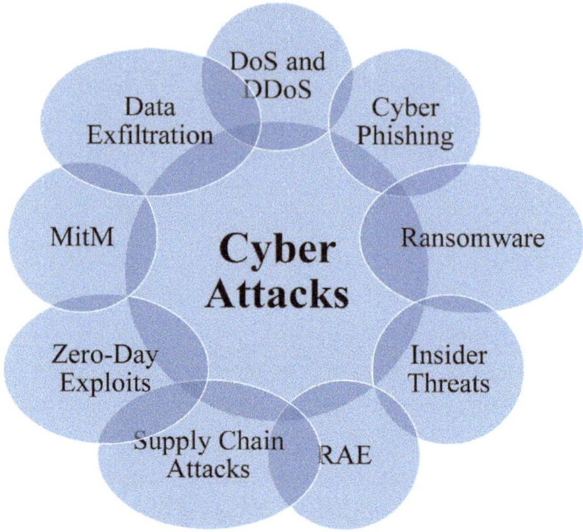

Figure 12.2 Different types of cyberattacks.

targeted in two separate incidents. These attacks, attributed to hackers, resulted in large-scale power outages. The attackers used malware to compromise critical systems, causing widespread disruptions. These attacks are some of the most well-documented instances of cyberattacks on critical infrastructure. The attacks mostly affected and disrupted the distribution grid, especially in the Kiev Oblast (Ukrainian province) [28].

In December 2015, a sophisticated cyberattack disrupted Ukraine's power grid, leading to power outages that affected hundreds of thousands of people in the Ivano-Frankivsk region [29]. The attack involved a coordinated effort to compromise the Supervisory Control and Data Acquisition (SCADA) systems that control the grid. It was attributed to a hacking group, specifically the SandWorm team, and marked one of the first instances of a cyberattack directly causing significant physical damage to critical infrastructure [30]. In December 2016, a similar attack occurred, again targeting Ukraine's power grid. The attackers used malware to gain access to and disrupt the control systems of the power distribution companies. This attack caused power outages in parts of Kiev and other regions of Ukraine, affecting hundreds of thousands of people [31]. These incidents underscored the vulnerability of critical infrastructure, including the power grid, to cyberattacks. They served as a wake-up call to governments and organizations worldwide about the potential for such attacks to disrupt essential services.

Cyber Security Infrastructure and Security Agency (CISA) of the U.S. Department of Homeland Security conducted multiple interviews with affected stakeholders and discovered that power disruptions stemmed from remote cyber intrusions affecting three regional electric power distribution companies, known as Oblenergos [32]. Such cyberattacks impacted more than 200,000 customers and for a very long time, the affected Oblenergos were still operating under limited capacity. As per the studies conducted, the cyberattack was well-orchestrated and synchronized due to thorough reconnaissance of the targeted networks. The cyberattacks on each organization unfolded in a remarkably coordinated manner, occurring within a 30-minute window, and impacting both central and regional facilities [33]. These attacks involved external individuals with malicious intent remotely manipulating circuit breakers. This was achieved by leveraging pre-existing remote administration tools at the operating system level or remote industrial control system (ICS) client software, accessed through virtual private network (VPN) connections. Studies suspected that the attackers had obtained legitimate credentials in advance to streamline their remote access during the cyberattack.

The perpetrators executed the KillDisk malware to wipe specific systems after the cyberattack. KillDisk is malicious software that selectively erases files on targeted systems and corrupts the master boot record, rendering those systems unusable [34]. Additionally, in at least one case, the cyber actors also overwrote Windows-based human-machine interfaces (HMIs) embedded in remote terminal units with the KillDisk malware. Furthermore, the attackers disabled Serial-to-Ethernet devices at substations by compromising their firmware. They also scheduled disconnections for server Uninterruptible Power Supplies (UPS) through the UPS remote management interface. Studies suggested that these actions were taken to disrupt the anticipated recovery efforts.

The presence of BlackEnergy malware [35] was detected in the systems, though it remains unclear if this malware had a direct impact on the cyberattacks. The malware is believed to have been delivered through spear-phishing emails containing malicious Microsoft Office attachments [36]. Few studies suspect that BlackEnergy might have served as an initial point of entry for acquiring valid credentials; however, this aspect was not assessed. It's essential to emphasize that any remote access Trojan could have been employed, and no specific features or capabilities of BlackEnergy were reportedly utilized in these incidents.

12.4.2 Israel's electricity authority attack (2016)

In January 2016, Israel experienced a significant cyber-attack on its power grid, shedding light on the growing threat of cyber warfare against critical infrastructure. While detailed information about the attack is limited due to the sensitive nature of such incidents and concerns for national security, available reports provide insights into this unsettling event.

The primary targets of the cyber-attack were Israel's electricity distribution companies, including the Israel Electric Corporation, which is the nation's main electricity provider [37]. The attackers employed a common and increasingly prevalent tactic in the world of cyber warfare: a phishing campaign. By sending malicious emails to employees within the targeted organizations, the attackers attempted to infiltrate the network and gain unauthorized access to sensitive systems [38].

The specifics of the phishing campaign remain somewhat obscure, but it's clear that the attackers were attempting to trick employees into downloading malware or clicking on malicious links. These tactics are often part of a broader strategy to compromise a network, gain a foothold within the infrastructure, and escalate their privileges to access critical systems. While the precise motivations behind the attack are not entirely clear, studies suggested that the attack was intended to disrupt the operation of Israel's power grid. Such disruptions can have significant consequences, impacting not only the regular lives of citizens but also critical services like hospitals, emergency response systems, and other essential infrastructure.

The response to the attack was swift and effective. The IEC's cybersecurity team was able to contain and mitigate the threat before it caused extensive damage [39]. The electricity supply to consumers was largely unaffected, which reflects the importance of preparedness and the dedication of security personnel in safeguarding critical infrastructure. This attack, however, served as a wake-up call for Israel and the broader international community. It underscored the vulnerabilities of essential infrastructure to cyber threats and the urgent need to enhance cybersecurity measures. Here are some key takeaways from this incident.

- **Increasing Sophistication of Cyber Attacks**: The attack on Israel's power grid in 2016 highlighted the increasing sophistication of cyber threats. The use of phishing campaigns is a common entry point for cybercriminals, and it can lead to significant consequences if not adequately addressed.

- **National Security Concerns**: As with many cyber-attacks on critical infrastructure, details are often kept confidential due to national security concerns. This secrecy can sometimes limit the public's understanding of the scope and impact of such attacks.
- **Need for Cybersecurity Investment**: Israel's response to this attack demonstrated the importance of investing in robust cybersecurity measures. It is essential for governments and organizations to continually enhance their cyber defences to protect against potential future attacks.
- **Preparedness and Vigilance**: Cybersecurity teams and professionals play a crucial role in detecting, responding to, and mitigating cyber threats. The timely and effective response to the 2016 attack on Israel's power grid highlights the importance of preparedness and vigilance.
- **Global Relevance**: While this incident occurred in Israel, it is a reminder that cyberattacks on critical infrastructure are a global concern. Nations around the world face similar threats and must work together to combat cyber threats effectively (Fig. 12.3).

12.4.3 Other significant cyber attacks

Dragonfly (Energetic Bear) Campaign is a long-running cyber-espionage campaign targeting energy and utility companies worldwide, including the U.S. and Europe. The attackers, suspected to be a Russian state-sponsored group, gained access to energy sector networks and could have potentially disrupted operations [40]. Stuxnet (2010) may not be a direct attack on a smart grid, but it targeted SCADA systems used in industrial control processes [40]. It was designed to specifically sabotage Iran's nuclear program, but it raised concerns about the vulnerability of critical infrastructure to cyberattacks [41].

CrashOverride (Industroyer) Attack (2016) was a malware which targeted Ukraine's power grid in December 2016. It was designed to disrupt the grid's operations, leading to a significant power outage. The attack demonstrated the potential for malware to disrupt critical infrastructure [42]. While initially appearing as ransomware, the NotPetya attack in 2017 was later determined to be a state-sponsored attack targeting Ukraine. It disrupted various critical systems, including energy, logistics, and financial institutions, highlighting the interconnectedness of these sectors [43]. Trisis (2017) was a sophisticated malware attack on a Saudi Arabian petrochemical plant [35]. While not a smart grid, it targeted industrial control systems and showcased the potential for attackers to disrupt critical infrastructure [35] (Table 12.6).

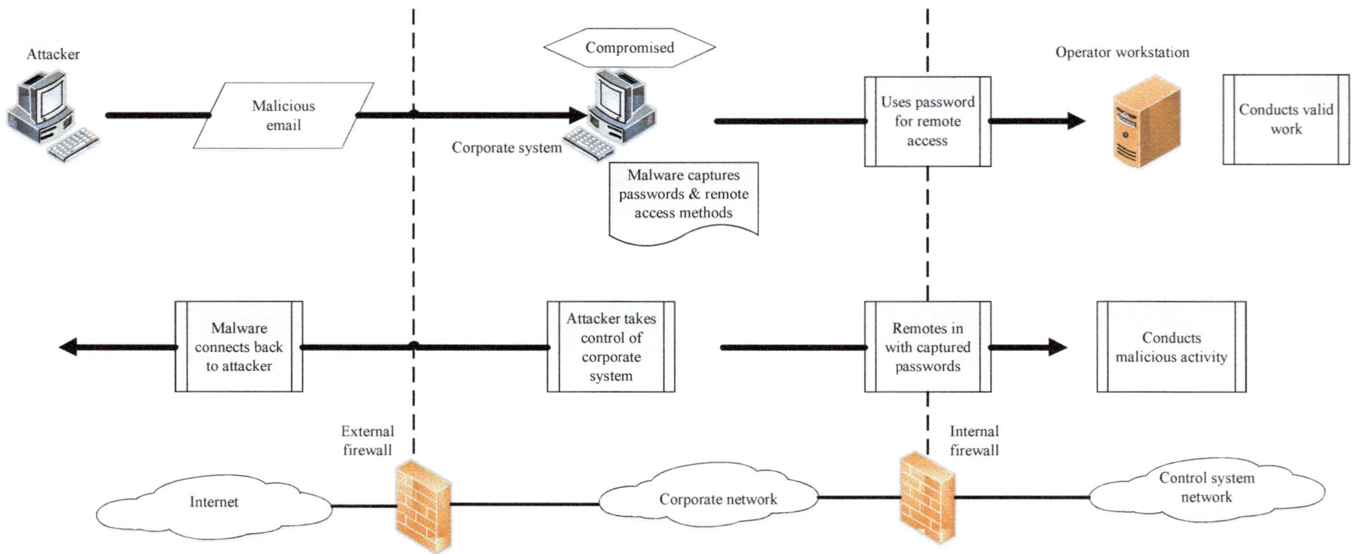

Figure 12.3 A typical cyberattack.

Table 12.6 Summary of a few significant cyber-attack events on power grids.

Attack example	Smart grid domain/area	Nature of the attack	Impact	Countermeasures	References
Stuxnet attack	SCADA systems	Malware	Disrupted operations, specifically targeting centrifuges in iranian nuclear facilities.	Regular system updates, network segmentation, whitelisting, strong access controls, robust security, security patches, and monitoring	[44]
Ukraine power grid attack	Distribution systems	Remote access and kill disk	Power outage, widespread disruption	Improved network security, strict authentication, measures, intrusion detection systems, regular employee training on cyber threats	[33]
Triton attack	Industrial control systems (ICS)	Malware (trisis)	Shut down of safety systems in a petrochemical plant.	Regular security audits, air-gapped networks for critical systems, continuous monitoring for unusual behavior	[45]
Dragonfly attack	Energy sector	Phishing	Data theft, compromising, of sensitive information.	Enhanced email security, employee training on phishing attacks, multi-factor authentication	[40]

Continued

Table 12.6 Summary of a few significant cyber-attack events on power grids.—cont'd

Attack example	Smart grid domain/area	Nature of the attack	Impact	Countermeasures	References
Havex attack	SCADA systems	Trojan horse	Compromised control systems, collecting sensitive data.	Updated antivirus software, regular system monitoring, network segmentation	[46]
Mirai botnet attack	IoT devices	Botnet attack	DDoS attacks on IoT devices, impacting grid operations.	Regular firmware updates, strong password protection, traffic monitoring and filtering	[47]
Aorora attack	Grid resilience	Physical attack	Damaged equipment, potential cascade	Physical security, grid design	[48]
NotPetya (petya) attack	IT infrastructure	Ransomware	Widespread disruption, financial losses	Patch management, data backup	[49]

12.5 Summary

In this chapter, a comprehensive exploration of the multifaceted landscape of cyber security in smart grids was presented. The significance of international standards, regulations, and best practices in safeguarding critical power grid infrastructures was illustrated, while also providing real-world insights through case studies of cyberattacks. The pivotal role of organizations such as IEC, NIST, IEEE, NERC, US DoE, in shaping the cyber security landscape for power grids, was studied. IEC standards covering diverse aspects such as risk management, access control, and incident response were described. The NIST standard takes center stage, offering a comprehensive framework for cyber security. It focuses on risk management, security controls, and continuous monitoring. NIST's guidelines are widely recognized and adopted in the smart grid domain. The ISO standards provide a holistic approach to managing information security. These standards cover a wide spectrum of topics, including risk assessment, security policies, and incident management, offering a globally recognized guide to cyber security in power grids. IEEE standards are discussed, particularly their relevance to data exchange, access control, and system architecture. They offer valuable insights for the smart grid community, focusing on key aspects of cyber security. NERC CIP standards play a pivotal role in safeguarding power infrastructure in North America. They emphasize critical infrastructure protection through a combination of physical and cyber security measures.

The chapter concluded with an examination of notable case studies involving cyberattacks on smart grids. The case study of the Ukraine Power Grid Attack serves as a stark reminder of the vulnerabilities in power grids and the imperative of robust cyber security measures. The Israel's Electricity Authority Attack illustrates how cyberattacks can disrupt the power supply and raises concerns about the security of critical infrastructure. Other significant cyberattacks on power grids, were also touched upon, demonstrating that cyber threats are a global concern and necessitate ongoing vigilance and preparedness. The chapter serves as an invaluable resource for those seeking to understand and strengthen the cyber security of power grids in an increasingly digital world.

References

[1] J. Bhatt, O. Jani, V.S.K.V. Harish, Correction to: an assessment of Indian smart grid pilot for selection of best-fit communication technology, in: Recent Advances in

Power Systems: Select Proceedings of EPREC-2021, Springer Nature Singapore, Singapore, 2022. C1-C1.

[2] M.D. Smith, M.E. Paté-Cornell, Cyber risk analysis for a smart grid: how smart is smart enough? A multiarmed bandit approach to cyber security investment, IEEE Transactions on Engineering Management 65 (3) (2018) 434−447.

[3] J. Bhatt, O. Jani, V.S.K.V. Harish, Weather-aware selection of wireless technologies for neighborhood area network of Indian smart grid, in: Computational Intelligence for Engineering and Management Applications: Select Proceedings of CIEMA 2022, Springer Nature Singapore, Singapore, 2023, pp. 83−96.

[4] M. Ghiasi, T. Niknam, Z. Wang, M. Mehrandezh, M. Dehghani, N. Ghadimi, A comprehensive review of cyber-attacks and defense mechanisms for improving security in smart grid energy systems: past, present and future, Electric Power Systems Research 215 (2023) 108975.

[5] M.K. Hasan, A. Alkhalifah, S. Islam, N.B. Babiker, A.A. Habib, A.H.M. Aman, M.A. Hossain, Blockchain technology on smart grid, energy trading, and big data: security issues, challenges, and recommendations, Wireless Communications and Mobile Computing 2022 (2022) 1−26.

[6] D. Tang, Y.P. Fang, E. Zio, Vulnerability analysis of demand-response with renewable energy integration in smart grids to cyber attacks and online detection methods, Reliability Engineering and System Safety 235 (2023) 109212.

[7] J. Bhatt, V.S.K.V. Harish, O. Jani, G. Saini, Performance based optimal selection of communication technologies for different smart microgrid applications, Sustainable Energy Technologies and Assessments 53 (2022) 102495.

[8] S.S. Hussain, T.S. Ustun, A. Kalam, A review of IEC 62351 security mechanisms for IEC 61850 message exchanges, IEEE Transactions on Industrial Informatics 16 (9) (2019) 5643−5654.

[9] P.P. Biswas, H.C. Tan, Q. Zhu, Y. Li, D. Mashima, B. Chen, A synthesized dataset for cybersecurity study of IEC 61850 based substation, in: 2019 IEEE International Conference on Communications, Control, and Computing Technologies for Smart Grids (SmartGridComm), IEEE, October 2019, pp. 1−7.

[10] D. Dolezilek, D. Gammel, W. Fernandes, Cybersecurity Based on IEC 62351 and IEC 62443 for IEC 61850 Systems, 2020.

[11] S. Hacks, M. Kaczmarek-Heß, S. de Kinderen, D. Töpel, A multi-level cyber-security reference model in support of vulnerability analysis, in: International Conference on Enterprise Design, Operations, and Computing, Springer International Publishing, Cham, September 2022, pp. 19−35.

[12] M. Podrecca, G. Culot, G. Nassimbeni, M. Sartor, Information security and value creation: the performance implications of ISO/IEC 27001, Computers in Industry 142 (2022) 103744.

[13] P. Bragatto, S.M. Ansaldi, Cyber physical systems for occupational safety at industrial sites: opportunities and challenges, Serbian Journal of Management 17 (2) (2022) 451−461.

[14] P.S. Sarker, V. Venkataramanan, D.S. Cardenas, A. Srivastava, A. Hahn, B. Miller, Cyber-physical security and resiliency analysis testbed for critical microgrids with ieee 2030.5, in: 2020 8th Workshop on Modeling and Simulation of Cyber-Physical Energy Systems, IEEE, April 2020, pp. 1−6.

[15] D. Christensen, M. Martin, E. Gantumur, B. Mendrick, Risk assessment at the edge: applying NERC CIP to aggregated grid-edge resources, The Electricity Journal 32 (2) (2019) 50−57.

[16] S. Madnick, New cybersecurity regulations are coming. Here's how to prepare, Harvard Business Review (2022). https://hbr.org/2022/08/new-cybersecurity-regulations-are-coming-heres-how-to-prepare.

[17] P.G. Chiara, The cyber resilience act: the EU Commission's proposal for a horizontal regulation on cybersecurity for products with digital elements: an introduction, International Cybersecurity Law Review 3 (2) (2022) 255–272.
[18] M. Boeding, K. Boswell, M. Hempel, H. Sharif, J. Lopez Jr, K. Perumalla, Survey of cybersecurity governance, threats, and countermeasures for the power grid, Energies 15 (22) (2022) 8692.
[19] S. Shrivastava, A.P. Mathur, Next generation cyber-physical architecture and training, in: Cyberdefense: The Next Generation, Springer International Publishing, Cham, 2023, pp. 195–204.
[20] CEA, CEA (Cyber Security in Power Sector) Guidelines, 2021, Central Electricity Authority, Ministry of Power, Government of India, 2021. Available Online: https://cea.nic.in/wp-content/uploads/notification/2021/10/Guidelines_on_Cyber_Security_in_Power_Sector_2021-2.pdf.
[21] Z. Zhang, Environmental review and case study: NERC's cybersecurity standards for the electric grid: fulfilling its reliability day job and moonlighting as a cybersecurity model, Environmental Practice 13 (3) (2011) 250–264.
[22] J. Srinivas, A.K. Das, N. Kumar, Government regulations in cyber security: framework, standards and recommendations, Future Generation Computer Systems 92 (2019) 178–188.
[23] Meity, Notification on National Cyber Security Policy – 2013, Ministry of Electronics and Information Technology, Government of India, 2013. Available Online: www.meity.gov.in/sites/upload_files/dit/files/National%20Cyber%20Security%20Policy%20%281%29.pdf.
[24] Meity, The Information Technology Act, 2000, Ministry of Electronics and Information Technology, Government of India, 2000. Available Online: www.meity.gov.in/writereaddata/files/itbill2000.pdf.
[25] CERC, Indian Electricity Grid Code 2010, Central Electricity Regulatory Commission (CERC), Government of India, 2010. Available Online: https://cercind.gov.in/2010/ORDER/February2010/IEGC_Review_Proposal.pdf.
[26] S. Snigdhha, V. Patel, V.S.K.V. Harish, A comprehensive study and assessment of electricity acts and power sector policies of India on social, technical, economic, and environmental fronts, Sustainable Energy Technologies and Assessments 57 (2023) 103299.
[27] CERC, Communication System for Inter-state Transmission of Electricity Regulations, 2016, Central Electricity Regulatory Commission (CERC), Government of India, 2017. Available Online: https://cercind.gov.in/2017/regulation/134.pdf.
[28] M. Toecker, Why power generators can't ignore the Ukraine cyberattack, Power magazine (2016). Available Online: https://www.powermag.com/power-generators-cant-ignore-ukraine-cyberattack/.
[29] T. Pultarova, Cyber security-Ukraine grid hack is wake-up call for network operators [news briefing], Engineering and Technology 11 (1) (2016) 12–13.
[30] J.E. Sullivan, D. Kamensky, How cyber-attacks in Ukraine show the vulnerability of the US power grid, The Electricity Journal 30 (3) (2017) 30–35.
[31] A. Bindra, Securing the power grid: protecting smart grids and connected power systems from cyberattacks, IEEE Power Electronics Magazine 4 (3) (2017) 20–27.
[32] CISA, Cyber-attack against Ukrainian critical infrastructure, in: Cyber Security Infrastructure and Security Agency, 2021. Available Online: https://www.cisa.gov/news-events/ics-alerts/ir-alert-h-16-056-01.
[33] D.U. Case, Analysis of the cyber attack on the Ukrainian power grid, Electricity Information Sharing and Analysis Center (E-ISAC) 388 (1–29) (2016) 3.

[34] D.E. Whitehead, K. Owens, D. Gammel, J. Smith, Ukraine cyber-induced power outage: analysis and practical mitigation strategies, in: 2017 70th Annual Conference for Protective Relay Engineers (CPRE), IEEE, April 2017, pp. 1–8.

[35] M. Geiger, J. Bauer, M. Masuch, J. Franke, An analysis of black energy 3, Crashoverride, and Trisis, three malware approaches targeting operational technology systems, in: 2020 25th IEEE International Conference on Emerging Technologies and Factory Automation (ETFA), vol. 1, IEEE, September 2020, pp. 1537–1543.

[36] M. Mesbah, M. Azer, Cyber threats and policies for industrial control systems, in: 2019 International Conference on Smart Applications, Communications and Networking (SmartNets), IEEE, December 2019, pp. 1–6.

[37] Y. Liu, R. Fan, V. Terzija, Power system restoration: a literature review from 2006 to 2016, Journal of Modern Power Systems and Clean Energy 4 (3) (2016) 332–341.

[38] M.N. Al-Mhiqani, R. Ahmad, W. Yassin, A. Hassan, Z.Z. Abidin, N.S. Ali, K.H. Abdulkareem, Cyber-security incidents: a review cases in cyber-physical systems, International Journal of Advanced Computer Science and Applications 9 (1) (2018).

[39] L. Michaels, Y. Parag, Motivations and barriers to integrating 'prosuming' services into the future decentralized electricity grid: findings from Israel, Energy Research and Social Science 21 (2016) 70–83.

[40] F.B. Khan, A. Asad, H. Durad, S.M. Mohsin, S.N. Kazmi, Dragonfly cyber threats: a case study of malware attacks targeting power grids, Journal of Computing and Biomedical Informatics 4 (02) (2023) 172–185.

[41] A. Patil, S. Shinde, S. Banerjee, Stuxnet-tool for zero-day attack, in: Handbook of Research on Cyber Crime and Information Privacy, IGI Global, 2021, pp. 652–675.

[42] M.D. Firoozjaei, N. Mahmoudyar, Y. Baseri, A.A. Ghorbani, An evaluation framework for industrial control system cyber incidents, International Journal of Critical Infrastructure Protection 36 (2022) 100487.

[43] M.A. Mos, M.M. Chowdhury, The growing influence of ransomware, in: 2020 IEEE International Conference on Electro Information Technology (EIT), IEEE, July 2020, pp. 643–647.

[44] R. Langner, Stuxnet: dissecting a cyberwarfare weapon, IEEE Security and Privacy 9 (3) (2011) 49–51.

[45] A. Di Pinto, Y. Dragoni, A. Carcano, TRITON: the first ICS cyber attack on safety instrument systems, Proceedings of Black Hat United States of America 2018 (2018) 1–26.

[46] B. Kang, P. Maynard, K. McLaughlin, S. Sezer, F. Andrén, C. Seitl, T. Strasser, Investigating cyber-physical attacks against IEC 61850 photovoltaic inverter installations, in: 2015 IEEE 20th Conference on Emerging Technologies and Factory Automation (ETFA), IEEE, September 2015, pp. 1–8.

[47] G. Kambourakis, C. Kolias, A. Stavrou, The mirai botnet and the iot zombie armies, in: MILCOM 2017-2017 IEEE Military Communications Conference (MILCOM), IEEE, October 2017, pp. 267–272.

[48] M.F. Arani, A.A. Jahromi, D. Kundur, M. Kassouf, Modeling and simulation of the aurora attack on microgrid point of common coupling, in: 2019 7th Workshop on Modeling and Simulation of Cyber-Physical Energy Systems (MSCPES), IEEE, April 2019, pp. 1–6.

[49] A. Ren, C. Liang, I. Hyug, S. Broh, N.Z. Jhanjhi, A three-level ransomware detection and prevention mechanism, EAI Endorsed Transactions on Energy Web 7 (26) (2020).

Index

Note: 'Page numbers followed by *f* indicate figures and *t* indicate tables.'

D

E

F

G

H

I

T